Robust Adaptive Control

Advances in Design and Control

SIAM's Advances in Design and Control series consists of texts and monographs dealing with all areas of design and control and their applications. Topics of interest include shape optimization, multidisciplinary design, trajectory optimization, feedback, and optimal control. The series focuses on the mathematical and computational aspects of engineering design and control that are usable in a wide variety of scientific and engineering disciplines.

Series Volumes

Karafyllis, Iasson and Krstic, Miroslav, *Robust Adaptive Control: Deadzone-Adapted Disturbance Suppression*
Ramírez, Adrián, Sipahi, Rifat, Mondié, Sabine, and Garrido, Rubén, *Design of Delay-Based Controllers for Linear Time-Invariant Systems*
Oliveira, Tiago Roux and Krstic, Miroslav, *Extremum Seeking through Delays and PDEs*
Kaszkurewicz, Eugenius and Bhaya, Amit, *Business Dynamics Models: Optimization-Based One Step Ahead Optimal Control*
Pedregal, Pablo, *A Variational Approach to Optimal Control of ODEs*
Garon, André and Delfour, Michel C., *Transfinite Interpolations and Eulerian/Lagrangian Dynamics*
Martinelli, Agostino, *Observability: A New Theory Based on the Group of Invariance*
Betts, John T., *Practical Methods for Optimal Control Using Nonlinear Programming, Third Edition*
Rodrigues, Luis, Samadi, Behzad, and Moarref, Miad, *Piecewise Affine Control: Continuous-Time, Sampled-Data, and Networked Systems*
Ferrara, A., Incremona, G. P., and Cucuzzella, C., *Advanced and Optimization Based Sliding Mode Control: Theory and Applications*
Morelli, A. and Smith, M., *Passive Network Synthesis: An Approach to Classification*
Özbay, Hitay, Gümüşsoy, Suat, Kashima, Kenji, and Yamamoto, Yutaka, *Frequency Domain Techniques for $H\infty$ Control of Distributed Parameter Systems*
Khalil, Hassan K., *High-Gain Observers in Nonlinear Feedback Control*
Bauso, Dario, *Game Theory with Engineering Applications*
Corless, M., King, C., Shorten, R., and Wirth, F., *AIMD Dynamics and Distributed Resource Allocation*
Walker, Shawn W., *The Shapes of Things: A Practical Guide to Differential Geometry and the Shape Derivative*
Michiels, Wim and Niculescu, Silviu-Iulian, *Stability, Control, and Computation for Time-Delay Systems: An Eigenvalue-Based Approach, Second Edition*
Narang-Siddarth, Anshu and Valasek, John, *Nonlinear Time Scale Systems in Standard and Nonstandard Forms: Analysis and Control*
Bekiaris-Liberis, Nikolaos and Krstic, Miroslav, *Nonlinear Control under Nonconstant Delays*
Osmolovskii, Nikolai P. and Maurer, Helmut, *Applications to Regular and Bang-Bang Control: Second-Order Necessary and Sufficient Optimality Conditions in Calculus of Variations and Optimal Control*
Biegler, Lorenz T., Campbell, Stephen L., and Mehrmann, Volker, eds., *Control and Optimization with Differential-Algebraic Constraints*
Delfour, M. C. and Zolésio, J.-P., *Shapes and Geometries: Metrics, Analysis, Differential Calculus, and Optimization, Second Edition*
Hovakimyan, Naira and Cao, Chengyu, *$\mathcal{L}_1$ Adaptive Control Theory: Guaranteed Robustness with Fast Adaptation*
Speyer, Jason L. and Jacobson, David H., *Primer on Optimal Control Theory*
Betts, John T., *Practical Methods for Optimal Control and Estimation Using Nonlinear Programming, Second Edition*
Shima, Tal and Rasmussen, Steven, eds., *UAV Cooperative Decision and Control: Challenges and Practical Approaches*
Speyer, Jason L. and Chung, Walter H., *Stochastic Processes, Estimation, and Control*
Krstic, Miroslav and Smyshlyaev, Andrey, *Boundary Control of PDEs: A Course on Backstepping Designs*
Ito, Kazufumi and Kunisch, Karl, *Lagrange Multiplier Approach to Variational Problems and Applications*
Xue, Dingyü, Chen, YangQuan, and Atherton, Derek P., *Linear Feedback Control: Analysis and Design with MATLAB*
Hanson, Floyd B., *Applied Stochastic Processes and Control for Jump-Diffusions: Modeling, Analysis, and Computation*
Michiels, Wim and Niculescu, Silviu-Iulian, *Stability and Stabilization of Time-Delay Systems: An Eigenvalue-Based Approach*
Ioannou, Petros and Fidan, Barış, *Adaptive Control Tutorial*
Bhaya, Amit and Kaszkurewicz, Eugenius, *Control Perspectives on Numerical Algorithms and Matrix Problems*
Robinett III, Rush D., Wilson, David G., Eisler, G. Richard, and Hurtado, John E., *Applied Dynamic Programming for Optimization of Dynamical Systems*
Huang, J., *Nonlinear Output Regulation: Theory and Applications*
Haslinger, J. and Mäkinen, R. A. E., *Introduction to Shape Optimization: Theory, Approximation, and Computation*
Antoulas, Athanasios C., *Approximation of Large-Scale Dynamical Systems*
Gunzburger, Max D., *Perspectives in Flow Control and Optimization*
Delfour, M. C. and Zolésio, J.-P., *Shapes and Geometries: Analysis, Differential Calculus, and Optimization*
Betts, John T., *Practical Methods for Optimal Control Using Nonlinear Programming*
El Ghaoui, Laurent and Niculescu, Silviu-Iulian, eds., *Advances in Linear Matrix Inequality Methods in Control*
Helton, J. William and James, Matthew R., *Extending H^∞ Control to Nonlinear Systems: Control of Nonlinear Systems to Achieve Performance Objectives*

Robust Adaptive Control

Deadzone-Adapted Disturbance Suppression

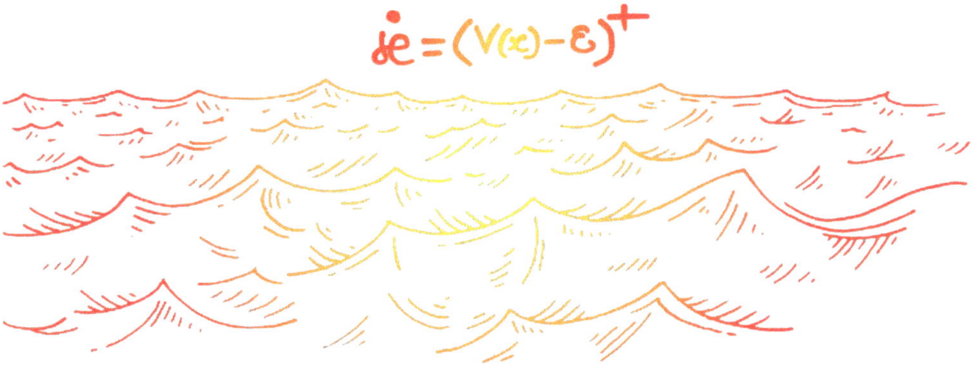

Iasson Karafyllis
National Technical University of Athens
Athens, Greece

Miroslav Krstic
University of California, San Diego
La Jolla, California

siam.
Society for Industrial and Applied Mathematics
Philadelphia

10 9 8 7 6 5 4 3 2 1

Publications Director	Kivmars H. Bowling
Executive Editor	Elizabeth Greenspan
Acquisitions Editor	Elizabeth Greenspan
Developmental Editor	Rose Kolassiba
Managing Editor	Kelly Thomas
Production Editor	David Riegelhaupt
Copy Editor	Paul Bernard
Production Manager	Rachel Ginder
Production Coordinator	Cally A. Shrader
Compositor	Cheryl Hufnagle
Graphic Designer	Doug Smock

Library of Congress Control Number: 2025935967

Contents

Preface

A brief history of Robust Adaptive Control. The purpose and power of feedback is to deal with uncertainty, in the system and its environment. *Robust* control offered the first systematic answer to attenuating the effect of uncertainty, starting in the 1960s. However, using a single, fixed controller, for a large uncertain family of systems and disturbances, and for all time, disregarding the information acquired in the course of the system operation, obviously falls short of what should be possible to achieve with feedback, combined with online learning about the system.

For this reason, a more ambitious alternative to robust control developed, called *adaptive* control, with general advances beginning in the 1970s. The idea that the system is being learned online, and that the uncertainty regarding the model would diminish over time, with the controller improved as a result, holds enormous appeal. Regardless of the challenges encountered, researchers will never stop aspiring to controller improvement in real time using adaptation. The current tide of incorporating machine learning into real-time control is a testament to that.

An exciting intellectual "clash" of the robust and adaptive control alternatives took place in the 1980s. Adaptive controllers, designed only for dealing with parametric uncertainties, were found out to lack robustness to disturbances and other non-parametric errors, with Charlie Rohrs and his coauthors blowing the whistle in a conference paper in 1982. It seemed, for a moment, as a vindication for using fixed ("robust"), non-adapting controllers for families of systems. But then, soon after, robustification tools for adaptive controllers were invented by Ioannou, Kokotovic, Narendra, Annaswamy, Praly, Morse, and others, accompanied by impressive theoretical results.

By about 1990, adaptive control had survived its robustness crisis and entered a new era of adaptive control development for nonlinear systems.

Yet, the 1980s-era robustification tools for adaptive controllers never fully overcame the challenges of non-parametric uncertainties. For robustified adaptive controllers to work, either a set in which the unknown parameters reside had to be known a priori (for the use of parameter projection) or a significant sacrifice in asymptotic performance had to be made with the sigma-modification ("leakage"), where the residual error depends on unknown parameters and, therefore, is unpredictably large when the parameters are fully unknown.

Four decades passed since the publication of the sigma-modification by Ioannou and Kokotovic in 1984. Modern concepts of disturbance-robustness came on the scene, such as Sontag's input-to-state stability (ISS) in 1989. Still, attaining ISS-like properties in adaptive control systems defied researchers for more than three additional decades. A folk belief developed that true disturbance-robustness, without a major sacrifice in asymptotic regulation, namely, without the residual error being unknown and unpredictably large, is not possible. This pessimistic belief took a deep root. With many unsuccessful attempts to go beyond what is achievable with the sigma-modification and other 1980s-era robustification tools, attempts to go further ceased.

This book: Deadzone-Adapted Disturbance Suppression (DADS). In the face of such a volume of unsucceeding efforts at stronger robustification of adaptive controllers, the results that we introduce in this book come as a big surprise, including to ourselves. Overcoming our own doubts that what we sought was possible was just as challenging as overcoming the technical obstacles. Yet, making first smaller steps, through examples, and then through generalizations, we identified, in parallel, the robust stability goals that can be achieved in adaptive control, and the methods to achieve them.

It is with great delight that we share our findings with research colleagues. With an appropriate combination of a deadzone in the adaptive law for the controller gain, and dynamic nonlinear damping, we achieve a set of stability properties which, inter alia, achieves the regulation of the state to an assignably small and known residual set, in the presence of unknown parameters for which no a priori knowledge is assumed, and in the presence of arbitrarily large disturbances. Several additional stability features are also achieved, understood to be important and necessary once one dives into the technical detail.

Since the robustness we seek chiefly deals with unknown disturbances of unlimited size, we refer to our robustification goal as "disturbance suppression." And since the key tool with which we achieve suppression to a residual set independent of the disturbance size is a deadzone in the update law, we refer to our method as "deadzone-adapted."

Hence, "deadzone-adapted disturbance suppression," or DADS.

The essence of the book is the following technical observation, not made before in the adaptive control field. The closed-loop system

$$\begin{array}{llcll} \text{plant:} & \dot{x} & = & \theta x + u + d & \\ \text{controller:} & u & = & -cx - \kappa\left[p(c) + qx^2\right] x\,, & c, q > 0 \\ \text{gain update:} & \dot{\kappa} & = & \underbrace{\gamma \max\left\{0, x^2 - \varepsilon\right\}}_{\text{deadzone}}\,, & \gamma, \varepsilon > 0\,,\ \kappa(0) > 0 \end{array} \tag{1}$$

where x is a scalar state, θ an arbitrary unknown parameter, u the control input, d an unmeasured bounded disturbance signal, $p(c)$ an appropriately chosen positive gain dependent on the gain c, and κ a positive dynamic gain updated with the help of a deadzone, yields the asymptotic property

$$\limsup_{t \to +\infty} |x(t)| \le \sqrt{\varepsilon} \tag{2}$$

as well as an explicit bound on $\kappa(t)$, which depends on $\sup_{t \ge 0} |d(t)|$. These properties are very complicated to prove and the book's key innovation is the tandem of a novel control design and a novel analysis methodology.

The essence of the design novelty is the following. The use of a deadzone alone in the update law is not new. And neither is the use of nonlinear damping qx^3. However, the addition of *dynamic nonlinear damping*, κx^3, along with the deadzone-equipped "pumping" of the gain κ, is what makes the result (2) possible. This result makes the asymptotic value of the state arbitrarily small and assignable, independently of both of the unknown quantities in the problem: the parameter θ and the disturbance bound $\sup_{t \ge 0} |d(t)|$.

The book's main subject is a generalization of the property (2) beyond the scalar case (1).

When a system is subject to time-varying disturbances and its model also has unknown parameters that are time varying, the resulting problem can be referred to as nonlinear robust adaptive control. Time-varying disturbances and parameters are not both to be regarded as disturbances, since, in the simplest of cases, such as linear time-varying plants under disturbances, the disturbances enter additively and the parameters multiplicatively to the state. The distinction between parameters and disturbances is retained in plants that are nonlinear in the state. Hence control that is robust *and* adaptive (not "robust adaptive" or "adaptive robust") is well defined.

The book's goals and contents. Nonlinear robust adaptive control is an inherently complicated and difficult topic and few research monographs tackle it, to some degree at best. This is because important technical difficulties jointly arise in nonlinear robust adaptive control. The need of using advanced stability properties for systems with inputs (disturbances), the need of dealing with complicated nonlinear systems obtained in closed loop, and the need to evaluate the controller and the performance of the closed-loop system with regards to several aspects (e.g., convergence properties, robustness, simplicity of the design, identification of the unknown parameters, etc.) are all simultaneously encountered in the area of nonlinear robust adaptive control.

This book aims to provide a systematic, mathematically rigorous approach that combines features from both adaptive control and robust control and guarantees many desired properties for the closed-loop system. More specifically, we present the following.

- We provide all stability notions that are useful in adaptive control (Chapter 2). Some of these stability notions (e.g., output stability notions) have been previously encountered in the literature on adaptive control, whereas other stability notions in Chapter 2 are new (e.g., the zero practical output asymptotic gain property). The latter notions are the stability notions that are actually needed to make adaptive control robust, without sacrificing asymptotic performance. We also provide sufficient Lyapunov-like criteria (some of them novel criteria) that can establish the validity of each stability notion.

- All robustness problems (e.g., parameter drift) that arise in adaptive control are illustrated by means of simple examples.

- Older and more recent approaches to robustification of adaptive controllers that have appeared in the literature are analyzed (Chapters 1–3).

- A novel adaptive control scheme is developed in detail: the Deadzone-Adapted Disturbance Suppression (DADS) control (Chapter 4–6). The scheme achieves an attenuation of the plant state to an *assignable* small level, despite the presence of disturbances and unknown parameters of arbitrary and unknown bounds and prevents gain and state drift regardless of the size of the disturbance and unknown parameter.

- A novel robust identifier that can work in parallel with every direct adaptive controller (not only DADS) is constructed (Chapter 7). The identifier can guarantee a prespecified level of robustness when there is sufficient excitation.

- An application of the DADS control scheme is presented: the wing rock instability problem is solved with the DADS controller (Chapter 8).

- Finally, the book gives ideas for the extension of the DADS control scheme to cases that are not studied in the book (Chapter 9).

We would like to be clear that we do not claim that this book is the ultimate answer to all robust adaptive control problems. Instead, we would like the reader to treat the book as a source of inspiration: many of the ideas described in the book can be extended in various directions and can be modified in order to deal with more complicated cases. The words of Winston Churchill—neither a feedback designer nor a mathematical authority—are nevertheless fitting here:

"Now this is not the end.
It is not even the beginning of the end.
But it is, perhaps, the end of the beginning."

We would like to accomplish the following: (i) to empower the fellow control researchers to use with rigor the stability notions provided in the book, and (ii) to help colleagues shed the urban myth which holds that a complete robustness analysis for adaptive control systems is not possible, as the book proves this myth invalid. We strongly believe that the use of the results contained in the book can open new theoretical research paths in adaptive control and can yield advances in control practice. The belief of some mathematicians that adaptive control is an engineering topic only, and not a sufficiently mathematical topic, should be corrected.

Another feature of the book is the absence of the phrase "persistency of excitation." We do not assume anything that is impossible to verify a priori. We do not assume that signals satisfy a persistency of excitation condition and we do not assume that disturbances have finite energy. We do not even assume that we know bounds for the unknown parameters. The book provides results that can be applied to realistic cases and can guarantee robustness without any unverifiable condition.

The users who will benefit from the book. The whole book deals with feedback stabilization problems for deterministic, nonlinear, time-invariant, uncertain systems (with some forays into time-varying parameters). Therefore, the book should be of interest to mathematicians working on feedback control theory. Moreover, the closed-loop systems in the book provide a unique combination of stability properties which are rarely met in other dynamical systems. This particular feature is of interest to all mathematicians working on stability theory.

All the designs in the book are given by explicit formulas and by applying systematic algorithms (backstepping). Thus, the book should be of interest to engineers working on control problems. The provided formulas of the controllers contain many parameters, and the role of each parameter is discussed in detail. Consequently, the control practitioner can understand what degrees of freedom are available and what he or she can do for the improvement of the behavior of the closed-loop system.

Graduate classes in engineering and applied mathematics could use this book. Parts of the book can become a source of homework exercises in classes such as nonlinear systems, nonlinear control, and adaptive control. The reader needs to have only a basic undergraduate-level background on control theory, differential equations, and calculus. Each chapter is written in a pedagogical style with many illustrative examples which are analyzed in detail. The proofs are also discussed so that the reader is not lost in the technical details but rather understands the core of the ideas behind the proofs. For the reader's convenience the Appendix of the book provides some notions and inequalities that are used in the book.

Acknowledgments. The authors would like to thank Dionysis Theodosis for his tremendous help in the writing process of the book and in the simulations of Chapter 7. We are also grateful to Antoine Chaillet, Petros Ioannou, Tamer Basar, and Markos Papageorgiou for the discussions that led to the inspiration of many ideas in this book.

We thank warmly our wives, Christina and Angela, and our daughters, Katerina, Olympia, Alexandra, and Victoria, for their support and encouragement.

Iasson Karafyllis, *Athens, Greece*
Miroslav Krstic, *La Jolla, California, U.S.A.*

Autumn, 2024

Notation

- Let $I \subseteq \mathbb{R}$ be an interval with non-empty interior and let $D \subseteq \mathbb{R}^n$. For every $a \geq 1$, $L^a(I; D)$ denotes the set of equivalence classes of Lebesgue measurable functions $f : I \to D$ for which $\|f\|_a = \left(\int_I |f(t)|^a \, dt\right)^{1/a} < +\infty$. $L^\infty(I; D)$ denotes the set of equivalence classes of Lebesgue measurable functions $f : I \to D$ for which $\|f\|_\infty := \operatorname{ess\,sup}_{t \in I} (|f(t)|) < +\infty$. When $D = \mathbb{R}$, we write $L^a(I)$ or $L^\infty(I)$. In many cases we use the simpler notation $\sup_{t \in I} (|f(t)|)$ to denote the essential supremum $\operatorname{ess\,sup}_{t \in I} (|f(t)|)$; the reader can understand from the context if the function $f : I \to D$ is measurable (in such a case $\sup_{t \in I} (|f(t)|)$ stands for $\operatorname{ess\,sup}_{t \in I} (|f(t)|)$).

- $\mathbb{R}_+ = [0, +\infty)$.

- For a vector $x \in \mathbb{R}^n$, $|x|$ denotes its standard Euclidean norm and x' denotes its transpose. For a matrix $A \in \mathbb{R}^{n \times m}$, A' denotes its transpose matrix. Unless stated otherwise, $|A|$ denotes the 2-norm of a matrix, i.e., the matrix norm induced by the standard Euclidean norm in $\mathbb{R}^n$.

- $x^+ = \max(x, 0)$ for $x \in \mathbb{R}$. In the neural network and machine learning literature, this function is often denoted as $x^+ = \mathrm{ReLU}(s)$, where ReLU is the notation common in neural networks, representing the "rectifier linear unit."

- $L^\infty_{loc}(\mathbb{R}_+; D)$ is the space of equivalence classes of Lebesgue measurable functions $f : \mathbb{R}_+ \to D$ for which their restriction on every bounded interval $I \subseteq \mathbb{R}_+$ with non-empty interior is of class $L^\infty(I; D)$.

- Let $D \subseteq \mathbb{R}^n$ be an open set and let $S \subseteq \mathbb{R}^n$ be a set that satisfies $D \subseteq S \subseteq \bar{D}$, where $\bar{D}$ is the closure of D. By $C^0(S\ ;\ \Omega)$ we denote the class of continuous functions on S, which take values in $\Omega \subseteq \mathbb{R}^m$. By $C^k(S\ ;\ \Omega)$, where $k \geq 1$ is an integer, we denote the class of functions on $S \subseteq \mathbb{R}^n$, which take values in $\Omega \subseteq \mathbb{R}^m$ and have continuous derivatives of order k. When $\Omega = \mathbb{R}$ then we write $C^0(S\ )$ or $C^k(S\ )$.

- By K we denote the class of increasing continuous functions $a : \mathbb{R}_+ \to \mathbb{R}_+$ with $a(0) = 0$. By K_∞ we denote the class of increasing continuous functions $a : \mathbb{R}_+ \to \mathbb{R}_+$ with $a(0) = 0$ and $\lim_{s \to +\infty} (a(s)) = +\infty$. By KL we denote the set of all continuous functions $\sigma : \mathbb{R}_+ \times \mathbb{R}_+ \to \mathbb{R}_+$ with the properties (i) for each $t \geq 0$ the mapping $\sigma(\cdot, t)$ is of class K; (ii) for each $s \geq 0$, the mapping $\sigma(s, \cdot)$ is non-increasing with $\lim_{t \to +\infty} (\sigma(s, t)) = 0$.

- $\limsup_{t \to +\infty} (f(t)) = \lim_{T \to +\infty} \left(\sup_{t \geq T} (f(t)) \right)$ for all bounded functions $f : \mathbb{R}_+ \to \mathbb{R}$ and for all $f \in L^\infty(\mathbb{R}_+)$. In the latter case $\sup_{t \geq T} (f(t))$ stands for $\operatorname{ess\,sup}_{t \geq T} (f(t))$.

- The Sobolev space $W^{1,\infty}(I)$, where $I \subseteq \mathbb{R}$ is an interval with non-empty interior, is the set of absolutely continuous functions $f : \bar{I} \to \mathbb{R}$ with $f, \dot{f} \in L^{\infty}(I; \mathbb{R})$, where $\bar{I}$ denotes the closure of $I \subset \mathbb{R}$.

Chapter 1
Prologue

1.1 ▪ Introduction

The present chapter assumes that the reader has already read the Preface, and it provides a review of the adaptive control methodologies that have been proposed in the literature. Qualitative comments that explain the characteristics of each described methodology are also given. Since the literature on adaptive control is huge, spanning over sixty years, our review cannot be exhaustive. Instead, we present either the most popular approaches or the most relevant ones to the topic of the book.

Chapter 1 also provides a preview of the book. The contents of each chapter of the book are presented in a brief way and the reader can understand what follows next.

1.2 ▪ Adaptive Control Approaches in the Literature

The problem of adaptive control arises naturally as it is rare (or even impossible) to know exactly the system parameters. Thus, it is of great interest to regulate a system in the presence of unknown parameters.

The design of controllers that can adapt to a specific system has attracted the attention of researchers and practitioners of adaptive control since the emergence of the subject in the form of the "MIT rule" in the late 1950s, namely, for about 65 years now. Standard tuning-based adaptive control theory went through many spurts of methodological advances from the late 1950s onward, remaining focused on linear systems until about 1990 [36], [58], [115], [130], and culminating in the designs for nonlinear systems in the mid-1990s [85]. Extensions of classes of systems have followed but paradigm-changing methodological advances—namely, fundamentally new approaches to adaptive control—have been nearly non-existent in the last two decades.

To understand the merits or shortfalls of any adaptive control design, it is crucial to understand the categorizations of adaptive control approaches. Classical and comprehensive references such as [58], [87], [125] are helpful for this purpose.

First, adaptive controllers can be categorized into indirect (updating plant parameters) and direct (updating controller parameters). The indirect approaches have the advantages of not requiring overparameterization, being more readily applicable to nonlinear systems, and yielding estimates of physical parameters, whereas the direct approaches avoid the online solution of "design equations" (such as Bezout, Riccati, etc.).

Second, adaptive controllers are categorized into certainty equivalence controllers, namely those designed for the "known parameter" situation, and those that are not of the certainty equivalence type.

Third, non-certainty equivalence adaptive controllers generally come in two classes. One class explicitly incorporates the functional form of the parameter update law, in order to compensate for the rapidly time-varying character of the parameter estimates (such adaptive controllers are usually designed using complete Lyapunov functions and are referred to as "Lyapunov-based" [85], [89], [126]). The other class allows off-the-shelf gradient and least-squares parameter estimators and does not incorporate the update law into the control law but, in order to compensate for the disturbing effects of the parameter estimation error and its rate of variation, which affects the validity of the control law, employs strengthening of the control law against such disturbances in some form of nonlinear damping (such adaptive controllers are referred to as "modular" [88], [90] because the controller and the parameter estimator are designed and analyzed separately, or "estimation-based" [125] because the estimator is not based on the regulation or stabilization objective but merely on the control-unrelated objective of estimating the parameters).

Event-triggered forms of adaptive control have existed for over two decades now. Direct adaptive control approaches for linear systems have been proposed in [105], [106], [107], [108]; the proposed schemes either employ event-triggering or sampled-data techniques. Direct adaptive control design methodologies with logic-based switching for linear and nonlinear control systems have been developed in [44], [45], [46], [47], [112], [113] (see also the references therein); the proposed supervisory adaptive control schemes employ multi-model based estimators of the performance of the "current" controller in conjunction with hierarchical hysteresis switching logic (which is the event-triggered element in the design). This approach is akin to estimation error-triggered controller scheduling (in a direct adaptive setting), as opposed to the regulation-triggered identification (in an indirect setting) in [9], [70], [72]. The advantage of logic-based switching [44], [45], [46], [47] is that it can also deal with systems with disturbances.

Least-squares (LS) identification is attractive because of the ability of the Riccati equation to adjust the adaptation gain in real time to the actual signal content of the regressor matrix so that all the channels of the parameter estimation vector evolve at comparable speed, rather than at vastly different time scales (with the slowest one being dominant). Recent advances and applications in LS-based adaptive control are [23], [84], [116], [151], [156]. However, in most LS adaptive control methods the parameter convergence is not guaranteed without persistency of excitation (PE), which is seldom verifiable a priori. A different LS estimator, which uses a hybrid dead-beat observer with delays was proposed in [68], [69]. The estimator in [68] does not require PE but only the weaker assumption of strong observability and was shown to be robust with respect to measurement noise. Dead-beat identifiers were first proposed in [1], [2], [43]. They require a PE assumption and are not of a LS type.

Robust adaptive control. The robustness problems in adaptive control are well known to the control community. Researchers are aware of the difficulty of studying robustness in nonlinear systems under adaptive control and the inherent limitations that may be hidden; see the relevant discussions in [58], [59], [75], [87], [115], [130] as well as the limitations for robustness that appear in discrete-time systems under adaptive control in [40], [41]. On the other hand, control practitioners are aware that some adaptive control schemes do not guarantee that the state of the closed-loop system is bounded when a disturbance is present. Particularly, even when the plant state is bounded, the phenomenon of parameter (or gain) drift may appear for certain systems under adaptive control and under the presence of disturbances. The following chapters of the book analyze the phenomenon of parameter drift and explain its origin: the lack of robustness with respect to additive and persistent disturbances. In 1982, "Rohr's counterexamples" exposed, in simulation, the absence of robustness of the nominal adaptive designs when unmodeled effects, including disturbances are present.

There are adaptive schemes that can guarantee that the state of the closed loop is bounded when a disturbance of unknown bound is present: see [58], [59], [115], [130] (for example, the

σ-modification and other leakage schemes). The leakage schemes are relatively simple modifications of the nonlinear adaptive controllers that were used in [87]. However, in such adaptive control schemes the radius of the residual set for the output is not known (and not assignable). The next chapters of the book present and analyze systematically the advantages and disadvantages of the σ-modification.

We would be misleading the reader if we were not to mention that one reference towers above the robust adaptive control literature that it precedes: the 800-page comprehensive 1996 textbook [58] by Ioannou and Sun. This book systematizes the numerous results on robust adaptive control that precede it, and advances them. Since our goal is to take robustness of adaptive control beyond what was achieved in [58], it is crucial to point out that our achievement is limited to robustness to bounded disturbances. The book [58] tackles a far broader class of uncertainties, including unmodeled dynamics. Handling unmodeled dynamics may become a new research direction for DADS but is not tackled here. What we do achieve is much stronger robustness to disturbances—particularly an asymptotic error independent of the disturbance size and the unknown parameter. What we achieve is not just with novel, more complex analysis, but also with a strengthened nonlinear dynamic control design, and with the use of a deadzone, rather than the σ-modification.

Another major textbook on adaptive control and its robustification is Tao's [143]. Published in 2003, six years after Ioannou and Sun's [58], it incorporates, additionally, discrete-time counterparts of the algorithms, extensions to multivariable systems, and refinements in the stability proofs. The use of deadzone in robust identification, combined with normalization, in both continuous and discrete time, is in pages 129–136. In the same textbook, the reader finds the deadzone in another role, not as a tool for robustification, but as an undesirable unknown nonlinearity affecting the actuator, overcome with adaptive control on pages 507–511.

A step even further in accessibility of adaptive control for linear systems is made in the tutorial textbook by Ioannou and Fidan [59]. Among many valuable tools, this textbook presents indirect adaptive pole placement, as well as adaptive linear quadratic (LQ) control, illustrated through a case study for an F-16 fighter jet.

Adaptive LQ control, whose early expositions one finds in [58, 59], has since drawn many authors, and resulted in dozens of papers and several books, dealing with adaptive optimal linear control supported by neural networks, reinforcement learning, and dynamic programming principles. Since all these adaptive LQ/optimal approaches employ approximations, robustifications of adaptation play a crucial role in them.

The book by Tsakalis and Ioannou [146] employs robustification in the adaptive control of time-varying linear systems.

Early accomplishments in robustification of *adaptive backstepping* controllers include [160], [161], [50], [51] for linear systems and [124], [154] for nonlinear systems. Some of these references employ the σ-modification and some employ projection.

A robust adaptive control methodology developed by T. Başar and coauthors in [7], [28], [29], [30], [121], [122], [145] is based on the use of robust identifiers, which are designed by means of game-theoretic arguments, and can handle tracking control problems with disturbances of finite energy. The methodology, in some cases, leverages persistency of excitation and achieves attenuation of the effect of disturbances that are not necessarily of finite energy.

In the supervisory framework of direct adaptive control, practical stability in the presence of external disturbances for plants belonging to known bounded sets has been studied in [45], [46], [106].

The approach of high-gain adaptive control (see [52] and the references therein) is a direct adaptive control methodology that involves controller gains that are updated in such a way that the gains always increase until convergence is achieved. The approach was modified and gave the adaptive λ-tracking methodology (see [3], [21], [35], [53], [54], [55], [57]): the desired convergence property is relaxed to a practical convergence property and the controller gain update law involves a deadzone. In this way, the gains stop to increase when convergence is achieved in a practical sense. The adaptive λ-tracking methodology has been applied to systems of relative degree 1 with polynomial nonlinearities, to systems of higher relative degree with bounded nonlinearities, as well as to certain engineering control problems. The methodology guarantees important robustness properties for the closed-loop system, although no KL stability estimates are provided and the effect of the disturbances/uncertainties is not quantified.

Adaptive control with performance guarantees. Before the introduction of adaptive backstepping, it was believed that the transient performance of adaptive controllers is destined to be unpredictable, i.e., arbitrarily bad [159], depending on the initial conditions and the unknown parameters. However, with strengthened adaptive controllers, the first transient bounds were introduced in [86], for linear systems, in both L_2 and L_∞ norms of the tracking error, which can be systematically reduced with the control gains. Such bounds were generalized for nonlinear systems in [87]. Some of the approaches from [87] to the shaping of transient performance of adaptive controllers are currently used in the aerospace industry.

Nevertheless, the results in [86, 87] do not guarantee uniform attractivity and/or KL estimates for the solution of the closed-loop system, when no PE assumption holds. Indeed, as shown in [91], the equilibrium set in the absence of PE is not uniformly attractive (from an open set containing the equilibrium set); hence, asymptotic stability does not hold and KL estimates are unavailable for the full state. The lack of KL estimates is strongly related to the robustness problems mentioned above (parameter drift under the presence of disturbances).

In the literature, one can find a number of different approaches for the problem of constructing adaptive feedback laws that can provide some performance guarantees for the closed-loop system. For tracking control problems, researchers have managed to establish Uniform Global Asymptotic Stability under the assumption of PE (see [97], [102], [103], [129], [155]). Using the Immersion and Invariance approach, introduced in [8], the authors of [150] managed to construct a strict Lyapunov function for a class of nonlinearly parameterized systems under an input-to-state stability (ISS) small-gain condition and a PE assumption, and guarantee global uniform asymptotic tracking. The immersion and invariance approach is also used in [96], [142]. Recent results assume less demanding excitation notions (e.g., excitation on an interval; see [24]). In [132], [133], an unbounded-in-time, time-varying adaptive control scheme is proposed by using a time-depended state transformation in order to guarantee uniform attractivity properties in the absence of PE.

It should also be noticed that the recent work [73] showed that minor modifications of the adaptive controller design procedures introduced in [87] can also guarantee semiglobal uniform output stability for nonlinear time-invariant systems under a smooth, time-invariant, and adaptive feedback law. Different modifications of the adaptive methodologies that are proposed in [87] are presented in [75]. These modifications employ strong nonlinear damping that depends both on the plant state and the parameter estimates and guarantee global KL estimates for the closed-loop system under adaptive control. It should be noticed here that the use of nonlinear damping (depending only on the plant state) was also suggested and used in [87] for better performance (providing smaller overshoots). In [9], [70], [72], an event-triggered adaptive control scheme is proposed in order to guarantee specific KL estimates for the solutions of the closed-loop system: the KL estimates that are obtained for a nominal controller that uses the exact values of the unknown parameters. In [162], an adaptive scheme is proposed that guarantees regulation of the state in finite time uniformly, assuming that a bound for the parameter vector is known; see also [153] and [158] for finite time regulation using state feedback gains. An unbounded time-scale transformation is used in [83] in order to regulate the state to the origin in finite time and, similarly, time-varying gains are used in [48] to guarantee global prescribed time convergence of the state to zero.

More recently, in [74] it was shown that the use of delays in adaptive control can ensure features that cannot be guaranteed by delay-free adaptive control for a scalar problem.

The methodology of funnel control was introduced in the early 2000s in [56] and was later developed in [15], [16], [42]. A similar approach to funnel control has been developed in [10], which is called prescribed performance control (see also [11], [12], [13], [14] for subsequent developments). Both methodologies aim at the attainment of prescribed transient and asymptotic behavior and are direct adaptive control schemes. The methodologies deal with tracking control problems and for certain classes of systems provide simple feedback formulas. The proposed feedback control is nonlinear and time-varying (even if the original control system is time-invariant) and is based on a "high-gain" approach. The design methodology can be described as follows: a time-varying set (the funnel) $F(t) \subseteq \mathbb{R}^n$ is prescribed so that we can guarantee all control objectives (performance and asymptotic objectives) when the state belongs to $F(t) \subseteq \mathbb{R}^n$ for all $t \geq 0$. The feedback design assigns nonlinear and time-varying controller gains which blow up when the state approaches the boundary of the set $F(t) \subseteq \mathbb{R}^n$. In this way, the controller does not allow the state to exit at time t the set $F(t) \subseteq \mathbb{R}^n$ for all $t \geq 0$ provided that the initial state is in the set $F(0)$. The funnel control methodology has been applied to various control problems (see the references in [17]) and both methodologies can give global results.

There are two major points that must be emphasized both for the funnel control methodology and the prescribed performance control methodology. The first point is the following fact: since the initial state must be in the set $F(0)$, it follows that the time-varying set $F(t) \subseteq \mathbb{R}^n$ must depend on the initial condition. Consequently, the proposed feedback law depends on the initial state. Therefore, the feedback law depends on time (time-varying feedback), on the current value of the state, and on the initial value of the state. The closed-loop system (even in the disturbance-free case) is not a standard dynamical system (e.g., the semigroup property does not hold). From a control-theoretical (and mathematical) point of view, this feature has important consequences: one cannot use stability notions in order to understand the properties of the resulting closed-loop systems.

The second point is subtler and is related to the first point. Both the funnel control methodology and the prescribed performance control methodology guarantee a desired behavior of the state, even when disturbances are present. Since the effect of the disturbances is not visible in the state behavior, it follows that the effect of the disturbances must be present in the control effort. Since we cannot use stability notions (e.g., ISS-like or IOS-like notions) and since no quantitative estimates are provided for the control input (which exploits time-varying controller gains that blow-up when the state approaches the boundary of the set $F(t) \subseteq \mathbb{R}^n$), there is no way to exclude the possibility of excessive control effort when disturbances are present.

The book [94], whose original edition was in 2012, and whose expanded 700-page edition is dated 2024, provides a thorough exposition on a model reference adaptive control (MRAC) augmentation to baseline robust controllers of the linear quadratic regulator (LQR) type with observer-based loop transfer recovery (OBLTR). A feedback relation underlies this methodology: OBLTR enforces "almost strictly positive real" (ASPR) properties of a transfer function matrix in the design of an output feedback MRAC, which, in turn, augments the OBLTR itself. The collection of tools in this book has been validated by the authors through numerous applications in aerial vehicle products of the Boeing company. Their work may constitute the broadest body of applications of adaptive control in industry. A strong emphasis on transient performance shaping is present in [94] and the associated successes in the applications of adaptive control to Boeing's aerial vehicles.

1.3 ▪ Preview of the Book

This book is devoted to the development of a novel robust adaptive control scheme, which was first introduced in [76], [78]. We call this control scheme the Deadzone-Adapted Disturbance Suppression (DADS) controller. However, the aim of the book is also the establishment of the area of robust adaptive control on solid mathematical foundations. The reader of the book will be able to express the objectives of robust adaptive control by using strict stability notions and will acquire the knowledge of methodologies that will allow the validation of many useful properties. For pedagogical reasons, we focus our attention on time-invariant systems. Therefore, only sporadically we provide some hints for the solution of tracking control problems.

Chapter 2 of the book describes the global stability notions needed in (robust) adaptive control and develops the mathematical tools that are used for the proof of such stability properties. While global stability notions and properties are presented in detail in standard textbooks like [80], usually these notions deal with the evolution of the full state of the closed-loop system, which in the case of adaptive control incorporates both the state of the plant and the state of the parameter estimator (as well as possibly the states of a state observer and of filters for the parameter estimator). In adaptive control, the goal of asymptotic performance is the regulation of the full state (or of an output mapping) of the plant, without having additional requirements (except for boundedness) of the parameter estimator. The state of the plant is thus to be regarded as an output of the overall closed-loop adaptive system. Therefore, all the stability notions given in [80] have to be extended to the case of a system with an output, and such extensions are provided in Chapter 2. Another issue that plays important role in adaptive control is the fact that we need to have stability properties that do not show convergence to a point or to a manifold but rather show convergence to a neighborhood of a point or a manifold. Furthermore, it is also necessary to know how large this neighborhood is. This implies the use of practical stability properties. The task of the development of practical output stability notions has been performed by many researchers in control theory (see notes and comments in Chapter 2). Chapter 2 provides and analyzes a collection of global output stability properties for input-free systems and systems with disturbances. Moreover, we show how the need of the output stability properties arises in adaptive control theory, and for each stability notion we provide verifiable sufficient conditions together with simple academic examples that illustrate the use of the provided sufficient conditions for the proof of each stability property.

Chapter 3 sets the objectives for a robust adaptive controller:

Robustness-Performance-Simplicity-Identification

In Chapter 3 we provide three different adaptive feedback design methodologies that have been introduced in recent years, and which cover some of the objectives in the set "Robustness-Performance-Simplicity-Identification". However, none of these three design methods achieves all of the four properties. These three different design methodologies are applied to the simplest scalar benchmark problem and are discussed in detail.

The development of the DADS scheme does not require knowledge of bounds for the unknown parameters and/or the disturbances and does not assume disturbances with finite energy. For pedagogical reasons, the development of the DADS adaptive control scheme in this book unfolds gradually:

- we start from Chapter 4 that is devoted to the simplest scalar benchmark problem,
- we continue with Chapter 5 where we add more complexity: in Chapter 5 we deal with systems with matched uncertainties; and, finally,
- in Chapter 6 we show how the DADS controller is designed for strict-feedback systems.

While Chapter 4 and Chapter 5 provide explicit formulas for the DADS controller, this is no longer possible to do in the case of strict-feedback systems with unmatched uncertainties. Instead, we present a backstepping methodology that allows the step-by-step construction of the DADS controller and the Lyapunov function for the closed-loop system. In these three chapters (Chapter 4, Chapter 5, and Chapter 6) we provide many simple academic examples that illustrate the design of the DADS controller and we discuss every design step.

The DADS controller combines two elements: (a) nonlinear damping with a scalar dynamic gain, and (b) a Lyapunov-based deadzone in the update law. The adaptive scheme is direct; no identification is performed. That does not mean that we cannot design identifiers for the unknown parameters. On the contrary, the design of the identifier becomes easier because the identifier is not involved in the control scheme. This task is performed in Chapter 7, where we propose robust identifiers that can work with persistent disturbances (of infinite energy). The robust identifiers of Chapter 7 can work in parallel with every direct adaptive controller (not only the DADS controller) and can provide reliable estimations of the unknown parameters when there is enough excitation.

Chapter 8 is devoted to the application of the DADS controller to a well-known technical problem: the aircraft wing rock instability. The aircraft wing rock phenomenon is an oscillation of the roll angle in an aircraft and is essentially a self-sustained oscillatory motion around the longitudinal aircraft axis. The wing rock is a substantially nonlinear and undesired phenomenon, which may become dangerous when the amplitude of the oscillation is large. Engineers have worked intensely on the suppression of the wing rock phenomenon without degrading the maneuverability of the aircraft. In Chapter 8 we show how the DADS controller can achieve the control objectives for the wing rock problem under the presence of unknown persistent disturbances (which inevitably are present to almost every mechanical system).

The last chapter of the book, Chapter 9, shows that some of the ideas presented in this book can give remarkable new results when applied to open/unsolved problems. In Chapter 9, we focus on two different issues that need further study:

- the effect of measurement errors, and
- the possible application of the DADS control scheme to tracking control problems for strict-feedback systems.

The book provides a useful Appendix, where we provide the necessary mathematical background for a complete understanding of the book (including the proofs of all results). A student/researcher who studies adaptive control usually has some knowledge obtained by the books [80] and [87]. In this book we do not assume that the reader has such knowledge although prior knowledge obtained by the books [80] and [87] will certainly help the reader. The only things that are required to be known are basic notions from a calculus course, like open and closed sets, intervals, *sup* and *inf*, (piecewise) continuous functions, limits, partial derivatives, and the chain rule. Based on these basic notions, the material that is given in the Appendix will help the reader to reach the point of understanding everything.

Chapter 2

Stability Notions Needed in Adaptive Control

2.1 ▪ Introduction

It is well known that stability theory plays a crucial role in feedback control theory. However, it is also true that adaptive control theory requires advanced stability notions that are not frequently used in standard feedback control theory. The present chapter aims at the description of the global stability notions needed in (robust) adaptive control and develops the mathematical tools that are used for the proof of such stability properties. Moreover, the present chapter shows why and how these stability properties arise in adaptive control.

It should be mentioned that it is possible that some of the readers will skip this chapter and will read the following chapters without familiarizing themselves to the stability notions that are described in the present chapter. We urge the reader not to do such a thing. Unfortunately, in the adaptive control literature many researchers use terms like “global asymptotic stability” in a very loose way and one needs to really “dig” into their papers in order to understand what the researchers actually prove/mean. It is far more preferable and far more promising for adaptive control theory to use an exact terminology that describes the obtained stability properties in a rigorous (and scientifically correct) way. That is why we strongly believe that the present chapter can inspire and help young researchers in adaptive control.

Finally, we would like to note that although the proofs of the results given in the present chapter are sometimes quite demanding, we try to present the material in the least technical way. Emphasis is placed on the detailed presentation of simple and pedagogical examples. For the interested and mathematically oriented reader we give the precise references where some of the results are proved. Moreover, we try to present some important issues in adaptive control by means of easy-to-grasp examples.

2.2 ▪ Stability Notions for Input-Free Systems

Let $f : \mathbb{R}^n \to \mathbb{R}^n$ be a locally Lipschitz vector field with $f(0) = 0$ and $h : \mathbb{R}^n \to \mathbb{R}^p$ be a continuous mapping with $h(0) = 0$. Consider the dynamical system

$$\dot{x} = f(x) \; , \; x \in \mathbb{R}^n \tag{2.1}$$

with output

$$y = h(x) \tag{2.2}$$

We assume that the dynamical system (2.1) is forward complete, i.e., for every $x_0 \in \mathbb{R}^n$ the unique solution $x(t) = \phi(t, x_0)$ of the initial-value problem (2.1) with initial condition $x(0) = x_0$ exists for all $t \geq 0$. We use the notation $y(t, x_0) = h(\phi(t, x_0))$ for all $t \geq 0$, $x_0 \in \mathbb{R}^n$ and $B_R = \{ x \in \mathbb{R}^n : |x| < R \}$ for all $R > 0$. We say that system (2.1), (2.2) is

- *Lagrange output stable* if for every $R > 0$ the set $\{ |y(t, x_0)| : x_0 \in B_R, t \geq 0 \}$ is bounded.
- *Lyapunov output stable* if for every $\varepsilon > 0$ there exists $\delta(\varepsilon) > 0$ such that for all $x_0 \in B_{\delta(\varepsilon)}$, it holds that $|y(t, x_0)| \leq \varepsilon$ for all $t \geq 0$.
- *Globally Asymptotically Output Stable (GAOS)* if system (2.1), (2.2) is Lagrange and Lyapunov output stable and $\lim\limits_{t \to +\infty} (y(t, x_0)) = 0$ for all $x_0 \in \mathbb{R}^n$.
- *Uniformly Globally Asymptotically Output Stable (UGAOS)* if system (2.1), (2.2) is Lagrange and Lyapunov output stable and for every $\varepsilon, R > 0$ there exists $T(\varepsilon, R) > 0$ such that for all $x_0 \in B_R$, it holds that $|y(t, x_0)| \leq \varepsilon$ for all $t \geq T(\varepsilon, R)$.

We say that system (2.1), (2.2) is *Globally Output Attractive (GOA)* if $\lim\limits_{t \to +\infty} (y(t, x_0)) = 0$ for all $x_0 \in \mathbb{R}^n$. We say that system (2.1), (2.2) is *practically Globally Output Attractive (p-GOA)* if there exists a constant $\tilde{\alpha} > 0$ such that $\limsup\limits_{t \to +\infty} (|y(t, x_0)|) \leq \tilde{\alpha}$ for all $x_0 \in \mathbb{R}^n$. The constant $\tilde{\alpha} > 0$ is called the *asymptotic residual constant.*

When $h(x) = x$ then the word "output" in the above properties is omitted (e.g., Lagrange stability, Lyapunov stability, GAS, UGAS, p-UGAS, GA, p-GA).

It should be noted that (see Theorem 2.2 on page 62 in [66]) UGAOS for system (2.1), (2.2) is equivalent to the existence of a function $\beta \in KL$ such that the following estimate holds for all $x_0 \in \mathbb{R}^n$ and $t \geq 0$:

$$|y(t, x_0)| \leq \beta(|x_0|, t) \tag{2.3}$$

UGAOS is a much stronger stability property than GAOS. UGAOS guarantees that there are no initial conditions for which convergence of the output to zero happens at a rate that may slow towards zero rate.

We say that system (2.1), (2.2) is *practically Uniformly Globally Asymptotically Output Stability (p-UGAOS)* if there exists a function $\beta \in KL$ and a constant $\alpha > 0$ such that the following estimate holds for all $x_0 \in \mathbb{R}^n$ and $t \geq 0$:

$$|y(t, x_0)| \leq \beta(|x_0|, t) + \alpha \tag{2.4}$$

The constant $\alpha > 0$ is called the *residual constant.* Notice that p-UGAOS with residual constant $\alpha > 0$ is equivalent to UGAOS for system (2.1) with output $y = (|h(x)| - \alpha)^+$. If system (2.1), (2.2) is p-UGAOS with residual constant $\alpha > 0$ then system (2.1), (2.2) is p-GOA with asymptotic residual constant $\tilde{\alpha} > 0$ less than or equal to the residual constant $\alpha > 0$, i.e., $\limsup\limits_{t \to +\infty} (|y(t, x_0)|) \leq \tilde{\alpha}$ for certain $\tilde{\alpha} \in (0, \alpha]$.

Lemma 1 in [73] shows that system (2.1), (2.2) is Lagrange output stable and Lyapunov output stable if and only if there exists a function $\zeta \in K_\infty$ such that the following estimate holds for all $x_0 \in \mathbb{R}^n$ and $t \geq 0$:

$$|y(t, x_0)| \leq \zeta(|x_0|) \tag{2.5}$$

None of the above stability properties guarantees or assumes that the set $\{ x \in \mathbb{R}^n : h(x) = 0 \}$ is invariant for (2.1). Invariance of the set $\{ x \in \mathbb{R}^n : h(x) = 0 \}$ for (2.1) is an additional property that can control the overshoot of the output (but there is no guarantee for this).

2.3 ▪ Proving Stability Properties for Input-Free Systems

The most convenient way of proving stability properties is by using Lyapunov-like functions. In this section, we focus on providing sufficient Lyapunov-like conditions for proving the output stability properties that were described above.

A requirement of all stability properties is the property of forward completeness. The following theorem provides the means of proving forward completeness for system (2.1). It is a special case of Theorem 1 in [77].

Theorem 2.1: *Let $H \in C^1(\mathbb{R}^n;\mathbb{R}_+)$ be a function for which there exist constants $c, \sigma \geq 0$ such that the following inequality holds for all $x \in \mathbb{R}^n$:*

$$\dot{H}(x) = \nabla H(x) f(x) \leq cH(x) + \sigma \tag{2.6}$$

Furthermore, for each $r > \inf\{\, H(x) \,:\, x \in \mathbb{R}^n \,\}$ define the set

$$S_r = \{\, x \in \mathbb{R}^n \,:\, H(x) \leq r \,\} \tag{2.7}$$

and suppose that $f(S_r)$ is bounded for every $r > \inf\{\, H(x) \,:\, x \in \mathbb{R}^n \,\}$. Then system (2.1) is forward complete.

Discussion of Theorem 2.1: Theorem 2.1 in the same spirit with the results in [6] for systems defined on $\mathbb{R}^n$ (it is a generalization of certain results in [6]) and with the results in [110] for abstract systems defined on normed linear spaces. However, in contrast with the results in [6], Theorem 2.1 does not require the set $S_r = \{\, x \in \mathbb{R}^n \,:\, H(x) \leq r \,\}$ to be bounded for all $r > \inf\{\, H(x) \,:\, x \in \mathbb{R}^n \,\}$, i.e., Theorem 2.1 does not require H to be radially unbounded. Instead, Theorem 2.1 requires boundedness of the set $f(S_r) = \{\, f(x) \,:\, x \in \mathbb{R}^n\,,\, H(x) \leq r \,\}$ for all $r > \inf\{\, H(x) \,:\, x \in \mathbb{R}^n \,\}$; this condition is automatically satisfied if $S_r = \{\, x \in \mathbb{R}^n \,:\, H(x) \leq r \,\}$ is bounded for all $r > \inf\{\, H(x) \,:\, x \in \mathbb{R}^n \,\}$, i.e., when H is radially unbounded, but can also be satisfied in other cases. The following example illustrates this point.

Now we digress to an example that illustrates Theorem 2.1 but is not inspired by adaptive control.

Example 2.1: Consider the following system:

$$\begin{aligned} \dot{z}_1 &= v_1 \\ \dot{z}_2 &= v_2 \\ \dot{v}_1 &= -g(z_1 - z_2) \\ \dot{v}_2 &= g(z_1 - z_2) \\ x &= (z_1, z_2, v_1, v_2)' \in \mathbb{R}^4 \end{aligned} \tag{2.8}$$

where $g \in C^1(\mathbb{R})$ is a function that satisfies

$$sg(s) \geq 0 \text{ for all } s \in \mathbb{R} \text{ and } \int_0^{+\infty} g(s)ds = \int_0^{-\infty} g(s)ds = +\infty \tag{2.9}$$

Define the function

$$H(x) = \frac{1}{2}v_1^2 + \frac{1}{2}v_2^2 + \int_0^{z_1 - z_2} g(s)ds \tag{2.10}$$

Definition (2.10) and equations (2.8) give us for all $x = (z_1, z_2, v_1, v_2)' \in \mathbb{R}^4$:

$$\dot{H}(x) = -v_1 g(z_1 - z_2) + v_2 g(z_1 - z_2) + (v_1 - v_2) g(z_1 - z_2) = 0 \tag{2.11}$$

Notice that the function H defined by (2.10) is not radially unbounded. Therefore, we cannot exploit the results in [6] in order to prove forward completeness for system (2.8). On the other hand, the function H satisfies all requirements of Theorem 2.1. Indeed, exploiting (2.9) and (2.10) we get $\inf\left\{ H(x) : x \in \mathbb{R}^4 \right\} = 0$. Moreover, (2.9) implies that for every $r > 0$ there exists $M(r) > 0$ such that the following implication holds:

$$\int_0^{z_1-z_2} g(s)ds \le r \Rightarrow |z_1 - z_2| \le M(r) \tag{2.12}$$

Thus, since $f(x) = (v_1, v_2, -g(z_1 - z_2), g(z_1 - z_2))'$ (recall (2.8)), we get for every $r > 0$

$$\begin{aligned} S_r &\subseteq \left\{ (z_1, z_2, v_1, v_2)' \in \mathbb{R}^4 : |v_1| \le \sqrt{2r}, |v_2| \le \sqrt{2r}, |z_1 - z_2| \le M(r) \right\} \\ f(S_r) &\subseteq \left\{ w \in \mathbb{R}^4 : |w| \le 2\sqrt{2r} + 2 \max_{|s| \le M(r)} (|g(s)|) \right\} \end{aligned} \tag{2.13}$$

where S_r is defined by (2.7). Since (2.11) implies that (2.6) holds with $c = \sigma = 0$, we conclude from Theorem 2.1 that system (2.8) is forward complete. ◁

The following proposition provides sufficient Lyapunov-like conditions for Lagrange output stability and Lyapunov output stability of system (2.1), (2.2). Its proof is provided in [73].

Proposition 2.1 (Lyapunov conditions for Lagrange and Lyapunov output stability): *Suppose that* (2.1) *is forward complete and there exist functions* $a \in K_\infty$, $W \in C^1(\mathbb{R}^n; \mathbb{R}_+)$ *with* $W(0) = 0$, *such that the following inequalities hold for all* $x \in \mathbb{R}^n$*:*

$$a(|h(x)|) \le W(x) \tag{2.14}$$

$$\nabla W(x) f(x) \le 0 \tag{2.15}$$

Then system (2.1), (2.2) *is Lagrange output stable and Lyapunov output stable.*

Example 2.2: Consider again the forward complete system (2.8) with output

$$y = h(x) = (v_1, v_2)' \in \mathbb{R}^2 \tag{2.16}$$

Notice that the function H defined by (2.10) satisfies the inequality $\frac{1}{2}|h(x)|^2 \le H(x)$ for all $x \in \mathbb{R}^4$. Therefore, (2.14) holds with $W = H$ and $a(s) = \frac{1}{2}s^2$. Moreover, (2.10) implies that $H(0) = 0$ and (2.11) shows that (2.15) holds. Therefore, by virtue of Proposition 2.1 we conclude that system (2.8), (2.16) is Lagrange output stable and Lyapunov output stable. ◁

Additional assumptions are needed in order to be able to assert GAOS. The following theorem provides sufficient conditions for GAOS. Its proof is provided in [73].

Theorem 2.2 (Lyapunov conditions for GAOS): *Suppose that* (2.1) *is forward complete and there exist functions* $a \in K_\infty$, $V, W \in C^1(\mathbb{R}^n; \mathbb{R}_+)$ *with* $V(0) = W(0) = 0$ *and a continuous, positive definite function* $\rho : \mathbb{R}_+ \to \mathbb{R}_+$ *for which the following inequalities hold for all* $x \in \mathbb{R}^n$*:*

$$a(|h(x)|) \le V(x) \tag{2.17}$$

$$\dot{V}(x) = \nabla V(x) f(x) \le -\rho(W(x)) \tag{2.18}$$

Moreover, suppose that (2.14) *holds and that there exists a continuous function* $\gamma : \mathbb{R}_+ \to \mathbb{R}_+$ *such that one of the following inequalities holds for all* $x \in \mathbb{R}^n$*:*

$$\dot{W}(x) = \nabla W(x) f(x) \le \gamma(V(x)) \tag{2.19}$$

or

$$\dot{W}(x) = \nabla W(x) f(x) \ge -\gamma(V(x)) \tag{2.20}$$

Finally, suppose that either ρ is non-decreasing or that there exists a continuous function $\zeta : \mathbb{R}_+ \to \mathbb{R}_+$ such that the following inequality holds for all $x \in \mathbb{R}^n$:

$$W(x) \leq \zeta(V(x)) \tag{2.21}$$

Then system (2.1), (2.2) *is GAOS.*

Remark: If $V \in C^1(\mathbb{R}^n;\mathbb{R}_+)$ is a radially unbounded function, then there exist continuous functions $\gamma, \zeta : \mathbb{R}_+ \to \mathbb{R}_+$ such that (2.19), (2.20), and (2.21) hold. Indeed, we obtain in this case the following corollary of Theorem 2.2, which can also be obtained as a direct consequence of Barbashin-Krasovskii-LaSalle's theorem.

Corollary 2.1: *Suppose that there exist functions $a \in K_\infty$, $V, W \in C^1(\mathbb{R}^n;\mathbb{R}_+)$ with $V(0) = W(0) = 0$ and a continuous, positive definite function $\rho : \mathbb{R}_+ \to \mathbb{R}_+$ for which* (2.14), (2.18) *hold for all $x \in \mathbb{R}^n$. Moreover, suppose that $V \in C^1(\mathbb{R}^n;\mathbb{R}_+)$ is a radially unbounded function. Then system* (2.1), (2.2) *is GAOS.*

Theorem 2.2 is less demanding than Barbashin-Krasovskii-LaSalle's theorem. The following example illustrates this point. Both the example and Theorem 2.2 itself draw some inspiration from adaptive control but go beyond adaptive control.

Example 2.3 (Verification of GAOS): Consider the following three-dimensional system:

$$\begin{aligned} &\dot{y} = -\left(1+w^2\right)y + \tfrac{2zg(z,w)}{(1+z^2)^2} \\ &\dot{z} = -g(z,w)y \\ &\dot{w} = w + |y| \\ &x = (y,z,w) \in \mathbb{R}^3 \end{aligned} \tag{2.22}$$

where $g : \mathbb{R}^2 \to \mathbb{R}$ is a bounded locally Lipschitz function. We start by showing forward completeness of system (2.22). This can be shown by using the function $H(x) = \left(y^2+z^2+w^2\right)/2$, which satisfies the following inequalities for all $x = (y,z,w) \in \mathbb{R}^3$:

$$\begin{aligned} &\dot{H}(x) = \nabla H(x)\dot{x} = -\left(1+w^2\right)y^2 + \tfrac{2zg(z,w)}{(1+z^2)^2}y \\ &-zg(z,w)y + w^2 + w\,|y| \\ &\leq -y^2 + 3\,|g(z,w)|\,|z|\,|y| + w^2 + |w|\,|y| \\ &\leq 9R^2z^2 + 3w^2/2 \leq \left(18R^2+3\right)H(x) \end{aligned} \tag{2.23}$$

where $R := \sup\left\{\, |g(z,w)| \;:\; (z,w) \in \mathbb{R}^2 \,\right\}$. It follows from (2.23) and Theorem 2.1 (notice that H is radially unbounded) that system (2.22) is forward complete. Next, we define the function $V(x) = \frac{1}{2}y^2 + \frac{z^2}{1+z^2}$, which satisfies (2.17) with $a(s) = s^2/2$ for all $s \geq 0$. Moreover, V satisfies the following inequality for all $x = (y,z,w) \in \mathbb{R}^3$:

$$\dot{V}(x) = \nabla V(x)\dot{x} = -\left(1+w^2\right)y^2 \leq -y^2 \tag{2.24}$$

Inequality (2.24) shows that (2.18) holds with $\rho(s) = 2s$ for all $s \geq 0$ (a non-decreasing function) and $W(x) = y^2/2$. Furthermore, (2.14) holds with $a(s) = s^2/2$ for all $s \geq 0$. Finally, we show that (2.19) holds. Indeed, by using the fact that $R := \sup\left\{\, |g(z,w)| \;:\; (z,w) \in \mathbb{R}^2 \,\right\}$, we obtain for all $x = (y,z,w) \in \mathbb{R}^3$

$$\begin{aligned} &\dot{W}(x) = \nabla W(x)\dot{x} = -\left(1+w^2\right)y^2 + \tfrac{2zg(z,w)}{(1+z^2)^2}y \\ &\leq -y^2 + R\,|y| \leq R^2/4 \end{aligned} \tag{2.25}$$

Inequality (2.25) shows that (2.19) holds with $\gamma(s) \equiv R^2/4$. Notice that since ρ is non-decreasing, we do not have to show (2.21) (although it happens to hold with $\zeta(s) = s$ for all $s \geq 0$). Therefore Theorem 2.2 guarantees that system (2.22) is GAOS.

Barbashin-Krasovskii-LaSalle's theorem could not have been used in order to prove GAOS for (2.22), because it admits unbounded solutions. To see this notice that the differential inequality $\dot{w} \geq w$ holds for all $x = (y, z, w) \in \mathbb{R}^3$. Thus, we get $w(t) \geq \exp(t)w(0)$ for all $t \geq 0$, and consequently, when $w(0) > 0$ the solutions of (2.22) are unbounded.

In the same way, Barbălat's Lemma could not have been easily used in order to prove GAOS for (2.22), because (2.24) shows that $\dot{W}(x)$ is bounded from above but not necessarily bounded from below. In other words, we cannot (easily) prove that the mapping $t \mapsto W(\phi(t, x))$ is uniformly continuous. ◁

The following example draws inspiration from adaptive control.

Example 2.4 (Verification of GAOS): Consider the system

$$\begin{aligned} \dot{\xi} &= -\left(z + k + \sigma\,(\theta + z)^2\right)\xi \\ \dot{z} &= \xi^2 \\ x &= (\xi, z)' \in \mathbb{R}^2 \end{aligned} \tag{2.26}$$

where $\theta \in \mathbb{R}$, $k > 0$, $\sigma \geq 0$ are constants, with output $y = \xi$. Using the radially unbounded function

$$V(x) = \frac{1}{2}\xi^2 + \frac{1}{2}z^2 \tag{2.27}$$

that satisfies for all $x \in \mathbb{R}^2$ the equation

$$\dot{V}(x) = -\left(k + \sigma\,(\theta + z)^2\right)\xi^2 \leq -k\xi^2 \leq 0 \tag{2.28}$$

we are in a position to conclude the following:

- by virtue of Theorem 2.1, system (2.26) is forward complete,
- the set $\left\{(0, z)' \in \mathbb{R}^2, z \in \mathbb{R}\right\}$ is invariant for (2.26); it is the set of equilibrium points of system (2.26),
- by virtue of Proposition 2.1, system (2.26) is Lagrange stable and Lyapunov stable,
- by using Corollary 2.1 with $W(x) = \frac{1}{2}\xi^2$ and $\rho(s) = 2ks$, system (2.26) is GAOS.

However, we notice that the convergence rate of $y = \xi \in \mathbb{R}$ to zero is not necessarily uniform with respect to the initial condition, i.e., UGAOS is not necessarily valid for (2.26). To see this, consider the case $\sigma = 0$ for which we are able to write a formula for the solution of (2.26) with initial condition $(\xi(0), z(0)) = (\xi_0, z_0) \in \mathbb{R}^2$:

$$\begin{aligned} \xi(t) &= \tfrac{K\xi_0}{K\cosh(Kt) + (z_0+k)\sinh(Kt)} \\ z(t) &= -k - K + \tfrac{2K(z_0+k+K)\exp(2Kt)}{(z_0+k+K)(\exp(2Kt)-1)+2K} \end{aligned} \tag{2.29}$$

where $K := \sqrt{\xi_0^2 + (k + z_0)^2}$. It is clearly shown by the above formulas that the convergence rate is state-dependent. Notice that convergence can happen at a rate that may slow towards zero rate. Indeed, when $z_0 = -k$ then $\xi(t) = \frac{2\xi_0\exp(-|\xi_0|t)}{1+\exp(-2|\xi_0|t)}$, which shows that the convergence rate tends to zero as ξ_0 tends to zero. Moreover, following the arguments in [118], we are able to show that system (2.26) with $\sigma = 0$ is only GAOS, and not UGAOS. We repeat the argument here: when $\xi_0 > 0$ and $z_0 + k < 0$, it holds that $\xi(t^*) = \max_{t\geq 0}(\xi(t)) = \sqrt{\xi_0^2 + (k + z_0)^2}$ for $t^* = \frac{1}{2\sqrt{\xi_0^2+(k+z_0)^2}}\ln\left(\frac{\sqrt{\xi_0^2+(k+z_0)^2}-z_0-k}{\sqrt{\xi_0^2+(k+z_0)^2}+z_0+k}\right) > 0$. If system (2.26) with $\sigma = 0$ were

UGAOS, then there would exist $T > 0$ such that for all $x_0 = (\xi_0, z_0) \in \mathbb{R}^2$ with $|x_0| \leq \sqrt{1+(k+1)^2}$ it holds that $|y(t, x_0)| \leq \frac{1}{2}$ for all $t \geq T$. Clearly, the sequence of initial conditions $x_{0,i} = \left(\frac{1}{i}, -k-1\right) \in \mathbb{R}^2$ with $i \geq 1$ being an integer satisfies $|x_{0,i}| \leq \sqrt{1+(k+1)^2}$ and $y\left(t_i^*, x_{0,i}\right) = \xi\left(t_i^*\right) = \sqrt{1+i^{-2}} > \frac{1}{2}$ for $t_i^* = \frac{1}{2\sqrt{1+i^{-2}}} \ln\left(1+2i^2\left(\sqrt{1+i^{-2}}+1\right)\right) \geq \frac{\ln\left(1+4i^2\right)}{2\sqrt{2}}$. Consequently, we would get $T \geq t_i^* \geq \frac{\ln\left(1+4i^2\right)}{2\sqrt{2}}$ for all integers $i \geq 1$; a contradiction. Thus, system (2.26) with $\sigma = 0$ is not UGAOS.

The example is continued below. $\triangleleft$

The counterexample in [118] showed that the existence of functions $V \in C^1(\mathbb{R}^n; \mathbb{R}_+)$ with $V(0) = 0$, $a, c \in K_\infty$ for which (2.17) holds and for which the following inequality holds for all $x \in \mathbb{R}^n$:

$$\dot{V}(x) = \nabla V(x) f(x) \leq -c\left(|h(x)|\right) \tag{2.30}$$

cannot guarantee UGAOS. On the other hand, it is known that the existence of functions $V \in C^1(\mathbb{R}^n; \mathbb{R}_+)$ with $V(0) = 0$, $a \in K_\infty$ and a continuous, positive definite function $\rho : \mathbb{R}_+ \to \mathbb{R}_+$ for which (2.17) holds and for which the following inequality holds for all $x \in \mathbb{R}^n$:

$$\dot{V}(x) = \nabla V(x) f(x) \leq -\rho\left(V(x)\right) \tag{2.31}$$

does guarantee UGAOS without any further assumption (see for instance the discussion in [139], [140] as well as Theorem 2.4 on page 90 in [66]). The key difference between (2.30) and (2.31) is that the dissipation rate in the former involves only the output norm, whereas the latter involves the Lyapunov function itself. However, in practice it is difficult to obtain a dissipation rate, like in (2.31), that depends on the Lyapunov function itself.

The following theorem provides Lyapunov-like conditions for UGAOS of system (2.1), (2.2) without requiring a dissipation rate that involves the Lyapunov function and shows the missing link between GAOS and UGAOS in terms of the dissipation rate. Its proof is provided in [73].

Theorem 2.3 (Lyapunov conditions for UGAOS): *Suppose that* (2.1) *is forward complete and there exist functions* $V, W \in C^1(\mathbb{R}^n; \mathbb{R}_+)$, $a \in K_\infty$ *and a continuous, positive definite function* $\rho : \mathbb{R}_+ \to \mathbb{R}_+$ *such that inequalities* (2.14), (2.15), (2.18) *hold.*
Then system (2.1), (2.2) *is UGAOS.*

The above statement shows that a sufficient additional requirement in order for (2.18) and (2.14) to ensure UAOS is that the dissipation rate W does not increase along the system's solutions.

Remark: Theorem 2.3 does not assume that (2.17) holds. Moreover, it is not assumed in Theorem 2.3 that $V(0) = 0$. Notice that if (2.17) and (2.31) hold then all assumptions of Theorem 2.3 are valid with $W = V$, in which case we recover a classical sufficient condition for UGAOS; see [66], [139], [140].

Example 2.4 (continued—verification of UGAOS): Consider again system (2.26) with $\sigma, k > 0$. We next show that system (2.26) with $k \geq \frac{1}{4\sigma} + \theta$ and output $y = \xi$ is UGAOS. Using the radially unbounded function defined by (2.27) that satisfies for all $x \in \mathbb{R}^2$ equation (2.28), we notice that (2.28) implies (2.18) with $W(x) = \xi^2/2$ and $\rho(s) = 2ks$. It is also clear that (2.14) holds with $a(s) = s^2/2$. Finally, we obtain for all $x \in \mathbb{R}^2$

$$\dot{W}(x) = -\left(z + k + \sigma\left(\theta + z\right)^2\right)\xi^2 \tag{2.32}$$

Since the inequality $z + k + \sigma\left(\theta + z\right)^2 \geq k - \frac{1}{4\sigma} - \theta$ holds for all $z \in \mathbb{R}$, we conclude from (2.32) that (2.15) holds when $k \geq \frac{1}{4\sigma} + \theta$. Therefore, Theorem 2.3 implies that system (2.26) is UGAOS when $k \geq \frac{1}{4\sigma} + \theta$.

We also notice that equation (2.32) implies the differential inequality

$$\dot{W}(x) \leq -2\left(k - \frac{1}{4\sigma} - \theta\right) W(x) \text{ for all } x \in \mathbb{R}^2. \tag{2.33}$$

Integrating the above differential inequality, we obtain the estimate

$$|y(t, x_0)| = |\xi(t)| \leq \exp\left(-\left(k - \frac{1}{4\sigma} - \theta\right) t\right) |\xi_0| \tag{2.34}$$

that holds for all $x_0 = (\xi_0, z_0)' \in \mathbb{R}^2$, $t \geq 0$. Therefore, when $k > \frac{1}{4\sigma} + \theta$ we are able to conclude that the output converges to zero at an exponential rate (notice that uniform attractivity does not necessarily imply an exponential convergence rate). The above estimate is independent of z_0 and that is the reason that such a strong stability property is termed in [140] as "State-Independent Uniform Output Stability". However, the reader should notice that a state-independent stability estimate (i.e., an estimate independent of z_0) cannot be easily shown in the case $k = \frac{1}{4\sigma} + \theta$, where we know that UGAOS holds.

Estimates like the ones provided above can be combined with other estimates in order to prove stability properties. Indeed, using (2.28) we can derive the estimate $\xi^2(t) + z^2(t) \leq \xi^2(0) + z^2(0)$ for all $t \geq 0$. Combining this estimate with the estimate

$$|\xi(t)| \leq \exp\left(-\left(k - \frac{1}{4\sigma} - \theta\right) t\right) |\xi(0)| \tag{2.35}$$

that holds when $k > \frac{1}{4\sigma} + \theta$, we can guarantee the following estimate for all $t \geq 0$ when $k > \frac{1}{4\sigma} + \theta$:

$$\begin{aligned} |\xi(t)z(t)| &\leq \exp\left(-\left(k - \tfrac{1}{4\sigma} - \theta\right) t\right) |\xi(0)|\, |x(0)| \\ &\leq \exp\left(-\left(k - \tfrac{1}{4\sigma} - \theta\right) t\right) |x(0)|^2 \end{aligned} \tag{2.36}$$

The above estimate is an estimate of the form (2.3) and shows UGAOS for system (2.26) with $k > \frac{1}{4\sigma} + \theta$ and output $y = \xi z$. $\triangleleft$

2.4 ▪ The Need of Output Stability Properties for Disturbance-Free Systems in Adaptive Control

Adaptive control is a particular case in which output stability notions are useful. Adaptive control strategies rely on a state extension to provide a dynamical estimate of unknown parameters. In this context, it is not always necessary to precisely estimate these unknown parameters to get a satisfactory behavior of the system's solutions. In other words, only part of the state variables of the closed-loop system are requested to converge to zero, which can be rephrased as a GAOS objective. To show this, consider the simplest of cases: the so-called case of a system that satisfies the "matching condition". Consider the following control system:

$$\begin{gathered} \dot{y} = f(y) + g(y)\left(u + (\varphi(y))'\,\theta\right) \\ y \in \mathbb{R}^n, \theta \in \mathbb{R}^p, u \in \mathbb{R} \end{gathered} \tag{2.37}$$

where $f, g : \mathbb{R}^n \to \mathbb{R}^n$, $\varphi : \mathbb{R}^n \to \mathbb{R}^p$ are locally Lipschitz vector fields with $f(0) = 0$ and $\varphi(0) = 0$, $y \in \mathbb{R}^n$ denotes the state, $u \in \mathbb{R}$ is the control input, and $\theta \in \mathbb{R}^p$ is the constant vector of unknown parameters. The control system (2.37) satisfies the "matching condition" (following the terminology in [87]) since the uncertain parameters are in the span of the control.

The following assumption is used for the control system (2.37).

(H) *There exist functions $P \in C^2(\mathbb{R}^n;\mathbb{R}_+)$ being positive definite and radially unbounded, $Q \in C^1(\mathbb{R}^n;\mathbb{R}_+)$ being positive definite, and a locally Lipschitz function $k \in C^0(\mathbb{R}^n;\mathbb{R})$ with $k(0)=0$ such that*

$$\nabla P(y)f(y) + \nabla P(y)g(y)k(y) \leq -Q(y) \text{ for all } y \in \mathbb{R}^n \tag{2.38}$$

When Assumption (H) holds then it is possible to regulate the state to zero by applying a standard adaptive control scheme. To see this, consider the adaptive controller

$$\begin{aligned} u &= k(y) - (\varphi(y))' \hat{\theta} \\ \tfrac{d\hat{\theta}}{dt} &= \Gamma^{-1}\nabla P(y)g(y)\varphi(y) \end{aligned} \tag{2.39}$$

where $\Gamma \in \mathbb{R}^{p\times p}$ is a constant, positive definite matrix (parameters of the controller). System (2.37) in closed loop with (2.39) then reads

$$\begin{aligned} \dot{y} &= f(y) + g(y)\left(k(y) - (\varphi(y))' z\right) \\ \dot{z} &= \Gamma^{-1}\nabla P(y)g(y)\varphi(y) \\ x &= (y,z) \in \mathbb{R}^n \times \mathbb{R}^p \end{aligned} \tag{2.40}$$

with $z = \hat{\theta} - \theta$ and output map defined by

$$h(x) = y \tag{2.41}$$

Using the radially unbounded function $V(x) := P(y) + \frac{1}{2}z'\Gamma z$, we obtain (by virtue of (2.31), (2.40)) the following differential inequality for all $x = (y,z) \in \mathbb{R}^n \times \mathbb{R}^p$:

$$\dot{V}(x) \leq -Q(y) \tag{2.42}$$

Using (2.42) we are in a position to conclude the following:

- by virtue of Theorem 2.1, system (2.40) is forward complete,
- the set $\{(0,z)' \in \mathbb{R}^n \times \mathbb{R}^p, z \in \mathbb{R}^p\}$ is invariant for (2.40); it is a set of equilibrium points for system (2.40),
- by virtue of Proposition 2.1, system (2.40) is Lagrange stable and Lyapunov stable,
- by using Corollary 2.1 with $W(x) = P(y)$, system (2.40) is GAOS.

The reader should notice the precision that the stability properties of system (2.40) are described by using the output stability notions that we introduced above. The above set of stability properties provides important performance guarantees for the closed-loop system; this is the reason for the success of adaptive control.

However, the convergence rate of $y \in \mathbb{R}^n$ to zero is not necessarily uniform with respect to the initial condition, i.e., UGAOS is not necessarily valid for (2.40). This is a source of various problems that can arise for the closed-loop system (2.40). The following sections will shed light to the problems that can appear for (2.40).

2.5 ▪ Stability Notions for Systems with (Disturbance) Inputs

Let $D \subseteq \mathbb{R}^p$ be a given closed set with $0 \in D$ and $f : \mathbb{R}^n \times D \to \mathbb{R}^n$ be a locally Lipschitz with respect to $x \in \mathbb{R}^n$ mapping with $f(0,0)=0$. Consider the control system

$$\dot{x} = f(x,d)\ ,\ x \in \mathbb{R}^n\ ,\ d \in D \tag{2.43}$$

We assume that system (2.43) is forward complete, i.e., for every $x_0 \in \mathbb{R}^n$ and for every Lebesgue measurable and locally essentially bounded input $d : \mathbb{R}_+ \to D$ the unique solution $x(t) = \phi(t, x_0; d)$ of the initial-value problem (2.43) with initial condition $x(0) = x_0$ corresponding to input $d : \mathbb{R}_+ \to D$ exists for all $t \geq 0$. We use the notation $y(t, x_0; d) = h(\phi(t, x_0; d))$ for all $t \geq 0$, $x_0 \in \mathbb{R}^n$ and for every Lebesgue measurable and locally essentially bounded input $d : \mathbb{R}_+ \to D$.

We say that system (2.43), (2.2) is *Input-to-Output Stable (IOS)* if there exist a function $\beta \in KL$ and a non-decreasing, continuous function $\gamma : \mathbb{R}_+ \to \mathbb{R}_+$ with $\gamma(0) = 0$ such that the following estimate holds for all $x_0 \in \mathbb{R}^n$, $t \geq 0$ and for every $d \in L^\infty(\mathbb{R}_+; D)$:

$$|y(t, x_0; d)| \leq \beta(|x_0|, t) + \gamma(\|d\|_\infty) \tag{2.44}$$

We say that system (2.43), (2.2) is *practically Input-to-Output Stable (p-IOS)* if there exist a function $\beta \in KL$, a non-decreasing, continuous function $\gamma : \mathbb{R}_+ \to \mathbb{R}_+$ with $\gamma(0) = 0$ and a constant $\alpha > 0$ such that the following estimate holds for all $x_0 \in \mathbb{R}^n$, $t \geq 0$ and for every $d \in L^\infty(\mathbb{R}_+; D)$:

$$|y(t, x_0; d)| \leq \beta(|x_0|, t) + \gamma(\|d\|_\infty) + \alpha \tag{2.45}$$

The constant $\alpha > 0$ is called the *residual constant* while the function γ is called the *gain function of the input* $d \in D$ *to the output* y. It is clear that the gain function is a measure of the sensitivity of system (2.43), (2.2) to the disturbance $d \in D$. Notice that p-IOS with residual constant $\alpha > 0$ is equivalent to IOS for system (2.43) with output $y = (|h(x)| - \alpha)^+$. In the disturbance-free case, the p-IOS property coincides with the p-UGAOS property.

We say that system (2.43), (2.2) satisfies the *practical Output Asymptotic Gain (p-OAG)* property if there exists a non-decreasing, continuous function $\tilde{\gamma} : \mathbb{R}_+ \to \mathbb{R}_+$ with $\tilde{\gamma}(0) = 0$ and a constant $\tilde{\alpha} > 0$ such that the following estimate holds for all $x_0 \in \mathbb{R}^n$ and for every $d \in L^\infty(\mathbb{R}_+; D)$:

$$\limsup_{t \to +\infty} (|y(t, x_0; d)|) \leq \tilde{\gamma}(\|d\|_\infty) + \tilde{\alpha} \tag{2.46}$$

The constant $\tilde{\alpha} > 0$ is called the *asymptotic residual constant* while the non-decreasing, continuous function $\tilde{\gamma}$ is called the *asymptotic gain function of the input* $d \in D$ *to the output* y. It is clear that the asymptotic gain function and the asymptotic residual constant are additional measures of the sensitivity of system (2.43), (2.2) to the disturbance $d \in D$. It is a consequence of the semigroup property that if system (2.43), (2.2) satisfies the p-IOS property then system (2.43), (2.2) also satisfies the p-OAG property with asymptotic gain function being less than or equal to the gain function (i.e., $\tilde{\gamma}(s) \leq \gamma(s)$ for all $s \geq 0$) and asymptotic residual constant being less than or equal to the residual constant (i.e., $\tilde{\alpha} \leq \alpha$). In the disturbance-free case, the p-OAG property coincides with the p-GOA property.

When $\tilde{\gamma} \equiv 0$ we say that system (2.43), (2.2) satisfies the *zero practical Output Asymptotic Gain property (zero p-OAG)*.

When $h(x) = x$ then the word "output" in the above properties is either replaced by the word "state" (e.g., ISS, p-ISS) or is omitted (e.g., p-AG, zero p-AG).

We say that system (2.43) satisfies the *practical Uniform Bounded-Input-Bounded-State (p-UBIBS)* property if there exists a function $\bar{\gamma} \in K_\infty$ and a constant $\bar{\alpha} > 0$ such that the following estimate holds for all $x_0 \in \mathbb{R}^n$ and for every $d \in L^\infty(\mathbb{R}_+; D)$:

$$\sup_{t \geq 0} (|\phi(t, x_0; d)|) \leq \bar{\gamma}(|x_0|) + \bar{\gamma}(\|d\|_\infty) + \bar{\alpha} \tag{2.47}$$

The p-UBIBS property with $\bar{\alpha} = 0$ is called the UBIBS property.

Clearly, the p-UBIBS property is equivalent to the existence of a continuous function $B : \mathbb{R}^n \times \mathbb{R}_+ \to \mathbb{R}_+$ for which the following estimate holds for all $x_0 \in \mathbb{R}^n$ and for every $d \in L^\infty(\mathbb{R}_+; D)$:

$$\sup_{t \geq 0} \left(|\phi(t, x_0; d)| \right) \leq B\left(x_0, \|d\|_\infty\right) \tag{2.48}$$

In the disturbance-free case, the p-UBIBS property coincides with the Lagrange stability property. Moreover, in the disturbance-free case, the UBIBS property coincides with the combination of Lagrange and Lyapunov stability.

2.6 ▪ Proving Stability Properties for Systems with (Disturbance) Inputs

A requirement of all stability properties for systems with inputs is the property of forward completeness. The following theorem is an extension of Theorem 2.1 and provides the means of proving forward completeness for system (2.43).

Theorem 2.4: *Let $H \in C^1(\mathbb{R}^n; \mathbb{R}_+)$ be a function for which there exist non-decreasing functions $c, \sigma \in C^0(\mathbb{R}_+; \mathbb{R}_+)$ such that the following inequality holds for all $(x, d) \in \mathbb{R}^n \times D$:*

$$\dot{H}(x, d) = \nabla H(x) f(x, d) \leq c\left(|d|\right) H(x) + \sigma\left(|d|\right) \tag{2.49}$$

Furthermore, for each $r > \inf\{ H(x) : x \in \mathbb{R}^n \}$ define by means of (2.7) the set S_r as well as the set

$$D_r = \{ d \in D : |d| \leq r \} \tag{2.50}$$

and suppose that $f(S_r \times D_r)$ is bounded for every $r > \inf\{ H(x) : x \in \mathbb{R}^n \}$. Then system (2.43) is forward complete.

Proof: Let $x_0 \in \mathbb{R}^n$ and $d \in L^\infty_{loc}(\mathbb{R}_+; D)$ be given. Since f is locally Lipschitz with respect to $x \in \mathbb{R}^n$, the initial-value problem $\dot{x} = f(x, d)$ with initial condition $x(0) = x_0$ has a unique solution $x(t) \in \mathbb{R}^n$ defined for $t \in [0, t_{\max})$, where $t_{\max} \in (0, +\infty]$. Furthermore, if $t_{\max} < +\infty$ then $\limsup\limits_{t \to t_{\max}^-} \left(|x(t)|\right) = +\infty$. We show next by means of a contradiction argument that $t_{\max} = +\infty$.

Without loss of generality, we may assume that (2.49) holds with $\sigma \equiv 0$ (if not then we can replace H by $1 + H$ and c by $c + \sigma$). Suppose that $t_{\max} < +\infty$ and define $\|d\| = \sup\limits_{0 \leq t \leq t_{\max}} \left(|d(t)|\right)$.

Inequality (2.49) and definition $\|d\| = \sup\limits_{0 \leq t \leq t_{\max}} \left(|d(t)|\right)$ as well as the fact that c is non-decreasing give the following differential inequality for $t \in [0, t_{\max})$ a.e.:

$$\frac{d}{dt}\left(H(x(t))\right) \leq c\left(\|d\|\right) H(x(t)) \tag{2.51}$$

Integrating (2.51), we obtain that

$$H(x(t)) \leq \exp\left(c\left(\|d\|\right) t\right) H(x_0) \text{ for all } t \in [0, t_{\max}) \tag{2.52}$$

which implies that

$$H(x(t)) \leq \exp\left(c\left(\|d\|\right) t_{\max}\right) H(x_0) \text{ for all } t \in [0, t_{\max}) \tag{2.53}$$

Let $r = \exp\left(c\left(\|d\|\right) t_{\max}\right) H(x_0) + \|d\|$ and notice that definitions $\|d\| = \sup\limits_{0 \leq t \leq t_{\max}} \left(|d(t)|\right)$, (2.7), (2.50) and estimate (2.53) imply that $(x(t), d(t)) \in S_r \times D_r$ for $t \in [0, t_{\max})$ a.e.. Since by assumption, $f(S_r \times D_r)$ is bounded, it follows that there exists a constant $M > 0$, such that

$$|\dot{x}(t)| \leq M \text{ for } t \in [0, t_{\max}) \text{ a.e.} \tag{2.54}$$

Inequality (2.54) implies that

$$|x(t)-x_0| \le M\,t \text{ for all } t \in [0,t_{\max}) \tag{2.55}$$

and consequently (using the triangle inequality)

$$|x(t)| \le |x_0| + M\,t_{\max} \text{ for all } t \in [0,t_{\max}) \tag{2.56}$$

Therefore, estimate (2.56) shows that $\limsup\limits_{t\to t_{\max}^-}(|x(t)|) = +\infty$ cannot hold. Consequently, $t_{\max} = +\infty$. The proof is complete. ◁

Discussion of Theorem 2.4: Theorem 2.4 in the same spirit with the results in [6] for systems defined on $\mathbb{R}^n$ (it is a generalization of certain results in [6]) and with the results in [110] for abstract systems defined on normed linear spaces. However, in contrast with the results in [6], Theorem 2.4 does not require the set $S_r = \{\, x \in \mathbb{R}^n : H(x) \le r \,\}$ to be bounded for all $r > \inf\{\, H(x) : x \in \mathbb{R}^n \,\}$, i.e., Theorem 2.4 does not require H to be radially unbounded. Instead, Theorem 2.4 requires boundedness of the set $f(S_r \times D_r)$ for all $r > \inf\{\, H(x) : x \in \mathbb{R}^n \,\}$; this condition is automatically satisfied if H is radially unbounded (notice that in this case $S_r = \{\, x \in \mathbb{R}^n : H(x) \le r \,\}$ is compact for all $r > \inf\{\, H(x) : x \in \mathbb{R}^n \,\}$ and since $D \subseteq \mathbb{R}^p$ is closed it follows that $D_r = \{\, d \in D : |d| \le r \,\}$ is compact for all $r > \inf\{\, H(x) : x \in \mathbb{R}^n \,\}$).

Example 2.5: Consider the system

$$\begin{aligned} &\dot{x} = f(x,d) \\ &x \in \mathbb{R}^n, d \in \mathbb{R}^p \end{aligned} \tag{2.57}$$

where $f \in C^1(\mathbb{R}^n \times \mathbb{R}^p)$ is a mapping that satisfies the following estimate for all $x \in \mathbb{R}^n, d \in \mathbb{R}^p$:

$$|f(x,d)| \le (L + \kappa(|d|))\,|x| + \kappa(|d|) \tag{2.58}$$

where $\kappa \in K_\infty$ and $L > 0$ is a constant. Define the radially unbounded function

$$H(x) = \frac{1}{2}|x|^2 \tag{2.59}$$

for which we get (using (2.50) and completing the squares) for all $x \in \mathbb{R}^n, d \in \mathbb{R}^p$

$$\begin{aligned} \dot{H}(x,d) &= \nabla H(x) f(x,d) = x' f(x,d) \\ &\le |x|\,|f(x,d)| \le (L+\kappa(|d|))\,|x|^2 + |x|\,\kappa(|d|) \\ &\le \left(L+\kappa(|d|)+\tfrac{1}{2}\right)|x|^2 + \tfrac{1}{2}\kappa^2(|d|) \\ &= (2L + 2\kappa(|d|)+1)\,H(x) + \tfrac{1}{2}\kappa^2(|d|) \end{aligned} \tag{2.60}$$

It follows from (2.60) that (2.49) holds for appropriate non-decreasing functions $c,\sigma \in C^0(\mathbb{R}_+;\mathbb{R}_+)$. We conclude from Theorem 2.4 that system (2.57) is forward complete. ◁

The p-UBIBS property is a more demanding property than simple forward completeness because the p-UBIBS property requires boundedness of solutions for all bounded inputs. Therefore, there are systems which are forward complete but do not satisfy the p-UBIBS property (e.g., the system $\dot{x} = d$ with $x \in \mathbb{R}, d \in \mathbb{R}$). The following result provides sufficient conditions for the p-UBIBS property.

Theorem 2.5 (Sufficient Lyapunov-like conditions for the p-UBIBS property): *Let $H_i \in C^1(\mathbb{R}^n;\mathbb{R}_+)$, $i = 1,\dots,m$ be a family of functions for which there exists a non-decreasing function $R \in C^0(\mathbb{R}_+;\mathbb{R}_+)$ such that the following implication holds for all $(x,d) \in \mathbb{R}^n \times D$ and $i = 1,\dots,m$:*

$$H_i(x) \ge R(|d|) \Rightarrow \dot{H}_i(x,d) = \nabla H_i(x) f(x,d) \le 0 \tag{2.61}$$

Furthermore, suppose that the function $H(x) = \max_{i=1,\dots,m}(H_i(x))$ *is a radially unbounded function. Then system (2.36) satisfies the p-UBIBS property. Moreover, if in addition the function* $H(x) = \max_{i=1,\dots,m}(H_i(x))$ *is positive definite and* $R(0) = 0$ *then system (2.36) satisfies the UBIBS property.*

Proof: Let arbitrary $x_0 \in \mathbb{R}^n$ and $d \in L^\infty(\mathbb{R}_+; D)$ be given. Since f is locally Lipschitz with respect to $x \in \mathbb{R}^n$, the initial-value problem $\dot{x} = f(x, d)$ with initial condition $x(0) = x_0$ has a unique solution $x(t) \in \mathbb{R}^n$ defined for $t \in [0, t_{\max})$, where $t_{\max} \in (0, +\infty]$. Furthermore, if $t_{\max} < +\infty$ then $\limsup_{t \to t_{\max}^-}(|x(t)|) = +\infty$.

We claim that the following estimate holds for all $t \in [0, t_{\max})$ and $i = 1, \dots, m$:

$$H_i(x(t)) \le \max\left(H_i(x_0), R\left(\|d\|_\infty\right)\right) \tag{2.62}$$

The proof is made by contradiction. Suppose that there exists $i = 1, \dots, m$ and $T \in [0, t_{\max})$ such that $H_i(x(T)) > \max\left(H_i(x_0), R\left(\|d\|_\infty\right)\right)$. Since $H_i(x(0)) \le \max\left(H_i(x_0), R\left(\|d\|_\infty\right)\right)$, by continuity of the mapping $t \mapsto H_i(x(t))$ the set

$$\left\{ t \in [0, T] : H_i(x(t)) = \max\left(H_i(x_0), R\left(\|d\|_\infty\right)\right) \right\} \tag{2.63}$$

is non-empty. Define $\tau = \sup\left\{ t \in [0, T] : H_i(x(t)) = \max\left(H_i(x_0), R\left(\|d\|_\infty\right)\right) \right\}$ and notice that continuity of the mapping $t \mapsto H_i(x(t))$ implies that $H_i(x(\tau)) = \max\left(H_i(x_0), R\left(\|d\|_\infty\right)\right)$ and $\tau < T$. Definition of τ guarantees that $H_i(x(t)) \ge \max\left(H_i(x_0), R\left(\|d\|_\infty\right)\right)$ for all $t \in [\tau, T]$. Thus, we obtain from (2.61) and the fact that $R \in C^0(\mathbb{R}_+; \mathbb{R}_+)$ is a non-decreasing function for $t \in [\tau, T]$ a.e.:

$$\frac{d}{dt}\left(H_i(x(t))\right) \le 0 \tag{2.64}$$

Inequality (2.64) implies that $H_i(x(T)) \le H_i(x(\tau)) = \max\left(H_i(x_0), R\left(\|d\|_\infty\right)\right)$, which contradicts the estimate $H_i(x(T)) > \max\left(H_i(x_0), R\left(\|d\|_\infty\right)\right)$.

The fact that $H(x) = \max_{i=1,\dots,m}(H_i(x))$ and (2.62) implies that the following estimate holds for all $t \in [0, t_{\max})$:

$$H(x(t)) \le \max\left(H(x_0), R\left(\|d\|_\infty\right)\right) \tag{2.65}$$

Since $H \in C^0(\mathbb{R}^n; \mathbb{R}_+)$ is a radially unbounded function we can guarantee from (2.65) that $t_{\max} = +\infty$.

Since H is a radially unbounded function the minimum $b = \min\left\{ H(x) : x \in \mathbb{R}^n \right\}$ is well-defined and satisfies the inequality $b \ge 0$. We define the non-decreasing function $a(s) := \max\left\{ |x| : H(x) \le s \right\}$ for $s \ge b$, which is well-defined by virtue of the fact that H is a radially unbounded function. Then, we get $|x| \le a\left(H(x)\right)$ for all $x \in \mathbb{R}^n$. The function $\tilde{a}(s) := \int_s^{s+1} a(p)dp$ is a continuous, non-decreasing function defined on $[b, +\infty)$ that satisfies $a(s) \le \tilde{a}(s)$ for all $s \ge b$. Consequently, we get $|x| \le \tilde{a}\left(H(x)\right)$ for all $x \in \mathbb{R}^n$. Using (2.65) we get the following estimate for all $t \ge 0$:

$$|x(t)| \le \tilde{a}\left(\max\left(H(x_0), R\left(\|d\|_\infty\right)\right)\right) \tag{2.66}$$

It follows from the above estimate that (2.48) holds with $B(x, s) = \tilde{a}\left(\max\left(H(x), R(s)\right)\right)$ for all $(x, s) \in \mathbb{R}^n \times \mathbb{R}_+$. Therefore, the p-UBIBS property holds for (2.43).

If in addition the function $H(x) = \max_{i=1,\dots,m}(H_i(x))$ is positive definite and $R(0) = 0$ then Lemma 3.5 on page 138 in [80] implies the existence of functions $a_1, a_2 \in K_\infty$ such that the following inequality holds for all $x \in \mathbb{R}^n$:

$$a_1(|x|) \leq H(x) \leq a_2(|x|) \tag{2.67}$$

Combining the above inequality with (2.65) we obtain the following estimate for all $t \geq 0$:

$$\begin{aligned}|x(t)| &\leq \max\left(a_1^{-1}(a_2(|x_0|)), a_1^{-1}(R(\|d\|_\infty))\right) \\ &\leq a_1^{-1}(a_2(|x_0|)) + a_1^{-1}(R(\|d\|_\infty))\end{aligned} \tag{2.68}$$

The above estimate shows the validity of (2.47) with $\bar{\alpha} = 0$ and $\bar{\gamma}(s) = a_1^{-1}(a_2(s)) + a_1^{-1}(R(s))$. Since $\bar{\gamma} \in K_\infty$ (recall that $R(0) = 0$), we conclude that the UBIBS property holds for (2.43). The proof is complete. $\triangleleft$

Example 2.6: Consider the following system:

$$\begin{aligned}\dot{x}_1 &= p(x,d)\left(\left(1 + d - x_1^2\right)x_1 + d\right) \\ \dot{x}_2 &= -q(x,d)(x_2 + d) \\ x &= (x_1, x_2) \in \mathbb{R}^2, d \in \mathbb{R}\end{aligned} \tag{2.69}$$

where $p, q : \mathbb{R}^2 \times \mathbb{R} \to \mathbb{R}_+$ are non-negative, locally Lipschitz functions. Define the functions

$$H_i(x) = \frac{1}{2}x_i^2, i = 1, 2 \tag{2.70}$$

for which we have for all $(x, d) \in \mathbb{R}^2 \times \mathbb{R}$

$$\begin{aligned}\dot{H}_1(x,d) &= \nabla H_1(x)\dot{x} = p(x,d)\left(\left(1 + d - x_1^2\right)x_1^2 + x_1 d\right) \\ &\leq p(x,d)\left((1 + |d|)x_1^2 - x_1^4 + |d||x_1|\right) \\ &\leq p(x,d)\left(2(1 + |d|)H_1(x) - 4H_1^2(x) + \tfrac{1}{2}|d|^2 + H_1(x)\right) \\ &\leq p(x,d)\left(2(|d| + 2 - 2H_1(x))H_1(x) + \tfrac{1}{2}|d|^2 - H_1(x)\right)\end{aligned} \tag{2.71}$$

It follows that when $H_1(x) \geq \frac{1}{2}|d|^2$ and $H_1(x) \geq \frac{|d|+2}{2}$ then $\nabla H_1(x)\dot{x} \leq 0$. Moreover, we also have for all $(x, d) \in \mathbb{R}^2 \times \mathbb{R}$

$$\dot{H}_2(x,d) = \nabla H_2(x)\dot{x} = -q(x,d)\left(x_2^2 + x_2 d\right) \leq -\frac{q(x,d)}{2}\left(2H_2(x) - d^2\right) \tag{2.72}$$

It follows that when $H_2(x) \geq |d|^2/2$ then $\nabla H_2(x)\dot{x} \leq 0$.

Therefore, (2.61) holds with $R(s) = \max\left(s + 2, s^2\right)/2$. Since $H(x) = \max_{i=1,2}(H_i(x)) = \max\left(x_1^2, x_2^2\right)/2$ is a radially unbounded function, Theorem 2.5 implies that the p-UBIBS property holds for system (2.69). $\triangleleft$

The following theorem provides sufficient conditions for the p-IOS property. It is a direct consequence of Lemma 2.14 on page 82 and Remark 2.4 on page 85 in [66].

Theorem 2.6 (Lyapunov conditions for p-IOS): *Suppose that* (2.43) *is forward complete and there exist functions $a \in K_\infty$, a non-decreasing, continuous function $\varphi : \mathbb{R}_+ \to \mathbb{R}_+$, $V_i \in C^1(\mathbb{R}^n; \mathbb{R}_+)$ $i = 1, ..., m$ with $V_i(0) = 0$ for $i = 1, ..., m$, and continuous, positive definite functions $\rho_i : \mathbb{R}_+ \to \mathbb{R}_+$, $i = 1, ..., m$ for which the following inequality holds for all $x \in \mathbb{R}^n$:*

$$a(|h(x)|) \leq \max_{i=1,...,m}(V_i(x)) \tag{2.73}$$

and the following implication holds for all $x \in \mathbb{R}^n$, $d \in D$, $i = 1, ..., m$:

$$V_i(x) \geq \varphi(|d|) \Rightarrow \dot{V}_i(x,d) = \nabla V_i(x) f(x,d) \leq -\rho_i(V_i(x)) \tag{2.74}$$

Then system (2.43), (2.2) *satisfies the p-IOS property with gain function* $\gamma(s) = a^{-1}(\varphi(s)) - a^{-1}(\varphi(0))$ *and residual constant* $\alpha = a^{-1}(\varphi(0))$. *Moreover, if* $\varphi(0) = 0$ *then system* (2.43), (2.2) *satisfies the IOS property with gain function* $\gamma = a^{-1} \circ \varphi$.

Example 2.7: Consider the following system:

$$\begin{aligned} \dot{y} &= -\left(z + k + \sigma(\theta + z)^2\right) y + d \\ \dot{z} &= y^2 \\ x &= (y, z)' \in \mathbb{R}^2, d \in \mathbb{R} \end{aligned} \tag{2.75}$$

where $\theta \in \mathbb{R}$, $\sigma, k > 0$ are constants that satisfy the inequality $k > \frac{1}{4\sigma} + \theta$. The disturbance-free version of system (2.75) was studied in Example 2.4. The radially unbounded function

$$H(x) = y^2/2 + z^2/2 \tag{2.76}$$

satisfies for all $x \in \mathbb{R}^2, d \in \mathbb{R}$ the equation

$$\dot{H}(x,d) = \nabla H(x) f(x,d) = -\left(k + \sigma(\theta + z)^2\right) y^2 + yd \leq -\frac{k}{2} y^2 + \frac{1}{2k} d^2 \tag{2.77}$$

Inequality (2.77) implies that all assumptions of Theorem 2.4 hold for system (2.75). Therefore, system (2.75) is forward complete. The function

$$V(x) = y^2/2 \tag{2.78}$$

satisfies for all $x \in \mathbb{R}^2, d \in \mathbb{R}$ the inequality (using the fact that $z + k + \sigma(\theta + z)^2 \geq k - \frac{1}{4\sigma} - \theta$ for all $z \in \mathbb{R}$)

$$\begin{aligned} &\dot{V}(x,d) = \nabla V(x) f(x,d) = -\left(z + k + \sigma(\theta + z)^2\right) y^2 + dy \\ &\leq -\left(k - \tfrac{1}{4\sigma} - \theta\right) y^2 + |d|\,|y| \end{aligned} \tag{2.79}$$

The function V defined by (2.78) satisfies inequality (2.73) with $a(s) = s^2/2$. Moreover, due to (2.79) we obtain the implication

$$\frac{1}{2\varepsilon^2} |d|^2 \leq V(x) \Rightarrow \dot{V}(x,d) = \nabla V(x) f(x,d) \leq -\varepsilon y^2 = -2\varepsilon V(x) \tag{2.80}$$

where

$$\varepsilon := \frac{1}{2}\left(k - \frac{1}{4\sigma} - \theta\right) > 0 \tag{2.81}$$

Consequently, implication (2.74) holds with $\varphi(s) = \frac{1}{2\varepsilon^2} s^2$, $\rho(s) = 2\varepsilon s$ and $r = 0$. We conclude from Theorem 2.6 that (2.75) satisfies the IOS property with gain function $\gamma(s) = s/\varepsilon^2$. ◁

Small-gain arguments can also be used for the derivation of IOS estimates (see for instance [66]). Here we prefer to give a (vector) Lyapunov-like version of the small-gain theorem instead of the usual trajectory-based version.

Theorem 2.7 (Small-gain conditions for IOS): *Suppose that there exist functions* $a \in K_\infty$, *non-decreasing, continuous functions* $\gamma_1, \gamma_2, \zeta, p : \mathbb{R}_+ \to \mathbb{R}_+$ *with* $\gamma_1(0) = \gamma_2(0) = \zeta(0) = p(0) = 0$, $V_1, V_2, U \in C^1(\mathbb{R}^n; \mathbb{R}_+)$ *with* $V_1(0) = V_2(0) = 0$ *and* U *being positive definite and radially unbounded, a constant* $c \geq 0$, *and a continuous, positive definite function* $\rho : \mathbb{R}_+ \to \mathbb{R}_+$ *for which the following inequalities hold for all* $x \in \mathbb{R}^n$, $d \in D$:

$$a(|h(x)|) \leq \max(V_1(x), V_2(x)) \tag{2.82}$$

$$\dot{U}(x,d) = \nabla U(x) f(x,d) \leq cU(x) + p(V_1(x)) + p(V_2(x)) + \zeta(|d|) \tag{2.83}$$

and the following implications hold for all $x \in \mathbb{R}^n$, $d \in D$:

$$V_1(x) \geq \max\left(\zeta\left(|d|\right), \gamma_1\left(V_2(x)\right)\right) \Rightarrow \dot{V}_1(x,d) = \nabla V_1(x) f(x,d) \leq -\rho\left(V_1(x)\right) \tag{2.84}$$

$$V_2(x) \geq \max\left(\zeta\left(|d|\right), \gamma_2\left(V_1(x)\right)\right) \Rightarrow \dot{V}_2(x,d) = \nabla V_2(x) f(x,d) \leq -\rho\left(V_2(x)\right) \tag{2.85}$$

Moreover, suppose that $\gamma_1\left(\gamma_2(s)\right) < s$ and $\gamma_2\left(\gamma_1(s)\right) < s$ for all $s > 0$. Then system (2.36), (2.2) satisfies the IOS property.

Proof: The proof is a simple application of Theorem 5.3 on page 219 in [66]. More, specifically we apply Theorem 5.3 in [66] with $W(t,x) = \exp(-(c+1)t)U(x)$. Due to the fact that U is positive definite and radially unbounded, there exist functions $a_1, a_2 \in K_\infty$ such that $a_1\left(|x|\right) \leq U(x) \leq a_2\left(|x|\right)$ for all $x \in \mathbb{R}^n$. Using Lemma 3.2 in [66], we obtain a function $q \in K_\infty$ such that $a_1^{-1}(rs) \leq q(r)\, q(s)$ for all $r, s \geq 0$. Defining $\mu(t) = \frac{1}{q(\exp((c+1)t))}$ we conclude that the inequality $q^{-1}\left(\mu(t)s\right) \leq \exp(-(c+1)t) a_1(s)$ holds for all $t, s \geq 0$. Putting everything together, we get the inequality $q^{-1}\left(\mu(t)\,|x|\right) \leq W(t,x) \leq a_2\left(|x|\right)$ for all $t \geq 0$, $x \in \mathbb{R}^n$. Using (2.83) we also obtain the inequality for all $t \geq 0$, $x \in \mathbb{R}^n$, and $d \in D$

$$\frac{\partial W}{\partial t}(t,x) + \frac{\partial W}{\partial x}(t,x) f(x,d) \leq -W(t,x) + p\left(V_1(x)\right) + p\left(V_2(x)\right) + \zeta\left(|d|\right) \tag{2.86}$$

The proof is complete. ◁

Example 2.8: Consider the following system:

$$\begin{aligned} \dot{x}_1 &= -x_1 + b x_2^m + d \\ \dot{x}_2 &= -x_2^m + x_1 \\ x &= (x_1, x_2)' \in \mathbb{R}^2, d \in \mathbb{R} \end{aligned} \tag{2.87}$$

where $b \in \mathbb{R}$ is a constant with $|b| < 1$ and m is an odd positive integer. Since $|b| < 1$ there exists $\lambda \in (0,1)$ with $|b| < \lambda^2$. The functions

$$V_i(x) = x_i^2/2, i = 1, 2 \tag{2.88}$$

satisfy inequality (2.82) with $h(x) = x$ and $a(s) = s^2/4$. Moreover, we have the following implications for all $x \in \mathbb{R}^2, d \in \mathbb{R}$:

$$\max\left(\frac{2}{(1-\lambda)^2}|d|^2, \frac{2^{m-1}b^2}{\lambda^2} V_2^m(x)\right) \leq V_1(x) \Rightarrow \dot{V}_1(x,d) = \nabla V_1(x)\dot{x} \leq -(1-\lambda)V_1(x) \tag{2.89}$$

$$\frac{2^{(1-m)/m}}{\lambda^{2/m}} V_1^{1/m}(x) \leq V_2(x) \Rightarrow \dot{V}_2(x,d) = \nabla V_2(x)\dot{x} \leq -2(1-\lambda)V_2(x) \tag{2.90}$$

It follows that implications (2.84), (2.85) hold with $\rho(s) = (1-\lambda)s$, $\gamma_1(s) = \frac{2^{m-1}b^2}{\lambda^2} s^m$, $\gamma_2(s) = \frac{2^{(1-m)/m}}{\lambda^{2/m}} s^{1/m}$, and $\zeta(s) = \frac{2}{(1-\lambda)^2} s^2$ for $s \geq 0$. The small-gain conditions $\gamma_1\left(\gamma_2(s)\right) < s$ and $\gamma_2\left(\gamma_1(s)\right) < s$ for all $s > 0$ hold when $|b| < \lambda^2$. Finally, the positive definite and radially unbounded function $U(x) = V_1(x) + V_2(x)$ satisfies for all $x \in \mathbb{R}^2, d \in \mathbb{R}$

$$\begin{aligned} &\dot{U}(x,d) = \nabla U(x) f(x,d) = -x_1^2 + b x_1 x_2^m + d x_1 - x_2^{m+1} + x_1 x_2 \\ &\leq \tfrac{b^2+1}{2} x_1^2 + \tfrac{1}{2} x_2^{2m} + \tfrac{1}{2} x_2^2 + \tfrac{2}{(1-\lambda)^2}|d|^2 + \tfrac{(1-\lambda)^2}{8} x_1^2 \end{aligned} \tag{2.91}$$

The above inequality shows that (2.83) holds for appropriate $p \in K_\infty$ and appropriate $c \geq 0$. We conclude from Theorem 2.7 that system (2.87) is ISS. ◁

The zero p-OAG property is a novel property that has not been studied in the literature. The only result that allows the proof of the zero p-OAG property deals with a specific class of systems:

systems with deadzone. The following theorem is the foundation upon which we build the robust adaptive designs in our book. It employs what at first might seem like an unusual form of a deadzone acting on the Lyapunov function, with an update gain $\exp(-z(t))$ that decays as the gain $z(t)$ being updated increases. With the analysis tool presented in this theorem we are able to construct robust adaptive designs which achieve the following properties: practical regulation to an arbitrarily small set for the plant state, regulation to an even smaller set if the asymptotic values of the disturbance and the unknown parameter are small, with the set size proportional to the small asymptotic value of the disturbance and parameter, an exponential transient decay of the plant state to a residual set proportional to the "historic" size of the disturbance and parameter, and the prevention of the unbounded drift of the adaptive gain $z(t)$.

Theorem 2.8 (Lyapunov conditions for systems with deadzone): *Let $\Theta \subseteq \mathbb{R}^p$, $D \subseteq \mathbb{R}^l$ be closed sets with $0 \in \Theta$ and $0 \in D$. Consider the system*

$$\begin{aligned} \dot{\xi} &= f(\xi, z, \theta, d) \\ \dot{z} &= \Gamma \exp(-z) \left(V(\xi, z) - \varepsilon\right)^{+} \end{aligned} \tag{2.92}$$

where $x = (\xi, z) \in \mathbb{R}^n \times \mathbb{R}$ is the state, $\theta \in \Theta \subseteq \mathbb{R}^p$, $d \in D \subseteq \mathbb{R}^l$ are disturbances, $\Gamma, \varepsilon > 0$ are constants, $f : \mathbb{R}^n \times \mathbb{R} \times \Theta \times D \to \mathbb{R}^n$ is a locally Lipschitz with respect to $(x, z) \in \mathbb{R}^n \times \mathbb{R}$ mapping with $f(0, z, \theta, 0) = 0$ for all $(z, \theta) \in \mathbb{R} \times \Theta$. Let $V \in C^1(\mathbb{R}^n \times \mathbb{R})$ with $V(0, z) = 0$ for all $z \in \mathbb{R}$ and $V(\xi, z) > 0$ for all $\xi \neq 0$, $z \in \mathbb{R}$ be a function that satisfies the following properties:

(i) *For every $M > 0$ the set $\{ (\xi, z) \in \mathbb{R}^n \times [-M, M] \, : \, V(\xi, z) \leq M \}$ is bounded.*

(ii) *There exist functions $\kappa, \lambda, \rho \in K_\infty$ and constants $a, c, \delta > 0$ $b \geq 0$, for which the inequality $\rho(s) \geq cs$ holds for all $s \in [0, \delta]$ and such that the following inequality holds for all $(\xi, z) \in \mathbb{R}^n \times \mathbb{R}, \theta \in \Theta, d \in D$:*

$$\begin{aligned} &\tfrac{\partial V}{\partial \xi}(\xi, z) f(\xi, z, \theta, d) + \Gamma \exp(-z) \tfrac{\partial V}{\partial z}(\xi, z) \left(V(\xi, z) - \varepsilon\right)^{+} \\ &\leq -\rho\left(V(\xi, z)\right) + \chi\left(|d|, |\theta|, \exp(z)\right) \end{aligned} \tag{2.93}$$

where $\chi(s_1, s_2, s_3) = a \frac{s_1^2 + \left((s_2 - b - \lambda(s_3))^{+}\right)^2}{1 + \kappa(s_3)}$ for all $s_1, s_2, s_3 \geq 0$.

Then there exist functions $B \in C^0(\mathbb{R}^n \times \mathbb{R} \times \mathbb{R}_+ \times \mathbb{R}_+)$, $\gamma \in K_\infty$, $\sigma \in KL$ such that for every $(\xi_0, z_0) \in \mathbb{R}^n \times \mathbb{R}$, $d \in L^\infty(\mathbb{R}_+; D)$, $\theta \in L^\infty(\mathbb{R}_+; \Theta)$ the unique solution of (2.92) *with $(\xi(0), z(0)) = (\xi_0, z_0)$ exists and satisfies the following estimates for all $t \geq 0$:*

$$\limsup_{t \to +\infty} \left(V(\xi(t), z(t))\right) \leq \varepsilon, \tag{2.94}$$

$$\limsup_{t \to +\infty} \left(V(\xi(t), z(t))\right) \leq \gamma\left(\chi\left(\limsup_{t \to +\infty}\left(|d(t)|\right), \limsup_{t \to +\infty}\left(|\theta(t)|\right), \exp\left(\lim_{t \to +\infty}\left(z(t)\right)\right)\right)\right) \tag{2.95}$$

$$V(\xi(t), z(t)) \leq \sigma\left(V(\xi_0, z_0), t\right) + \gamma\left(\chi\left(\|d\|_\infty, \|\theta\|_\infty, \exp(z_0)\right)\right) \tag{2.96}$$

$$|\xi(t)| \leq B\left(\xi_0, z_0, \|d\|_\infty, \|\theta\|_\infty\right) \tag{2.97}$$

$$z_0 \leq z(t) \leq \lim_{s \to +\infty}\left(z(s)\right) \leq B\left(\xi_0, z_0, \|d\|_\infty, \|\theta\|_\infty\right) \tag{2.98}$$

Moreover, if the inequality $\rho(s) \geq cs$, holds for all $s \geq 0$ then $\gamma(s) = c^{-1}s$ for all $s \geq 0$ and $\sigma(s, t) = \exp(-ct)s$ for all $t, s \geq 0$.

Remark: Estimate (2.95) shows that $\lim_{t \to +\infty}\left(|\xi(t)|\right) = 0$ whenever $\lim_{t \to +\infty}\left(d(t)\right) = 0$ and $\limsup_{t \to +\infty}\left(|\theta(t)|\right) \leq b + \lambda\left(\exp\left(\lim_{t \to +\infty}\left(z(t)\right)\right)\right)$. Inequality (2.94) shows the zero p-OAG property for system (2.92) with output $y = V(\xi, z)$ with asymptotic residual constant $\varepsilon > 0$. Inequalities (2.97), (2.98) show that system (2.92) satisfies the p-UBIBS property.

Proof: We start by noticing that there exist functions $\sigma \in KL$, $\gamma \in K_\infty$ with the following property: for every $T \in (0, +\infty]$, $\mu \geq 0$ and for every absolutely continuous function $y : [0, T) \to \mathbb{R}_+$ for which the following inequality holds for $t \in [0, T)$ a.e.:

$$\dot{y}(t) \leq -\rho\,(y(t)) + \mu \tag{2.99}$$

the following estimate holds for all $t \in [0, T)$:

$$y(t) \leq \sigma\,(y(0), t) + \gamma\,(\mu) \tag{2.100}$$

The existence of functions $\sigma \in KL$, $\gamma \in K_\infty$ follows from Lemma 2.14 on page 82 in [66] and by noticing that (2.108) shows that the implication

$$y(t) \geq \rho^{-1}\,(2\mu) \Rightarrow \dot{y}(t) \leq -\frac{1}{2}\rho\,(y(t)) \tag{2.101}$$

holds for $t \in [0, T)$ a.e. Moreover, if the inequality $\rho\,(s) \geq cs$, holds for all $s \geq 0$ then by directly integrating (2.99) we get (2.100) with $\gamma\,(s) = c^{-1}s$ for all $s \geq 0$ and $\sigma\,(s, t) = \exp(-ct)s$ for all $t, s \geq 0$.

Define $\bar{c}(r) := \inf\left\{ \frac{\rho(l)}{l} : 0 < l \leq r \right\}$ for all $r > 0$. Notice that continuity of $\rho \in K_\infty$ and the fact that the inequality $\rho\,(l) \geq cl$ holds for all $l \in [0, \delta]$ guarantee that $\bar{c} : (0, +\infty) \to (0, +\infty)$ is a well-defined, non-increasing function that satisfies

$$\rho\,(l) \geq \bar{c}\,(r)\,l \text{ for all } r > 0 \text{ and } l \in [0, r] \tag{2.102}$$

Without loss of generality, we may assume that $\bar{c} : (0, +\infty) \to (0, +\infty)$ is continuous (otherwise replace the function $\bar{c}$ by the function $\tilde{c}(r) := \int_r^{r+1} \bar{c}(l)dl$ defined for $r > 0$). Next define for $r \geq 0$ the non-increasing, continuous, and positive function

$$C(r) := \bar{c}\,(1 + \sigma\,(r, 0) + \gamma\,(ar)) \tag{2.103}$$

Moreover, define for $r \geq 0$

$$A(r) := \kappa^{-1}\left(\left(\frac{ar}{C(r)\varepsilon} - 1\right)^+\right) \tag{2.104}$$

$$G(r) := \ln\left(1 + \max\left(A(r), \exp(r)\right) + \frac{\Gamma}{C(r)}\left(1 + \frac{a}{C(r)}\right)r\right) \tag{2.105}$$

Notice that both A and G are non-decreasing, continuous functions.

Let arbitrary $(\xi_0, z_0) \in \mathbb{R}^n \times \mathbb{R}$, $d \in L^\infty\,(\mathbb{R}_+; D)$, $\theta \in L^\infty\,(\mathbb{R}_+; \Theta)$ be given. Since the functions $f : \mathbb{R}^n \times \mathbb{R} \times \Theta \times D \to \mathbb{R}^n$ and V are locally Lipschitz with respect to $(x, z) \in \mathbb{R}^n \times \mathbb{R}$ there exists an unique solution $x(t) = (\xi(t), z(t))$ of (2.92) with $x(0) = (\xi(0), z(0)) = (\xi_0, z_0)$ defined on a maximal interval $[0, t_{max})$, where $t_{max} \in (0, +\infty]$. For every $t \in [0, t_{max})$, we have from (2.92) that $\dot{z}(t) \geq 0$. Therefore, $z(t)$ is non-decreasing on $[0, t_{max})$. Moreover, the mapping

$$z \mapsto \chi\,(|d|\,, |\theta|\,, \exp(z)) \tag{2.106}$$

is non-increasing. Using (2.93) we conclude that the following inequalities hold almost everywhere in $[0, t_{max})$:

$$\frac{d}{dt}V(\xi(t), z(t)) \leq -\rho\,(V(\xi(t), z(t))) + \chi\,(|d(t)|\,, |\theta(t)|\,, \exp(z(t))) \tag{2.107}$$

which implies

$$\frac{d}{dt}V(\xi(t),z(t)) \leq -\rho\left(V(\xi(t),z(t))\right) + \chi\left(\|d\|_\infty, \|\theta\|_\infty, \exp(z_0)\right) \tag{2.108}$$

It follows from (2.99) and (2.100) that the following estimate holds for all $t \in [0, t_{max})$:

$$V(\xi(t),z(t)) \leq \sigma\left(V(\xi_0,z_0),t\right) + \gamma\left(\chi\left(\|d\|_\infty, \|\theta\|_\infty, \exp(z_0)\right)\right) \tag{2.109}$$

Moreover, if the inequality $\rho(s) \geq cs$ holds for all $s \geq 0$ then $\gamma(s) = c^{-1}s$ for all $s \geq 0$ and $\sigma(s,t) = \exp(-ct)s$ for all $t, s \geq 0$.

It follows from (2.107), (2.102), estimate (2.109) and definition (2.103) that the following inequality holds for $t \in [0, t_{max})$ a.e.:

$$\frac{d}{dt}V(\xi(t),z(t)) \leq -C(s)V(\xi(t),z(t)) + a\frac{\|d\|_\infty^2 + \left(\left(\|\theta\|_\infty - b - \lambda\left(\exp(z_0)\right)\right)^+\right)^2}{1+\kappa(\exp(z(t)))} \tag{2.110}$$

where

$$s := V(\xi_0,z_0) + |z_0| + \|\theta\|_\infty^2 + \|d\|_\infty^2 \tag{2.111}$$

Integrating (2.110) and using the fact that $z(t)$ is non-decreasing on $[0, t_{max})$, we obtain the following estimate for all $t \in [0, t_{max})$:

$$V(\xi(t),z(t)) \leq \exp\left(-C(s)t\right)V(\xi_0,z_0) + \frac{1}{C(s)}\chi\left(\|d\|_\infty, \|\theta\|_\infty, \exp(z_0)\right) \tag{2.112}$$

Moreover, from estimate (2.112) and definition (2.111), we get the following estimate for all $t \in [0, t_{max})$:

$$V(\xi(t),z(t)) \leq V(\xi_0,z_0) + \frac{a}{C(s)}\left(\|\theta\|_\infty^2 + \|d\|_\infty^2\right) \leq s + \frac{a}{C(s)}s \tag{2.113}$$

Next, we show that $z(t)$ is bounded. Since $z(t)$ is non-decreasing on $[0, t_{max})$, we have $z(t) \geq z_0$ for all $t \in [0, t_{max})$. We next distinguish the following cases:

<u>Case 1:</u> $\exp(z(t)) \leq 1 + \max\left(A(s), \exp(s)\right)$ for all $t \in [0, t_{max})$

In this case, it follows from (2.105) that

$$z(t) \leq G(s) \text{ for all } t \in [0, t_{max}) \tag{2.114}$$

<u>Case 2:</u> There exists $T \in (0, t_{max})$ such that $\exp(z(T)) > 1 + \max\left(A(s), \exp(s)\right)$. Since $\exp(z(0)) \leq \exp(s) < 1 + \max\left(A(s), \exp(s)\right)$, continuity of $z(t)$ guarantees that there exist $t_0 \in (0, T)$ such that

$$\exp(z(t_0)) = 1 + \max\left(A(s), \exp(s)\right) \tag{2.115}$$

Since $z(t)$ is non-decreasing, it follows that

$$\exp(z(t)) \geq 1 + \max\left(A(s), \exp(s)\right) \text{ for } t \in [t_0, t_{max}) \tag{2.116}$$

Inequality (2.116) in conjunction with (2.111) and (2.104) gives the following inequalities for $t \in [t_0, t_{max})$:

$$\begin{gathered}\exp(z(t)) \geq A(s) \geq \kappa^{-1}\left(\left(\tfrac{as}{C(s)\varepsilon} - 1\right)^+\right) \geq \kappa^{-1}\left(\left(a\tfrac{\|\theta\|_\infty^2 + \|d\|_\infty^2}{C(s)\varepsilon} - 1\right)^+\right) \\ \Rightarrow \kappa\left(\exp(z(t))\right) \geq \left(a\tfrac{\|\theta\|_\infty^2 + \|d\|_\infty^2}{C(s)\varepsilon} - 1\right)^+ \geq a\tfrac{\|\theta\|_\infty^2 + \|d\|_\infty^2}{C(s)\varepsilon} - 1 \Rightarrow C(s)\varepsilon \geq a\tfrac{\|\theta\|_\infty^2 + \|d\|_\infty^2}{1+\kappa(\exp(z(t)))}\end{gathered} \tag{2.117}$$

Consequently, we obtain from (2.110) for almost all $t \in [t_0, t_{max})$

$$\frac{d}{dt} V(\xi(t), z(t)) \leq -C(s) V(\xi(t), z(t)) + C(s)\varepsilon \tag{2.118}$$

Integrating the above differential inequality, we obtain for all $t \in [t_0, t_{max})$

$$V(\xi(t), z(t)) \leq \exp\left(-C(s)(t - t_0)\right) V(\xi(t_0), z(t_0)) + \varepsilon \tag{2.119}$$

Using (2.92) and inequality (2.119), we have for all $t \in [t_0, t_{max})$

$$\frac{d}{dt}\left(\exp(z(t))\right) \leq \Gamma \exp\left(-C(s)(t - t_0)\right) V(\xi(t_0), z(t_0)) \tag{2.120}$$

Integrating (2.120) and using (2.113) and (2.115), we obtain for all $t \in [t_0, t_{max})$

$$\begin{aligned} \exp(z(t)) &\leq \exp\left(z(t_0)\right) + \tfrac{\Gamma}{C(s)} V(\xi(t_0), z(t_0)) \\ &= 1 + \max\left(A(s), \exp(s)\right) + \tfrac{\Gamma}{C(s)} V(\xi(t_0), z(t_0)) \\ &\leq 1 + \max\left(A(s), \exp(s)\right) + \tfrac{\Gamma}{C(s)}\left(1 + \tfrac{a}{C(s)}\right) s \end{aligned} \tag{2.121}$$

Since $z(t)$ is non-decreasing, inequality (2.121) holds for all $t \in [0, t_{max})$.

It follows from (2.121) and definitions (2.111), (2.105) that in both cases the following estimates hold for all $t \in [0, t_{max})$:

$$z_0 \leq z(t) \leq G(s) \tag{2.122}$$

$$|z(t)| \leq G(s) \tag{2.123}$$

Estimate (2.123) implies that $z(t)$ is bounded. Therefore, assumption (i) in conjunction with (2.109) guarantees that $\xi(t)$ is bounded for all $t \in [0, t_{max})$. Consequently, $t_{max} = +\infty$. Moreover, estimate (2.109) implies directly the required estimate (2.96).

Since $z(t)$ is non-decreasing and bounded from above, $\lim_{\tau \to +\infty}(z(\tau))$ exists. Thus, we obtain from (2.122)

$$z_0 \leq z(t) \leq \lim_{\tau \to +\infty}(z(\tau)) \leq G(s) \tag{2.124}$$

Next, define the following function for $r \geq 0$:

$$H(r) := \max\left\{ |w| : V(w, \zeta) \leq \left(1 + \frac{a}{C(r)}\right) r \text{ and } |\zeta| \leq G(r) \right\} \tag{2.125}$$

Assumption (i) guarantees that H is well-defined. Since G is non-decreasing and C is non-increasing, it follows from (2.125) that H is non-decreasing. Moreover, it follows from estimates (2.113), (2.123) and definitions (2.111), (2.125) that

$$|x(t)| \leq H(s) \text{ for all } t \geq 0 \tag{2.126}$$

Define the continuous function $\tilde{H}(r) := G(r) + \int_r^{r+1} H(l)dl$ for $r \geq 0$ (notice that since H is non-decreasing, it is locally bounded and Riemann-integrable on every bounded interval of $\mathbb{R}_+$). We conclude from (2.126), (2.122), and (2.111) that estimates (2.98), (2.97) hold with $B\left(\xi_0, z_0, \|d\|_\infty, \|\theta\|_\infty\right) = \tilde{H}\left(V(\xi_0, z_0) + |z_0| + \|\theta\|_\infty^2 + \|d\|_\infty^2\right)$.

We next show estimate (2.94). Estimate (2.122) and the fact that $z(t)$ is non-decreasing guarantees that the function $z(t)$ has a finite limit as $t \to +\infty$. This implies that the function $\exp(z(t))$ has a finite limit as $t \to +\infty$. Moreover, the facts that $d \in L^\infty(\mathbb{R}_+; D)$, $\theta \in L^\infty(\mathbb{R}_+; \Theta)$ and (2.123), (2.126) imply that $\frac{d}{dt}\left(V(\xi(t), z(t))\right)$ is of class $L^\infty(\mathbb{R}_+)$. It follows that the function $\frac{d}{dt}(\exp(z(t))) = \Gamma\left(V(\xi(t), z(t)) - \varepsilon\right)^+$ is uniformly continuous, in addition to $\int_0^{+\infty} \frac{d}{dt}(\exp(z(t)))\,dt = \lim_{t\to+\infty}(\exp(z(t))) - \exp(z_0) \le \exp(G(s)) - \exp(z_0) < +\infty$. From Barbălat's Lemma (see [73], [80]), we have

$$\lim_{t\to+\infty}\left(\frac{d}{dt}(\exp(z(t)))\right) = \lim_{t\to+\infty}\left(\Gamma\left(V(\xi(t), z(t)) - \varepsilon\right)^+\right) = 0 \tag{2.127}$$

Therefore, estimate (2.94) holds.

Next, we show estimate (2.95). Let arbitrary $\epsilon > 0$ be given. Then there exists $\bar{T} > 0$ such that

$$|d(t)| \le \limsup_{\tau\to+\infty}(|d(\tau)|) + \epsilon \text{ for } t \ge \bar{T} \text{ a.e.} \tag{2.128}$$

$$|\theta(t)| \le \limsup_{\tau\to+\infty}(|\theta(\tau)|) + \epsilon \text{ for } t \ge \bar{T} \text{ a.e.} \tag{2.129}$$

$$z(t) \ge \lim_{\tau\to+\infty}(z(\tau)) - \epsilon \text{ for all } t \ge \bar{T} \tag{2.130}$$

Define $L_d = \limsup_{\tau\to+\infty}(|d(\tau)|)$, $L_\theta = \limsup_{\tau\to+\infty}(|\theta(\tau)|)$ and $L_z = \lim_{\tau\to+\infty}(z(\tau))$. It follows from (2.107) that the following differential inequality holds for $t \ge \bar{T}$ a.e.:

$$\frac{d}{dt}V(\xi(t), z(t)) \le -\rho(V(\xi(t), z(t))) + \chi\left(L_d + \epsilon, L_\theta + \epsilon, \exp\left(L_z - \epsilon\right)\right) \tag{2.131}$$

We get from (2.99), (2.100), and (2.131) for all $t \ge \bar{T}$

$$V(\xi(t), z(t)) \le \sigma\left(V\left(\xi(\bar{T}), z(\bar{T})\right), t - \bar{T}\right) + \gamma\left(\chi\left(L_d + \epsilon, L_\theta + \epsilon, \exp\left(L_z - \epsilon\right)\right)\right) \tag{2.132}$$

Consequently, we get from estimate (2.132)

$$\limsup_{t\to+\infty}\left(V(\xi(t), z(t))\right) \le \gamma\left(\chi\left(L_d + \epsilon, L_\theta + \epsilon, \exp\left(L_z - \epsilon\right)\right)\right) \tag{2.133}$$

Since (2.133) holds for arbitrary $\epsilon > 0$, we obtain the required estimate (2.95). The proof is complete. $\lhd$

Example 2.9: This example conveys the essence of the adaptive robust stability studies in this book. It covers the closed-loop dynamics for a scalar linear plant. Consider the following system:

$$\begin{aligned} &\dot{\xi} = \theta\xi - \left(c + \tfrac{(1+\exp(z))^2}{2c} + \tfrac{1+\exp(z)}{4a}\left(1+\xi^2\right)\right)\xi + d \\ &\dot{z} = \Gamma\exp(-z)\left(\tfrac{1}{2}\xi^2 - \varepsilon\right)^+ \\ &x = (\xi, z)' \in \mathbb{R}^2, d \in \mathbb{R}, \theta \in \mathbb{R} \end{aligned} \tag{2.134}$$

where $a, c, \Gamma, \varepsilon > 0$ are constants. The function $V(\xi, z) = \xi^2/2$ satisfies for all $(\xi, z) \in \mathbb{R}^2$, $\theta, d \in \mathbb{R}$

$$\begin{aligned} &\tfrac{\partial V}{\partial \xi}(\xi, z)\dot{\xi} + \tfrac{\partial V}{\partial z}(\xi, z)\dot{z} = \theta\xi^2 - \left(c + \tfrac{(1+\exp(z))^2}{2c} + \tfrac{1+\exp(z)}{4a}\left(1+\xi^2\right)\right)\xi^2 + \xi d \\ &\le |\theta|\,\xi^2 - \left(c + \tfrac{(1+\exp(z))^2}{2c} + \tfrac{1+\exp(z)}{4a}\left(1+\xi^2\right)\right)\xi^2 + \tfrac{a|d|^2}{1+\exp(z)} + \tfrac{1+\exp(z)}{4a}\xi^2 \\ &= \left(|\theta| - 1 - \exp(z)\right)\xi^2 + \left(1+\exp(z)\right)\xi^2 - \left(c + \tfrac{(1+\exp(z))^2}{2c} + \tfrac{1+\exp(z)}{4a}\xi^2\right)\xi^2 + \tfrac{a|d|^2}{1+\exp(z)} \\ &\le \left(|\theta| - 1 - \exp(z)\right)^+\xi^2 + \left(1+\exp(z)\right)\xi^2 - \left(c + \tfrac{(1+\exp(z))^2}{2c} + \tfrac{1+\exp(z)}{4a}\xi^2\right)\xi^2 + \tfrac{a|d|^2}{1+\exp(z)} \\ &\le a\tfrac{\left(\left(|\theta|-1-\exp(z)\right)^+\right)^2}{1+\exp(z)} + \tfrac{1+\exp(z)}{4a}\xi^4 + \tfrac{c}{2}\xi^2 + \tfrac{(1+\exp(z))^2}{2c}\xi^2 \\ &-\left(c + \tfrac{(1+\exp(z))^2}{2c} + \tfrac{1+\exp(z)}{4a}\xi^2\right)\xi^2 + \tfrac{a|d|^2}{1+\exp(z)} \\ &= -\tfrac{c}{2}\xi^2 + a\tfrac{|d|^2 + \left(\left(|\theta|-1-\exp(z)\right)^+\right)^2}{1+\exp(z)} = -cV(\xi, z) + a\tfrac{|d|^2 + \left(\left(|\theta|-1-\exp(z)\right)^+\right)^2}{1+\exp(z)} \end{aligned} \tag{2.135}$$

Consequently, inequality (2.93) holds with $b = 1$, $\kappa(s) = \lambda(s) = s$ for $s \geq 0$. It follows that system (2.134) satisfies all assumptions of Theorem 2.8. Therefore, there exists a function $B \in C^0(\mathbb{R} \times \mathbb{R} \times \mathbb{R}_+ \times \mathbb{R}_+)$ such that for every $(\xi_0, z_0) \in \mathbb{R} \times \mathbb{R}$, $d \in L^\infty(\mathbb{R}_+; \mathbb{R})$, $\theta \in L^\infty(\mathbb{R}_+; \mathbb{R})$ the unique solution of (2.134) with $(\xi(0), z(0)) = (\xi_0, z_0)$ that corresponds to $d \in L^\infty(\mathbb{R}_+; \mathbb{R})$, $\theta \in L^\infty(\mathbb{R}_+; \mathbb{R})$ exists for all $t \geq 0$ and satisfies the following estimates for all $t \geq 0$:

$$\limsup_{t \to +\infty} (|\xi(t)|) \leq \sqrt{2\varepsilon} \tag{2.136}$$

$$|\xi(t)| \leq \exp(-ct/2)\,|\xi_0| + \sqrt{\frac{2a}{c\,(1+\exp(z_0))}}\left(\|d\|_\infty + (\|\theta\|_\infty - 1 - \exp(z_0))^+\right) \tag{2.137}$$

$$\begin{aligned} &\limsup_{t \to +\infty} (|\xi(t)|) \\ &\leq \sqrt{\frac{2a}{c\left(1+\exp\left(\lim_{t\to+\infty}(z(t))\right)\right)}}\,\limsup_{t \to +\infty}(|d(t)|) \\ &\leq \sqrt{\frac{2a}{c\left(1+\exp\left(\lim_{t\to+\infty}(z(t))\right)\right)}}\left(\limsup_{t \to +\infty}(|\theta(t)|) - 1 - \exp\left(\lim_{t\to+\infty}(z(t))\right)\right)^+ \end{aligned} \tag{2.138}$$

as well as estimates (2.98), (2.97). When $\theta \in \mathbb{R}$ is a constant, the stability properties of system (2.134) with output $y = \xi$ can be stated with precision:

- estimates (2.98), (2.97) guarantee the p-UBIBS property,
- estimate (2.136) guarantees the zero p-OAG property with asymptotic residual constant equal to $\sqrt{2\varepsilon}$,
- estimate (2.137) guarantees the p-IOS property with gain function $\gamma(s) = s\sqrt{2a/c}$ and residual constant $\alpha = (|\theta| - 1)^+ \sqrt{2a/c}$.

When $\theta \in \mathbb{R}$ is a constant and in the disturbance-free case (i.e., when $d \equiv 0$), the stability properties of system (2.134) with output $y = \xi$ can also be stated with precision:

- estimates (2.98), (2.97) guarantee the Lagrange stability property,
- the set $\{(0,z)' \in \mathbb{R}^2, z \in \mathbb{R}\}$ is invariant for (2.134); however, it should be noticed that (depending on the value of $\theta \in \mathbb{R}$) there may be equilibrium points of system (2.134) out of the set $\{(0,z)' \in \mathbb{R}^2, z \in \mathbb{R}\}$,
- estimate (2.137) guarantees the p-UGAOS property and residual constant

$$\alpha = (|\theta| - 1)^+ \sqrt{2a/c} \tag{2.139}$$

- estimates (2.136), (2.137) show that system (2.134) with output $y = \xi$ is p-GOA with asymptotic residual constant equal to $\tilde{\alpha} = \min\left(\sqrt{2\varepsilon}, (|\theta| - 1)^+ \sqrt{2a/c}\right) \leq \alpha$,
- estimates (2.136) and (2.138) guarantee the estimate

$$\limsup_{t \to +\infty} (|\xi(t)|) \leq \min\left(\sqrt{2\varepsilon}, \sqrt{\frac{2a}{c\,(1+l)}}\,(|\theta| - 1 - l)^+\right) \tag{2.140}$$

where $l = \exp\left(\lim_{t\to+\infty}(z(t))\right)$; thus, we can obtain an explicit condition for the convergence of the output to zero, namely, $\lim_{t\to+\infty}(\xi(t)) = 0$ when $|\theta| \le 1 + l$. Consequently, we get the following result:

"At least one of the following holds:

1. $\lim_{t\to+\infty}(\xi(t)) = 0$,
2. $|\theta| > 1$ and $z(t) \le \lim_{s\to+\infty}(z(s)) < \ln(|\theta| - 1)$ for all $t \ge 0$."

It is clear that system (2.134) with output $y = \xi$ is rich of stability properties. ◁

We end this section by providing sufficient conditions for the p-OAG property. The following theorem is a novel result which is close in spirit to LaSalle's theorem (see [80]), which is not applicable to systems with inputs. In order to overcome the presence of inputs the following result exploits a family of functions (instead of a single function).

Theorem 2.9 (LaSalle-like conditions for the p-OAG property): *Suppose that for every $x_0 \in \mathbb{R}^n$ and $d \in L^\infty(\mathbb{R}_+; D)$ the solution of (2.43) with initial condition $x(0) = x_0$ is bounded. Moreover, suppose that there exists a non-decreasing function $b \in C^0(\mathbb{R}_+; \mathbb{R}_+)$ and families of functions $H_s \in C^1(\mathbb{R}^n)$, $Q_s \in C^0(\mathbb{R}^n; \mathbb{R}_+)$, parameterized by $s \ge 0$, such that the following properties hold for all $s \ge 0$, $x \in \mathbb{R}^n$:*

$$Q_s(x) = 0 \Rightarrow |h(x)| \le b(s) \tag{2.141}$$

$$\max\{\nabla H_s(x) f(x,d) \,:\, d \in D,\ |d| \le s\} \le -Q_s(x) \tag{2.142}$$

Then system (2.43), (2.2) satisfies the p-OAG property with asymptotic residual constant $\tilde{\alpha} = b(0)$ and asymptotic gain function $\tilde{\gamma}(s) = b(s) - b(0)$.

Before we proceed with the proof of Theorem 2.9 it is worth noticing some things about Theorem 2.9. First, the family of functions $H_s \in C^1(\mathbb{R}^n)$ parameterized by $s \ge 0$ is not a family of positive definite functions. Second, we say that $H_s \in C^1(\mathbb{R}^n)$, $Q_s \in C^0(\mathbb{R}^n; \mathbb{R}_+)$ are families of functions $H_s \in C^1(\mathbb{R}^n)$, $Q_s \in C^0(\mathbb{R}^n; \mathbb{R}_+)$, parameterized by $s \ge 0$, instead of saying that H, Q are functions of $(x,s) \in \mathbb{R}^n \times \mathbb{R}_+$ because there is no regularity requirement with respect to $s \ge 0$; the functions $H_s \in C^1(\mathbb{R}^n)$, $Q_s \in C^0(\mathbb{R}^n; \mathbb{R}_+)$ are not assumed to be continuous (or even measurable) with respect to $s \ge 0$. When (2.73) and

$$V(x) \ge \varphi(|d|) \Rightarrow \dot{V}(x,d) = \nabla V(x) f(x,d) \le -c(x,|d|) \tag{2.143}$$

hold for certain functions $a \in K_\infty$, $V \in C^1(\mathbb{R}^n; \mathbb{R}_+)$, a non-decreasing, continuous function $\varphi : \mathbb{R}_+ \to \mathbb{R}_+$, and a locally Lipschitz function $c : \mathbb{R}^n \times \mathbb{R}_+ \to \mathbb{R}_+$ with $c(x,s) > 0$ for all $(x,s) \in \mathbb{R}^n \times \mathbb{R}_+$ with $V(x) > \varphi(s)$, then (2.141), (2.142) hold with

$$H_s(x) = \tfrac{1}{2}\left((V(x) - \varphi(s))^+\right)^2,\ Q_s(x) = (V(x) - \varphi(s))^+ \min\{c(x,l) \,:\, 0 \le l \le s\}$$
$$\text{and } b(s) = a^{-1}(\varphi(s))$$

Proof of Theorem 2.9: Let arbitrary $x_0 \in \mathbb{R}^n$, $d \in L^\infty(\mathbb{R}_+; D)$ be given and consider the bounded solution $x(t) \in \mathbb{R}^n$ of (2.43) with initial condition $x(0) = x_0$. Define $s = \|d\|_\infty$. Since the solution is bounded, there exists $R \in \mathbb{R}$ such that $H_s(x(t)) \ge R$ for all $t \ge 0$. Exploiting (2.142) and the fact that $s = \|d\|_\infty$ we get for all $t \ge 0$

$$\int_0^t Q_s(x(\tau))\,d\tau \le H_s(x(0)) - H_s(x(t)) \le H_s(x(0)) - R \tag{2.144}$$

Moreover, the facts that $d \in L^{\infty}(\mathbb{R}_+; D)$, $x(t) \in \mathbb{R}^n$ is bounded and (2.43) imply that $\dot{x}(t)$ is of class $L^{\infty}(\mathbb{R}_+)$. Since $Q_s \in C^0(\mathbb{R}^n; \mathbb{R}_+)$ is continuous, it is also uniformly continuous on a compact set that contains the bounded trajectory $\{x(t) \in \mathbb{R}^n : t \geq 0\}$. It follows that the function $t \mapsto Q_s(x(t))$ is uniformly continuous. From Barbălat's Lemma (see [73], [80]), we have

$$\lim_{t \to +\infty} (Q_s(x(t))) = 0 \tag{2.145}$$

We next claim that

$$\limsup_{t \to +\infty} (|y(t)|) \leq b(\|d\|_{\infty}) \tag{2.146}$$

We prove (2.146) by contradiction. Suppose that $\limsup_{t \to +\infty} (|y(t)|) > b(\|d\|_{\infty})$. Then by virtue of (2.2) there exists $\varepsilon > 0$ and a sequence of times $\{t_i \geq 0 : i = 1, 2, ...\}$ with $\lim_{i \to +\infty} (t_i) = +\infty$ such that $|y(t_i)| = |h(x(t_i))| \geq b(\|d\|_{\infty}) + \varepsilon$. Since the solution $x(t) \in \mathbb{R}^n$ is bounded, it follows that the sequence $\{x(t_i) \in \mathbb{R}^n : i = 1, 2, ...\}$ is bounded. By taking a subsequence if necessary, we may assume that there exists $x^* \in \mathbb{R}^n$ such that $\lim_{i \to +\infty} (x(t_i)) = x^*$. Consequently, by exploiting continuity of h, Q_s, (2.145), and the fact that $s = \|d\|_{\infty}$ we get $|h(x^*)| \geq b(\|d\|_{\infty}) + \varepsilon = b(s) + \varepsilon > b(s)$ and $Q_s(x^*) = 0$. Clearly, this contradicts (2.141).

Inequality (2.146) and the fact that $x_0 \in \mathbb{R}^n$, $d \in L^{\infty}(\mathbb{R}_+; D)$ are arbitrary imply the p-OAG property with asymptotic residual constant $\tilde{\alpha} = b(0)$ and asymptotic gain function $\tilde{\gamma}(s) = b(s) - b(0)$. The proof is complete. $\triangleleft$

Example 2.10: Consider the following planar system:

$$\begin{aligned} \dot{\xi} &= 0 \\ \dot{y} &= -yp(\xi, y, d) + (1 - y^2)^+ q(\xi, y, d) \\ x &= (\xi, y)' \in \mathbb{R}^2, d \in \mathbb{R} \end{aligned} \tag{2.147}$$

where $p, q : \mathbb{R}^3 \to \mathbb{R}$ are locally Lipschitz functions that satisfy $q(0,0,0) = 0$ and $p(\xi, y, d) > 0$ for all $(\xi, y, d) \in \mathbb{R}^3$. System (2.147) is not necessarily a system that satisfies the IOS property (e.g., consider the input-free case $p(\xi, y, d) \equiv 1$ and $q(\xi, y, d) = 2y$). However, system (2.147) satisfies the p-UBIBS property (use Theorem 2.5 with $H_1(x) = \xi^2/2$, $H_2(x) = y^2/2$ and $R(s) \equiv 1/2$).

The function

$$V(\xi, y) = y^2/2 \tag{2.148}$$

gives for all $(x, d) \in \mathbb{R}^2 \times \mathbb{R}$

$$\dot{V}(\xi, y, d) = -y^2 p(\xi, y, d) + y(1 - y^2)^+ q(\xi, y, d) \tag{2.149}$$

It follows that (2.143) holds with $\varphi(s) = 1/2$ and

$$c(\xi, y, |d|) = y^2 (V(\xi, y) - 1/2)^+ \min(p(\xi, y, |d|), p(\xi, y, -|d|)). \tag{2.150}$$

Moreover, (2.73) holds with $a(s) = s^2/2$. Consequently, (2.141), (2.142) hold with $b(s) \equiv 1$ and

$$H_s(x) = \left((V(x) - \varphi(s))^+\right)^2/2,\ Q_s(x) = y^2 (V(\xi, y) - 1/2)^+ \min\{p(\xi, y, l) : |l| \leq s\} \tag{2.151}$$

We conclude from Theorem 2.9 that the OAG property

$$\limsup_{t\to+\infty}\left(|y(t)|\right)\leq 1 \tag{2.152}$$

holds for all $d\in L^{\infty}(\mathbb{R}_{+};\mathbb{R})$. ◁

Example 2.11: Consider again system (2.134), which was studied in Example 2.9 and was shown (by means of Theorem 2.8) to be a system that satisfies the p-UBIBS property. Here we show the validity of inequality (2.136) by means of Theorem 2.9. We consider system (2.134) with output $y=\xi$ and we apply Theorem 2.9 with

$$H_s(x)=-\exp(z),\ Q_s(x)=\Gamma\left(\frac{1}{2}\xi^2-\varepsilon\right)^{+}\ \text{and}\ b(s)\equiv\sqrt{2\varepsilon} \tag{2.153}$$

Inequality (2.136), i.e., the zero p-OAG property with asymptotic residual constant equal to $\sqrt{2\varepsilon}$, is a direct consequence of Theorem 2.9.

It should be noticed that in this case the function $H_s(x)$ has negative values. ◁

2.7 ▪ The Need of Output Stability Properties for Systems with Disturbances in Adaptive Control

Since disturbances are always present in every system and since disturbances appear as external inputs for a system in the standard mathematical formalism that is used in control theory, it becomes clear that there are no control systems without external inputs. When adaptive controllers are used, one has to study the stability properties of the closed-loop system with external inputs. This study is not easy. In this section we focus on the simplest system, namely the scalar system

$$\begin{aligned}&\dot{y}=\theta y+u+d\\&y\in\mathbb{R},\theta\in\mathbb{R},u\in\mathbb{R},d\in\mathbb{R}\end{aligned} \tag{2.154}$$

where $\theta\in\mathbb{R}$ is an unknown parameter of the system, $u\in\mathbb{R}$ is the control input, and $d\in\mathbb{R}$ is the (unknown and possibly time-varying) external disturbance.

A nonlinear adaptive controller may be designed based on the methodologies proposed in [87]. Indeed, the adaptive controller

$$\begin{aligned}&u=-cy-\hat{\theta}y-qy^3\\&\tfrac{d\hat{\theta}}{dt}=\Gamma y^2\end{aligned} \tag{2.155}$$

where $c,\Gamma>0$, $q\geq 0$ are the controller parameters (gains) guarantees a number of stability properties in the disturbance-free case (i.e., when $d\equiv 0$). Indeed, by defining $z=\hat{\theta}-\theta$, we obtain the equivalent version of the closed-loop system (2.154) with (2.155):

$$\begin{aligned}&\dot{y}=-cy-zy-qy^3+d\\&\dot{z}=\Gamma y^2\\&x=(y,z)'\in\mathbb{R}^2,d\in\mathbb{R}\end{aligned} \tag{2.156}$$

The radially unbounded function $V(x)=\frac{1}{2}y^2+\frac{1}{2\Gamma}z^2$ satisfies the following inequality for all $y,z,d\in\mathbb{R}$:

$$\dot{V}(x,d)=-cy^2-qy^4+yd\leq-\frac{c}{2}y^2-qy^4+\frac{1}{2c}d^2 \tag{2.157}$$

Consequently, the disturbance-free version of system (2.156)

- by virtue of Theorem 2.1 and Proposition 2.1 satisfies the Lagrange and Lyapunov stability properties,

- by the set $\{(0,z)' \in \mathbb{R}\times\mathbb{R}, z\in\mathbb{R}\}$ is invariant for (2.156) (it is a set of equilibrium points for system (2.156)),
- by virtue of Theorem 2.2 is GAOS.

However, we cannot show UGAOS for system (2.156). Still, the adaptive controller (2.155) works nicely when disturbances are absent. Figure 2.1 shows the plots of the state variables and the control input in the disturbance-free case.

Another advantage of the adaptive controller (2.155) is its simplicity. The adaptive controller (2.155) consists of two parts: the static part $u = -cy - \hat{\theta}y - qy^3$ and the dynamic part $\frac{d\hat{\theta}}{dt} = \Gamma y^2$. The dynamic part of the adaptive controller (2.155) is called "an identifier" that produces the parameter estimate $\hat{\theta}(t)$ of the true value of the parameter θ.

By virtue of Theorem 2.4 and inequality (2.157) we can also guarantee that system (2.156) is forward complete. However, we cannot guarantee boundedness of the state variables for arbitrary inputs $d \in L^\infty(\mathbb{R}_+;\mathbb{R})$. In order to guarantee state boundedness through estimate (2.157), one has to assume that the disturbance has only "finite energy", i.e., that $d \in L^2(\mathbb{R}_+;\mathbb{R}) \cap L^\infty(\mathbb{R}_+;\mathbb{R})$. The reason that we cannot prove boundedness of the state variables for arbitrary inputs $d \in L^\infty(\mathbb{R}_+;\mathbb{R})$ is that the states are not bounded. Figure 2.2 shows the problem in the case of a persistent disturbance $d \in L^\infty(\mathbb{R}_+;\mathbb{R})$ with $d \notin L^2(\mathbb{R}_+;\mathbb{R})$.

Figure 2.1 and Figure 2.2 show another aspect of adaptive control schemes which are similar to (2.155). The identifier fails to guarantee convergence of the parameter estimate $\hat{\theta}(t)$ to θ. The variable $\hat{\theta}$ works more or less as a dynamic controller gain which can only increase with respect to time. The increase of $\hat{\theta}(t)$ does not stop in the case of a persistent disturbance $d \in L^\infty(\mathbb{R}_+;\mathbb{R})$ with $d \notin L^2(\mathbb{R}_+;\mathbb{R})$ and this causes a slow increase of the magnitude of the control input $|u(t)|$.

The problems shown in Figure 2.2 have been known in the control community for a long time now and many modifications have been proposed for handling such situations (see [58], [59]). The most popular modification is the so-called "σ-modification" (see [119] for the application of σ-modification even to infinite-dimensional systems) where only the identifier is modified by adding an extra term of the form "$-\sigma\hat{\theta}$" where $\sigma > 0$ is a constant; this term tries to keep $\hat{\theta}(t)$ bounded. Therefore, the adaptive controller (2.155) with σ-modification reads

$$\begin{aligned} &u = -cy - \hat{\theta}y - qy^3 \\ &\tfrac{d\hat{\theta}}{dt} = \Gamma y^2 - \sigma\hat{\theta} \end{aligned} \tag{2.158}$$

By defining $z = \hat{\theta} - \theta$, we obtain the equivalent version of the closed-loop system (2.154) with (2.158):

$$\begin{aligned} &\dot{y} = -cy - zy - qy^3 + d \\ &\dot{z} = \Gamma y^2 - \sigma z - \sigma\theta \\ &x = (y,z)' \in \mathbb{R}^2, d \in \mathbb{R} \end{aligned} \tag{2.159}$$

Again, an advantage of the adaptive controller (2.158) is its simplicity. The adaptive controller (2.158) consists again of two parts: the static part $u = -cy - \hat{\theta}y - qy^3$ and the dynamic part $\frac{d\hat{\theta}}{dt} = \Gamma y^2 - \sigma\hat{\theta}$. Due to the σ-modification we cannot expect the dynamic part of the adaptive controller (2.158) to work as "an identifier" that produces a reliable parameter estimate $\hat{\theta}(t)$ of the true value of the parameter θ. Instead, as noted above, the variable $\hat{\theta}$ works more or less as a dynamic controller gain which cannot increase too much.

The radially unbounded function $V(x) = \frac{1}{2}y^2 + \frac{1}{2\Gamma}z^2$ for system (2.159) satisfies the following inequality for all $y, z, d \in \mathbb{R}$:

$$\begin{aligned} \dot{V}(x,d) &= -cy^2 - qy^4 - \tfrac{\sigma}{\Gamma}z^2 - \tfrac{\sigma}{\Gamma}z\theta + yd \\ &\le -\tfrac{c}{2}y^2 - qy^4 - \tfrac{\sigma}{2\Gamma}z^2 + \tfrac{\sigma}{2\Gamma}\theta^2 + \tfrac{1}{2c}d^2 \\ &\le -\min(c,\sigma)\,V(x) + \tfrac{\sigma}{2\Gamma}\theta^2 + \tfrac{1}{2c}d^2 \end{aligned} \tag{2.160}$$

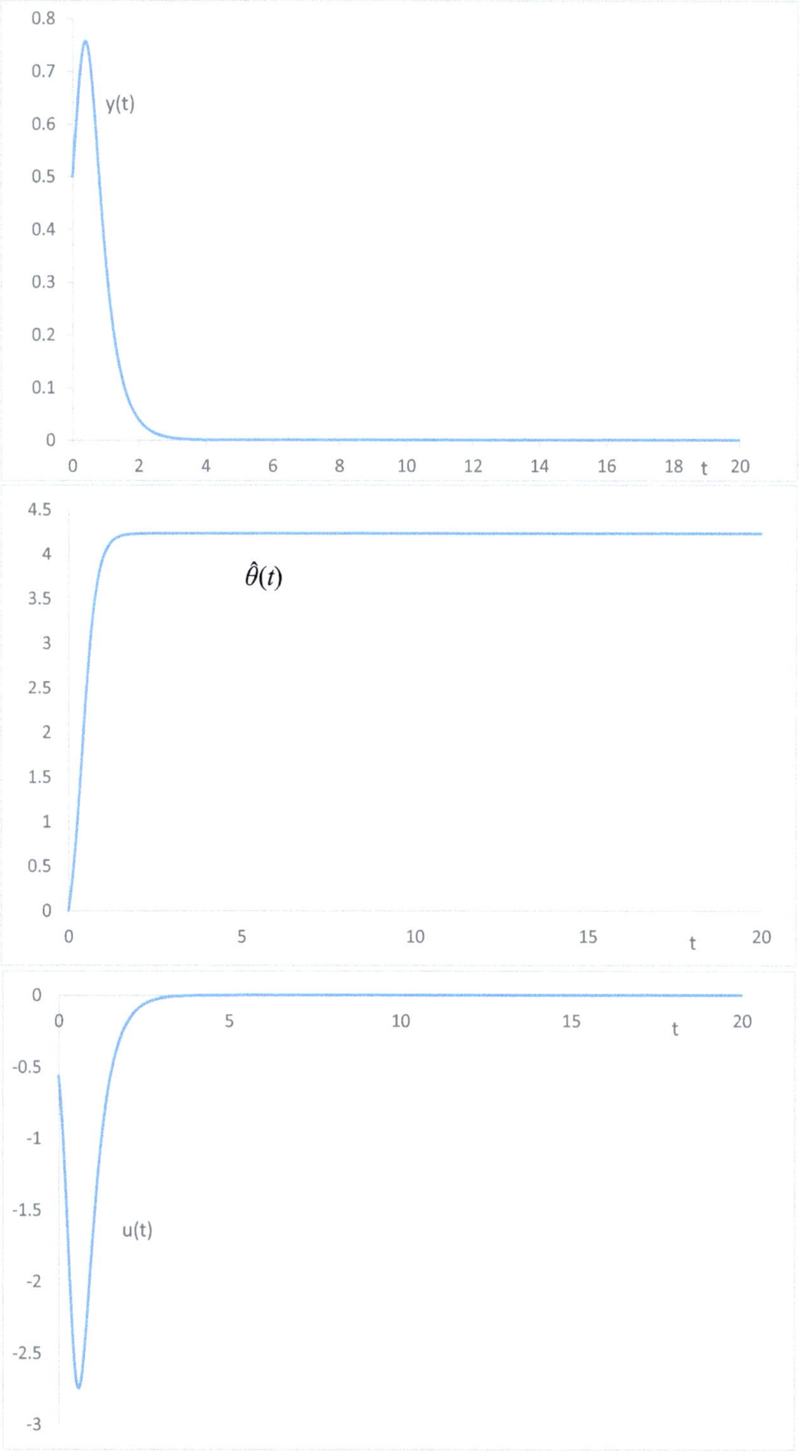

Figure 2.1. *The evolution of the state variables and the control input for the closed-loop system* (2.154), (2.155), *with* $d \equiv 0$, $\theta = 3$, $c = 1, \Gamma = 10, q = 0.5$ *and* $y(0) = 0.5$, $\hat{\theta}(0) = 0$.

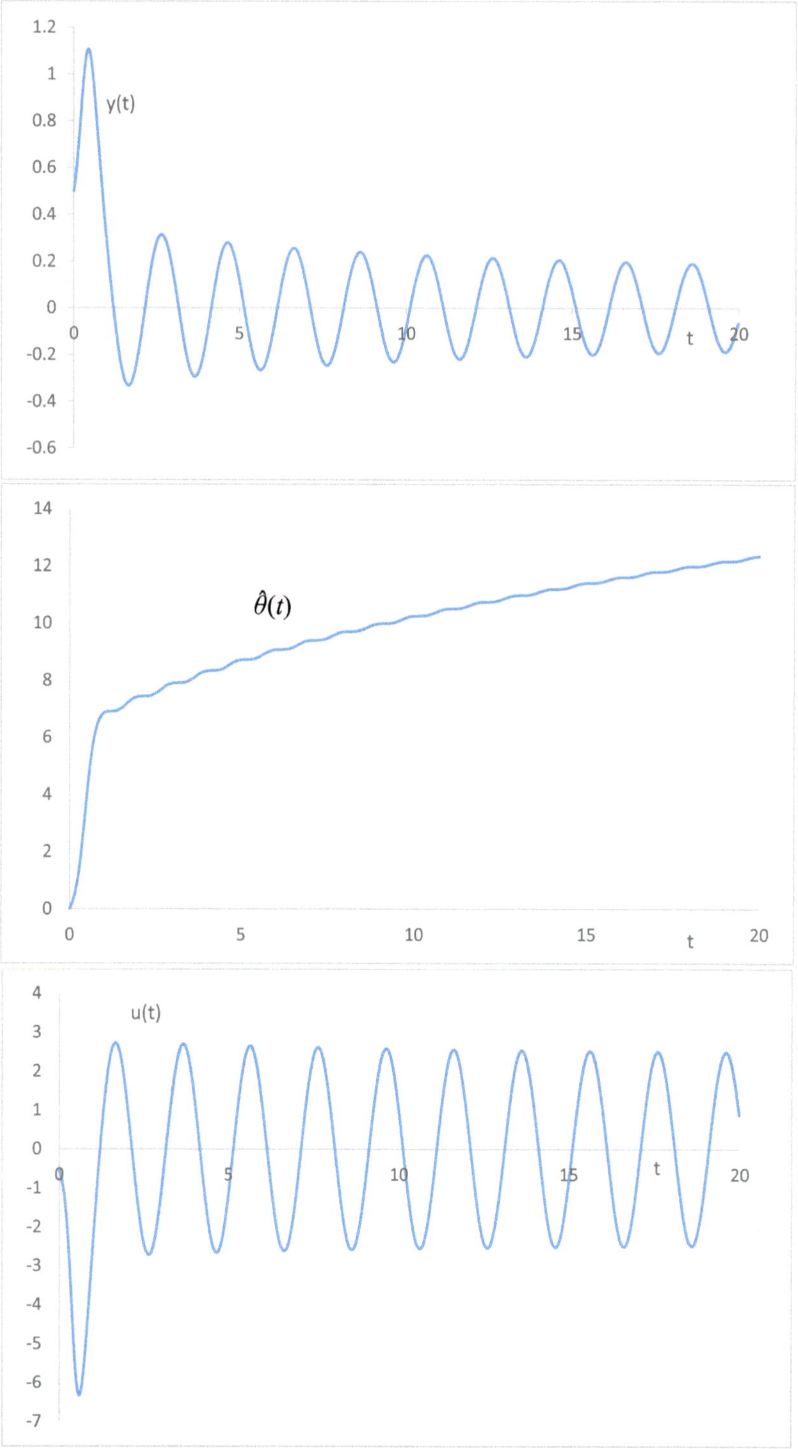

Figure 2.2. *The evolution of the state variables and the control input for the closed-loop system* (2.154), (2.155), *with* $d(t) = 2\sin(\pi t)$, $\theta = 3$, $c = 1, \Gamma = 10, q = 0.5$ *and* $y(0) = 0.5$, $\hat{\theta}(0) = 0$.

Integrating the above differential inequality, we obtain the following estimate for every $d \in L^{\infty}(\mathbb{R}_+;\mathbb{R})$:

$$
\begin{aligned}
|x(t)| &\leq \exp\left(-\min(c,\sigma)\,t/2\right)\sqrt{\max\left(\Gamma,1/\Gamma\right)}\,|x(0)| \\
&+\sqrt{\tfrac{\max(\Gamma,1)}{c\min(c,\sigma)}}\,\|d\|_{\infty} + |\theta|\sqrt{\tfrac{\sigma}{\min(\Gamma,1)\min(c,\sigma)}}
\end{aligned}
\tag{2.161}
$$

Estimate (2.161) guarantees the p-ISS property for system (2.159) with gain function $\gamma(s) = s\sqrt{\frac{\max(\Gamma,1)}{c\min(c,\sigma)}}$ and residual constant $\alpha = |\theta|\sqrt{\frac{\sigma}{\min(\Gamma,1)\min(c,\sigma)}}$. In the disturbance-free case we guarantee p-UGAS with residual constant $\alpha = |\theta|\sqrt{\frac{\sigma}{\min(\Gamma,1)\min(c,\sigma)}}$.

Although the p-ISS property is a very strong stability property and estimate (2.161) also guarantees exponential convergence, there is a problem here. The problem is that the residual constant is proportional to the absolute value of the unknown parameter, i.e., $\alpha \sim |\theta|$. One may say that the stability analysis that we have performed is very conservative and that is why we get such a result. Unfortunately, this is not the case. If we perform an analysis of the equilibrium points of system (2.159) with $d \equiv 0$ then we can easily see that there are equilibrium points of the form $x = (y,z) = \left(\pm\sqrt{\frac{\sigma(\theta-c)}{\Gamma+\sigma q}}, -\frac{\Gamma c+\theta\sigma q}{\Gamma+\sigma q}\right)$ when $\theta > c$. Consequently, the residual constant (which must be large enough so that the ball with radius the residual constant contains all equilibrium points) must be approximately proportional to the absolute value of the unknown parameter. The problem is real and it is shown in Figure 2.3 for the disturbance-free case and in Figure 2.4 for the case of a persistent disturbance.

The advantages of σ-modification are its simplicity, the fact that we can guarantee exponential convergence in the disturbance-free case with assignable convergence rate, the fact that we can guarantee the p-ISS property with assignable gain function. The disadvantage of σ-modification is the fact that the residual constant is not a priori known due to the dependence of the residual constant on the unknown parameter θ. Therefore, we cannot control the offset in the response of the closed-loop system—even in the disturbance-free case—and that can give a poor performance.

Having identified the weaknesses of the σ-modification (poor transients, regulation to an unknown neighborhood of the origin dependent on the unknown parameter), we remind the reader of Example 2.9, which introduced an adaptive system that overcomes these deficiencies of the σ-modification. This example, with its contrasting tools relative to the σ-modification (deadzone, update rate that decreases as the adaptive gain grows, several forms of "nonlinear damping") is generalized from the scalar linear case in the rest of the book.

We end the section by noticing the precision that the stability properties (and the problems) of the closed-loop systems under adaptive controllers are described by using the output stability notions that we introduced in the present chapter.

2.8 ▪ Notes and comments

1. There are many other stability properties beyond the ones that are discussed in the present chapter. We have focused mostly on the stability properties that are useful for adaptive control.

2. Output stability properties are very much related to the so-called "partial stability" properties. See for instance [147]. Output stability is a special case of stability with respect to two measures; see [92], [93], [114], [144].

3. Controller (2.155) with $q = 0$ is a "high-gain" adaptive controller (see [100], [111], [152]).

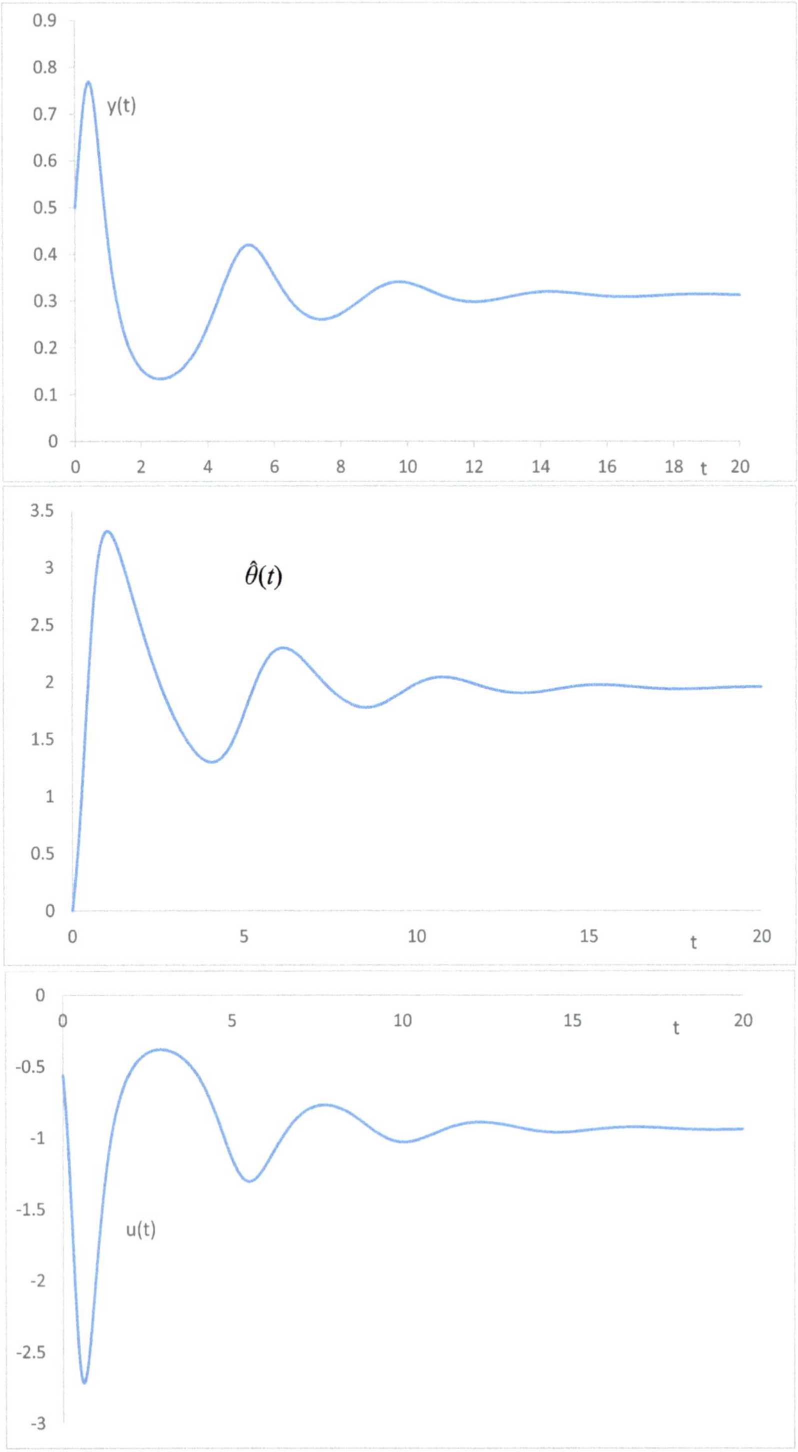

Figure 2.3. *The evolution of the state variables and the control input for the closed-loop system* (2.154), (2.158), *with* $d(t) \equiv 0$, $\theta = 3$, $c = 1, \Gamma = 10, \sigma = q = 0.5$ *and* $y(0) = 0.5$, $\hat{\theta}(0) = 0$.

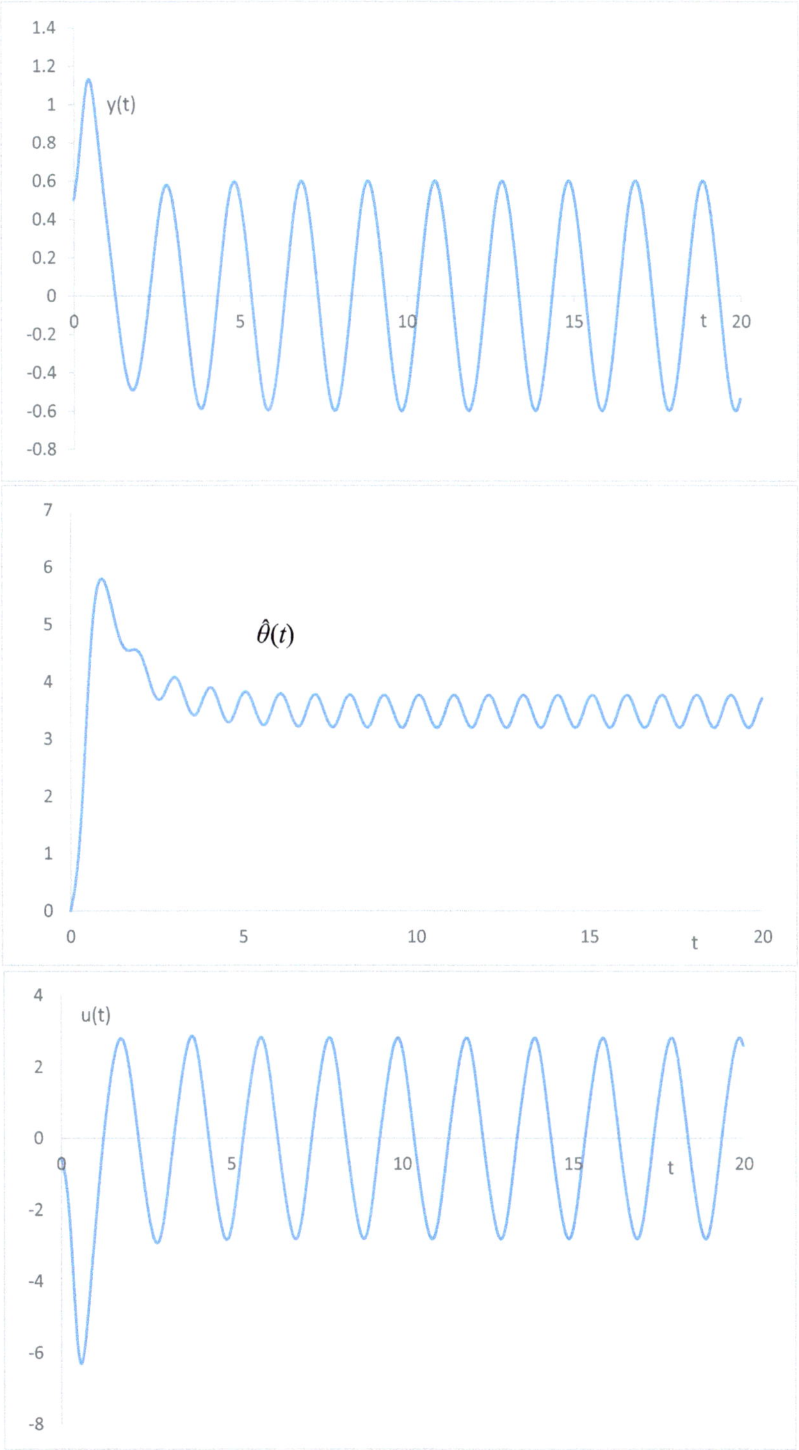

Figure 2.4. *The evolution of the state variables and the control input for the closed-loop system* (2.154), (2.158), *with* $d(t) = 2\sin(\pi t)$, $\theta = 3$, $c = 1, \Gamma = 10, \sigma = q = 0.5$ *and* $y(0) = 0.5$, $\hat{\theta}(0) = 0$.

4. ISS was introduced in control theory by E. D. Sontag in [134]. IOS was introduced in control theory in [61] and was studied by E. D. Sontag and Y. Wang in [139]. Both notions were studied by E. D. Sontag and Y. Wang; see [134], [135], [136], [138], [139], [140], [141]. The role that IOS and ISS played in modern feedback control theory is huge. The UBIBS property, i.e., property (2.47) with $\bar{\alpha} = 0$, was introduced in [140].

5. Some of the results that are given in the present chapter are not used in the subsequent chapters. However, we have given the results as well as examples for the corresponding stability properties because the results deal with stability notions that play a significant role in what follows. The reader can appreciate the clear presentation of this collection of stability properties, with a scope slightly expanded beyond adaptive control.

6. While writing this chapter, Prof. Antoine Chaillet mentioned to the authors that the same proof procedure that is used in the proof of Theorem 1.2 in [140] can be exploited in order to obtain the following sufficient condition for p-IOS:

   Theorem 2.10 (Lyapunov conditions for p-IOS): *Suppose that system* (2.43) *is p-UBIBS and there exist a function $a \in K_\infty$, a non-decreasing, continuous function $\varphi : \mathbb{R}_+ \to \mathbb{R}_+$, a function $V \in C^1(\mathbb{R}^n; \mathbb{R}_+)$ with $V(0) = 0$, and a function $\sigma \in KL$ for which the following inequality holds for all $x \in \mathbb{R}^n$:*

   $$a(|h(x)|) \leq V(x) \tag{2.162}$$

   and the following implication holds for all $x \in \mathbb{R}^n$, $d \in D$:

   $$V(x) \geq \varphi(|d|) \Rightarrow \dot{V}(x,d) = \nabla V(x) f(x,d) \leq -\sigma(V(x), |x|) \tag{2.163}$$

   Then system (2.43), (2.2) *satisfies the p-IOS property. Moreover, if $\varphi(0) = 0$ then system* (2.43), (2.2) *satisfies the IOS property.*

   In other words, it is possible to replace the UBIBS property in Theorem 1.2 in [140] by the less demanding p-UBIBS property.

7. Some of the results contained in the present chapter are novel results. For example, Theorem 2.4, Theorem 2.5, Theorem 2.8, and Theorem 2.9 have not appeared elsewhere (except of [79] which was a preliminary version of this chapter). A less general version of Theorem 2.8 first appeared in [78].

8. It is well known that, for time-invariant systems described by ODEs with locally Lipschitz right-hand sides, GAS and UGAS are actually equivalent properties; see [157]. However, GAOS is not equivalent to UGAOS for arbitrary output maps.

Chapter 3

Robustness-Performance-Simplicity-Identification: A Journey in Bringing Down Barriers in Adaptive Control

3.1 ▪ Introduction

What kind of properties should an adaptive controller guarantee, and what kind of properties *can* it guarantee? Such questions have occupied the researchers and practitioners of adaptive control since the emergence of the subject in the form of the "MIT rule" in the late 1950s, namely, for about 65 years now.

The first two decades of research in adaptive control dealt with the most basic question: when simultaneously performing control and the estimation of the plant or controller parameters, can one even guarantee that all the signals remain bounded and the state is regulated to a setpoint (or a reference trajectory)? These questions got settled affirmatively around 1980. But such "boundedness plus asymptotic regulation" results came without Lyapunov stability estimates and without an estimate of the convergence rate. Convergence, established using a conceptual result in real analysis called "Barbălat's Lemma," was established but left unquantified.

In 1982, "Rohr's counterexamples" exposed, in simulation, the absence of robustness of the nominal adaptive designs when unmodeled effects, including disturbances, are present. Various remedies were promptly devised, including the σ-modification/leakage, deadzone, and projection, but they all came with limitations on either the uncertainty or the initial condition or the disturbance size, or were accompanied with a loss of some of the basic properties, such as convergence, in the absence of the unmodeled effects. Remedies introduced tradeoffs but not full solutions to the problem of robustness.

Among the remedies studied was the reliance on Persistent Excitation (PE) which, in addition to guaranteeing convergence of parameter estimates, results in the exponential stability of the system in the absence of unmodeled effects. While PE adds robustness in the presence of unmodeled effects, as explored in many studies in the 1980s, this addition of robustness comes at a large cost. PE has to be induced by tracking persistently rich references, which may contradict the control objective of tracking a constant setpoint. Hence, robustness attained with PE, and the resultant perfect identification of the plant, often forfeits the performance objective sought with adaptive control.

Following these studies of robustness, PE, and identifiability in the 1980s, one major problem still remained unresolved with *linear* adaptive controllers: transient performance. While, in the absence of unmodeled effects, most adaptive controllers achieve the desired asymptotic performance (regulation or tracking), basic adaptive linear controllers are known to lead to transients that can be arbitrarily bad, before they ultimately lead to asymptotic convergence. By "transients" one means the transients of both the plant's state (or output) and the parameter estimate.

Such transients are coupled. Poor initial estimates lead to destabilizing tendencies, which result in the divergence of the plant's state, which in turn provides the excitation which drives the parameter estimates to stabilizing values and, finally, the plant state to small values. Linear adaptive controllers are unequipped to deal with these coupled transients of the plant and estimator. In the early 1990s, adaptive nonlinear controllers that use *backstepping* solved the problem of taming the transients and led to transient performance estimates that can be systematically shaped with various gains in the backstepping procedure and adaptation gains. However, such adaptive controllers with superior performance remained hampered by the presence of unmodeled effects, as they inherited only the robustification tools from the 1980s, whose capabilities were limited or entailed serious tradeoffs.

The progression from the linear adaptive designs of the 1960s–1980s to the nonlinear adaptive designs of the 1990s involved a progression in the complexity of such designs. Complex feedback laws are harder to implement, harder to debug, and less trusted by practitioners who have difficulty understanding them. A need exists for adaptive controllers that not only have strong robustness and performance but are also reasonably simple.

In summary, the history of adaptive control points to the following properties being desirable:

Robustness-Performance-Simplicity-Identification

A 65-year long struggle to achieve these properties has led to a widespread informal belief that such properties are not simultaneously achievable. While such a belief is not validated by formal results in *fundamental limitations* to the attainment of such results in adaptive control, the absence of such results was attributed to the nonlinear nature of adaptive control, and the belief of the existence of fundamental limitations was sustained, regardless of the absence of a proof of unachievability.

In this book we venture to actually break the barriers to the simultaneous achievement of these properties. In subsequent chapters we provide designs that simultaneously achieve these properties. And, hence, we disprove the beliefs that become entrenched through many decades of unsuccessful attempts.

Before we advance to the designs that successfully attain the above properties, it is important to review some of the alternatives that attain simultaneously some of the above properties but not all of them. This represents our journey of about a decade in making very disparate methodological attempts and making gradual progress towards the results presented in the remaining chapters of the book.

In the remainder of the present section, we get more specific regarding the meanings of the words *robustness*, *performance*, *simplicity*, and *identification* than we were in the historical narrative above.

I) Robustness

The closed-loop adaptive system must exhibit robustness in the presence of disturbances. More specifically, we want to be able to guarantee that every solution of the closed-loop system is bounded for bounded but (possibly) persistent inputs. Furthermore, we should be able to guarantee a p-IOS property for the closed-loop system, where the so-called "Input" (I) is the disturbance and the so-called "Output" (O) is the plant state. For example, the σ-modification method, which was explained in the previous chapter, guarantees the p-ISS property for systems that satisfy the matching condition (as well as other kinds of systems). Therefore, we may say that the σ-modification method can guarantee robustness.

Of course, as explained in the previous chapter, there are hidden issues here. The fact that the residual constant for the p-ISS property guaranteed by σ-modification for systems that satisfy the matching condition depends on the unknown parameters is problematic because this means that the residual constant is not a priori known. We regard this as a concession in performance made to attain robustness. Clearly, one would prefer to have a form of p-IOS robustness in which the residual constant is independent of unknown uncertainties.

II) Performance

At least in the disturbance-free case, the closed-loop system must guarantee a satisfactory performance for the plant state. We offer that such a performance requirement also extends to the case where disturbances are present. By "satisfactory performance" in the disturbance-free case we mean at least a property like GAOS, which captures only asymptotic performance (perfectly), or, preferably, p-UGAOS, or even UGAOS, in certain cases. It is, additionally, desirable not only to have the plant state's convergence be uniform in the initial condition but also have an arbitrarily assignable *rate* of convergence.

III) Simplicity

The feedback design methodology should, preferably, be simple enough so that it can be applied, with relative ease of coding, to two important classes of nonlinear systems: (a) systems that satisfy the matching condition, and (b) strict-feedback systems (and their generalizations, such as block-strict-feedback systems and their multi-input versions). Another aspect related to simplicity is the number of the controller states needed by the adaptive controller. While present-day computers can handle easily a large number of integrators, a large number of states introduces "more things that can go wrong" and makes the transients of a large number of states harder to shape.

IV) Identification

A basic fact of adaptive control is that a perfect identification of parameters—of either the plant or the controller—is not necessary for the attainment of perfect asymptotic regulation or trajectory tracking. The parameter estimates merely need to converge to values that are "good enough" (preferably stabilizing in a "frozen" sense). Perfect identification is not necessary in indirect adaptive control and the need for it is even less clear in direct adaptive control. However, one cannot disagree with the statement that, if an adaptive controller's parameter estimator can guarantee identification of the unknown parameters, this is certainly an advantage of the adaptive controller. With parameters that have converged, the adaptive controller implements a nominal non-adaptive controller, for which the desired behavior in closed loop is typically assignable and, additionally, with parameters that have converged, fewer states of the overall closed-loop system can behave unpredictably. For these reasons, we dedicate attention, in this chapter, to identifiability of the estimated parameters. We present some remarkably strong results on identification in closed loop, though we don't pursue them beyond this chapter.

More issues arise with identification: should we seek identification only in the disturbance-free case or should we somehow in addition guarantee some form of robustness of identification (i.e., of the convergence of the parameter estimates to the parameters' true values) when disturbances are present? Should we have a criterion that informs us when identification occurs? These are important issues that must not be neglected.

In this chapter we provide three different adaptive feedback design methodologies that we have introduced in recent years, and which cover some of the words in the set "Robustness-Performance-Simplicity-Identification". However, none of these three design methods achieves all of the four properties in the senses introduced above.

These three different design methodologies are applied to the simplest benchmark problem

$$\begin{aligned} &\dot{y} = \theta y + u + d \\ &y \in \mathbb{R}, u \in \mathbb{R}, \theta \in \mathbb{R}, d \in \mathbb{R} \end{aligned} \tag{3.1}$$

where $y \in \mathbb{R}$ is the plant state, $\theta \in \mathbb{R}$ is the (constant) unknown parameter, $u \in \mathbb{R}$ is the control input, and $d \in \mathbb{R}$ is the (unmeasured and possibly time-varying) disturbance. We assume that no bounds for the values of the unknown parameter and the disturbance are known. The simple problem (3.1) allows us to present the advantages and disadvantages of the design methodologies.

3.2 ▪ Performance + Identification: Batch Least-Squares Identifier

In this section we describe the use of the Batch Least-Squares Identifier (BaLSI) to system (3.1). The BaLSI appeared in [70], [72] and was later applied to various control problems (see [9]; even to the infinite-dimensional case in [71], [148], [149]). The design methodology of BaLSI consists of the following two steps:

1. design a nominal feedback law that stabilizes the system by assuming that the system parameters are known, and
2. achieve exact identification of the unknown parameters in finite time and use the parameter estimates in the nominal feedback law (certainty equivalence).

The nominal controller for system (3.1) can be the following feedback law:

$$u = -\left(c + \theta\right) y \tag{3.2}$$

where $c > 0$ is a constant. Feedback law (3.2) achieves global exponential stabilization of system (3.1) (with convergence rate $c > 0$) in the disturbance-free case, but it cannot be implemented because $\theta \in \mathbb{R}$ is unknown.

Next, we design a finite-time identifier for system (3.1) with $d = 0$. Our starting point is the equation

$$p(t,s) = \theta q(t,s) \tag{3.3}$$

where

$$\begin{aligned} p(t,s) &= y(t) - y(s) - \textstyle\int_s^t u(l)dl \\ q(t,s) &= \textstyle\int_s^t y(l)dl \end{aligned} \tag{3.4}$$

that holds for all $t, s \geq 0$ for which the solution of (3.1) exists. Notice that both $p(t,s)$ and $q(t,s)$ are available from direct measurement. Assuming that the solution of (3.1) exists on the interval $[0,\tau]$, where $\tau > 0$, equation (3.3) guarantees that the function

$$h(\vartheta) := \int_0^\tau \int_0^\tau \left(p(t,s) - \vartheta q(t,s)\right)^2 dsdt \tag{3.5}$$

attains a global minimum at $\vartheta = \theta$ (Gauss least-squares). By Fermat's theorem we get that

$$Y\left(\tau\right) = \theta Q\left(\tau\right) \tag{3.6}$$

where

$$\begin{aligned} Y\left(\tau\right) &:= \textstyle\int_0^\tau \int_0^\tau p(t,s)q(t,s)dsdt \\ Q\left(\tau\right) &:= \textstyle\int_0^\tau \int_0^\tau q^2(t,s)dsdt \end{aligned} \tag{3.7}$$

Consequently, when $Q\left(\tau\right) > 0$ (notice that $Q\left(\tau\right)$ is always non-negative) we obtain the least-squares estimate $\theta = Y\left(\tau\right)/Q\left(\tau\right)$. On the other hand, when $Q\left(\tau\right) = 0$ then using definitions (3.4) and (3.7), we conclude that $y(t) = 0$ for all $t \in [0,\tau]$. Therefore, when $Q\left(\tau\right) = 0$ the state is at the desired position.

When the data are fed to the least-squares identifier in batches and we perform identification at the discrete times τ_i, $i = 0, 1, 2, ...$ then we obtain the BaLSI. The selection of the discrete times τ_i, $i = 0, 1, 2, ...$ can be made in a regulation-triggered manner (see [70], [72]) but here we prefer to give the simple selection $\tau_i = iT$, $i = 0, 1, 2, ...$, where $T > 0$ is a constant. Thus, the update law with BaLSI becomes

$$\hat{\theta}((i+1)T) = \begin{cases} \hat{\theta}(iT) & \text{if} \quad Q((i+1)T) = 0 \\ \frac{Y((i+1)T)}{Q((i+1)T)} & \text{if} \quad Q((i+1)T) > 0 \end{cases} \tag{3.8}$$

The feedback law (3.2) is replaced by its certainty-equivalence version that exploits the parameter estimates provided by BaLSI:

$$u(t) = -\left(c + \hat{\theta}(iT)\right) y(t) \text{ for } t \in [iT, (i+1)T), \ i = 0, 1, 2, ... \tag{3.9}$$

Distinguishing the cases $y(0) \neq 0$, $y(0) = 0$ and working inductively, we can obtain the following estimates for the solution of the closed-loop system (3.1) with (3.8), (3.9), $d = 0$, and initial condition $\left(y(0), \hat{\theta}(0)\right) = \left(y_0, \hat{\theta}_0\right)$:

$$y(t) = \begin{cases} \exp\left(\left(\theta - c - \hat{\theta}_0\right) t\right) y_0 & \text{for} \quad t \in [0, T) \\ \exp\left(\left(\theta - \hat{\theta}_0\right) T\right) \exp(-ct) \, y_0 & \text{for} \quad t \geq T \end{cases} \tag{3.10}$$

$$\text{If } y_0 \neq 0 \text{ then } \hat{\theta}(t) = \begin{cases} \hat{\theta}_0 & \text{for} \quad t \in [0, T) \\ \theta & \text{for} \quad t \geq T \end{cases} \tag{3.11}$$

$$\text{If } y_0 = 0 \text{ then } \hat{\theta}(t) \equiv \hat{\theta}_0 \tag{3.12}$$

Figure 3.1 shows the plots for the closed-loop system (3.1) with (3.8), (3.9) of the state variables and the control input in the disturbance-free case.

It is clear that the performance shown in Figure 3.1 as well as in formulas (3.10), (3.11), (3.12) is satisfactory. Exponential convergence with assignable convergence rate is guaranteed. Moreover, exact identification of the unknown parameter is also guaranteed for almost all initial conditions.

The update law (3.8) uses the quantities $Y(t)$, $Q(t)$, which can be obtained by implementing the following system of ODEs:

$$\begin{aligned} \dot{a} &= y \\ \dot{b} &= a \\ \dot{\xi} &= a^2 \\ \dot{\gamma} &= u \\ \dot{\delta} &= \gamma \\ \dot{\zeta} &= a\gamma \end{aligned} \tag{3.13}$$

with

$$a(0) = b(0) = \xi(0) = \gamma(0) = \delta(0) = \zeta(0) = 0 \tag{3.14}$$

and the time-varying equations

$$\begin{aligned} Q(t) &= 2\left(t\xi(t) - b^2(t)\right) \\ Y(t) &= ta^2(t) - 2a(t)b(t) - 2t\zeta(t) + 2b(t)\delta(t) \end{aligned} \tag{3.15}$$

Consequently, the implementation of the adaptive feedback law (3.8), (3.9) is a feasible task.

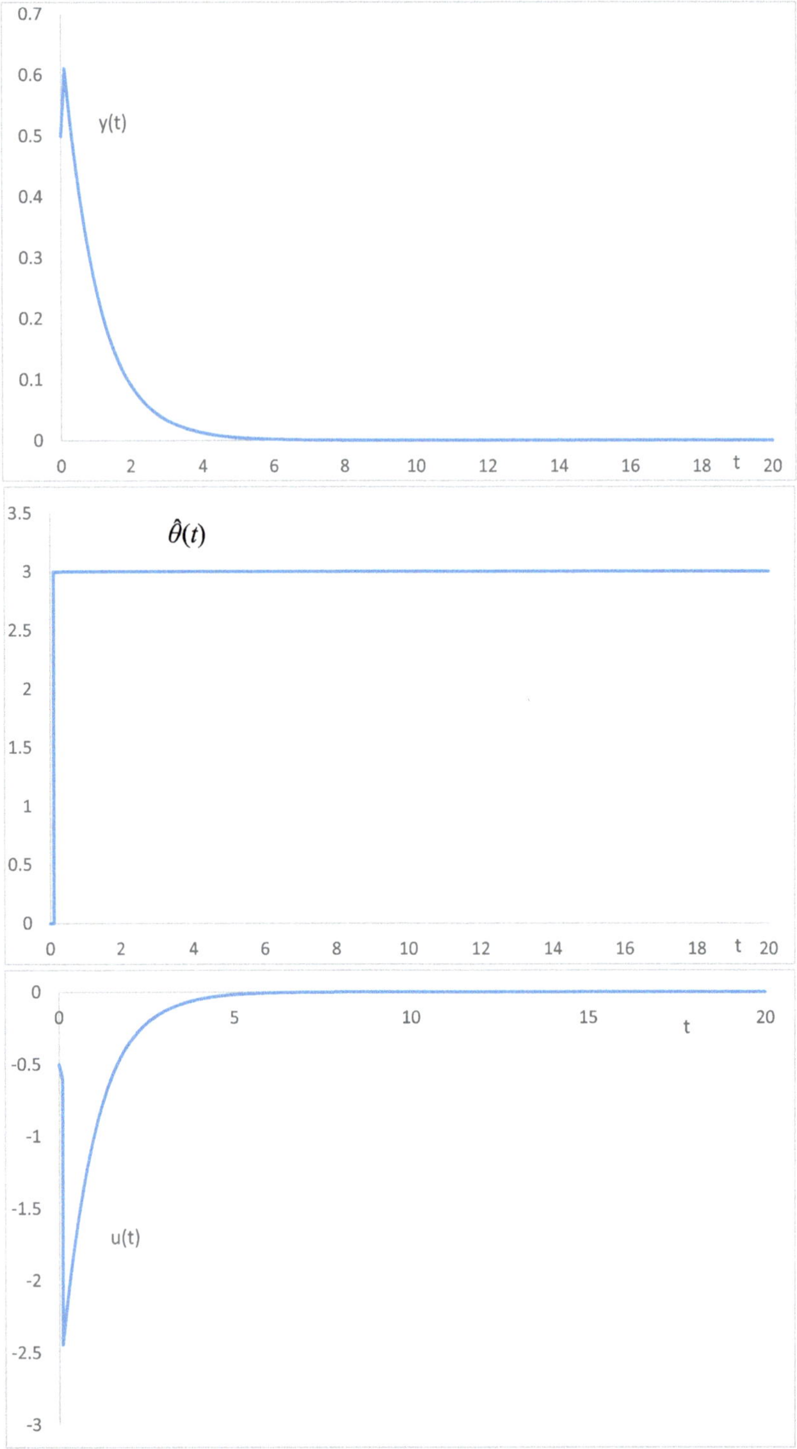

Figure 3.1. *The evolution of the state variables and the control input for the closed-loop system* (3.1) *with* (3.8), (3.9), *with* $d \equiv 0$, $\theta = 3$, $c = 1, T = 0.1$ *and* $y(0) = 0.5$, $\hat{\theta}(0) = 0$.

However, the BaLSI adaptive scheme has some shortcomings. First of all, it is not a simple scheme. The scheme cannot be generalized easily to systems that satisfy the matching condition or strict-feedback systems. The reason that explains this difficulty has to do with an essential feature of the BaLSI adaptive scheme: in order to be able to achieve exact parameter identification one needs a kind of an observability property for the unknown parameters. This observability property is also related to the nominal feedback (see the discussion in [70]) and it is not easily verified. Another aspect that has to be pointed out is the large number of additional ODEs that are required for the implementation of the BaLSI scheme even when the number of unknown parameters is not large.

The study of robustness of the closed-loop system (3.1) with (3.8), (3.9) is very difficult. Some numerical robustness tests were performed in [70], [72]. However, no theoretical robustness guarantee can be proved for the BaLSI adaptive scheme even in the simplest case (3.1). Figure 3.2 shows that there are robustness problems and that the performance in the case of persistent disturbances can be poor. This is expected since the discontinuous nature of the BaLSI adaptive scheme (due to switching times) can create robustness problems.

Nevertheless, the idea that the history of the solution can be used for the direct estimation of the unknown parameters, which is used in BaLSI, can be exploited as the following chapters show.

3.3 ▪ Performance + Simplicity: Obtaining KL Estimates

The book [87] provided KL estimates when the adaptive feedback controllers had a special characteristic: nonlinear damping terms. The paper [75] built on this idea and allowed the nonlinear damping term to depend not only on the plant state but also on the parameter estimates.

For system (3.1) the adaptive controller proposed in [75] is given by the equations

$$\begin{aligned} u &= -cy - \hat{\theta} y - \left(\tfrac{\delta}{2} y^2 + c^{-1} r + c^{-1} \hat{\theta}^2\right) y \\ \tfrac{d\hat{\theta}}{dt} &= \Gamma y^2 \end{aligned} \tag{3.16}$$

where $c, r, \Gamma, \delta > 0$ are parameters of the controller. The static part of the controller (3.16)

$$u = -cy - \hat{\theta} y - \left(\frac{\delta}{2} y^2 + c^{-1} r + c^{-1} \hat{\theta}^2\right) y \tag{3.17}$$

contains the nonlinear damping term $-\left(\frac{\delta}{2} y^2 + c^{-1} r + c^{-1} \hat{\theta}^2\right) y$, which depends both on the plant state $y \in \mathbb{R}$ and the parameter estimate $\hat{\theta} \in \mathbb{R}$.

We next describe the stability properties of the closed-loop system (3.1) with (3.16). Using the transformation $z = \hat{\theta} - \theta$ we obtain the following equivalent form of the closed-loop system (3.1), (3.16):

$$\begin{aligned} \dot{y} &= -\left(\tfrac{\delta}{2} y^2 + c + c^{-1} r + z + c^{-1} (\theta + z)^2\right) y + d \\ \dot{z} &= \Gamma y^2 \\ x &= (y, z)' \in \mathbb{R}^2, d \in \mathbb{R} \end{aligned} \tag{3.18}$$

Using the radially unbounded function $V(x) = \frac{1}{2} y^2 + \frac{1}{2\Gamma} z^2$ that satisfies the following inequality for all $x = (y, z)' \in \mathbb{R}^2, d \in \mathbb{R}$:

$$\dot{V} = -\left(\frac{\delta}{2} y^2 + c + c^{-1} r + c^{-1} (\theta + z)^2\right) y^2 + yd \leq -\frac{c}{2} y^2 + \frac{1}{2c} d^2 \tag{3.19}$$

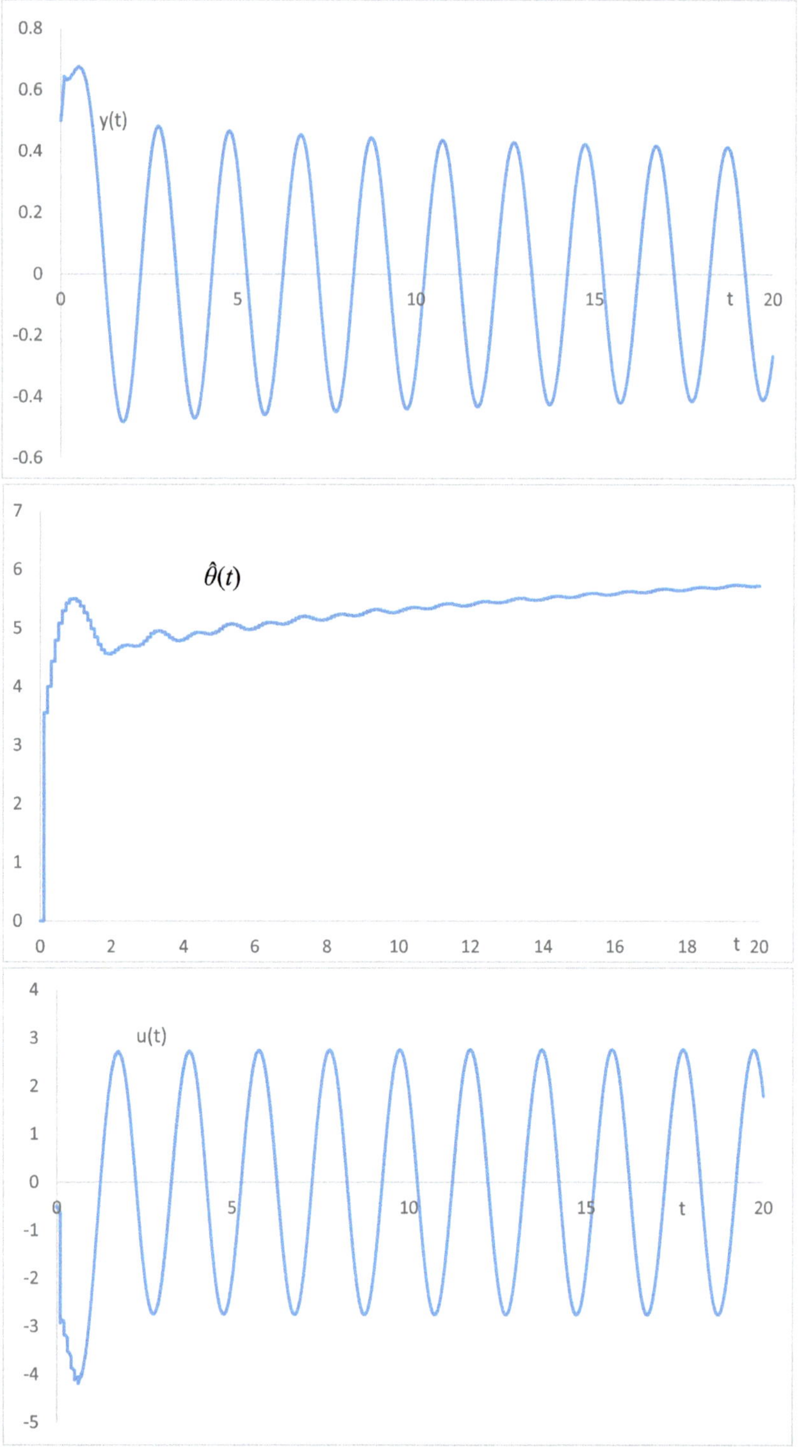

Figure 3.2. *The evolution of the state variables and the control input for the closed-loop system* (3.1) *with* (3.8), (3.9), *with* $d(t) = 2\sin(\pi t)$, $\theta = 3$, $c = 1, T = 0.1$ *and* $y(0) = 0.5$, $\hat{\theta}(0) = 0$.

we are in a position to conclude the following:

1. the set $\{(0,z)' \in \mathbb{R}^2, z \in \mathbb{R}\}$ is invariant for the disturbance-free version of (3.18); it is the set of equilibrium points of system (3.18),

2. by virtue of Proposition 1.1, the disturbance-free version of system (3.18) is Lagrange stable and Lyapunov stable,

3. by using Corollary 1.1 with $W(x) = \frac{1}{2}y^2$ and $\rho(s) = cs$, the disturbance-free version of system (3.18) is GAOS.

Inequality (3.19) implies that all assumptions of Theorem 1.4 hold for system (3.18). Therefore, system (3.18) is forward complete. The function $W(x) = y^2/2$ satisfies for all $x \in \mathbb{R}^2, d \in \mathbb{R}$ the inequality (using the fact that $z + c^{-1}(\theta + z)^2 \geq -\theta - c/4$ for all $z \in \mathbb{R}$)

$$\begin{aligned} \nabla W(x)\dot{x} &= -\left(\tfrac{\delta}{2}y^2 + c + c^{-1}r + z + c^{-1}(\theta+z)^2\right)y^2 + dy \\ &\leq -\left(\tfrac{\delta}{2}y^2 + c^{-1}r + \tfrac{3c}{4} - \theta\right)y^2 + dy \end{aligned} \tag{3.20}$$

Inequalities (3.19), (3.20) imply that all assumptions of Theorem 1.3 hold for the disturbance-free version of system (3.18) when $\theta \leq c^{-1}r + 3c/4$. Therefore, when $\theta \leq c^{-1}r + 3c/4$ then the disturbance-free version of system (3.18) is UGAOS. Moreover, when $\theta < c^{-1}r + 3c/4$ then we obtain from (3.20) the inequality $\dot{W} \leq -2\left(c^{-1}r + \frac{3c}{4} - \theta\right)W$ for the disturbance-free version of system (3.18), which allows us to derive the exponential convergence estimate $|y(t)| \leq \exp\left(-\left(c^{-1}r + \frac{3c}{4} - \theta\right)t\right)|y(0)|$ for all $t \geq 0$. The previous estimate shows exponential convergence of the output of the disturbance-free version of system (3.18) when $\theta < c^{-1}r + 3c/4$.

However, we can derive additional stability estimates for system (3.18). Rewriting the time derivative of W in the following way:

$$\begin{aligned} \nabla W(x)\dot{x} &= \left(\theta - \tfrac{\theta\sqrt{r}}{\max(|\theta|,\sqrt{r})}\right)y^2 + \tfrac{y^2\theta\sqrt{r}}{\max(|\theta|,\sqrt{r})} \\ &-(\theta+z)y^2 - \left(\tfrac{\delta}{2}y^2 + c + c^{-1}r + c^{-1}(\theta+z)^2\right)y^2 + dy \end{aligned} \tag{3.21}$$

and using the inequalities

$$dy \leq \frac{c}{4}y^2 + \frac{1}{c}d^2 \tag{3.22}$$

$$\left(\theta - \frac{\theta\sqrt{r}}{\max\left(|\theta|,\sqrt{r}\right)}\right)y^2 \leq \frac{1}{2\delta}\left(\theta - \frac{\theta\sqrt{r}}{\max\left(|\theta|,\sqrt{r}\right)}\right)^2 + \frac{\delta}{2}y^4 \leq \frac{1}{2\delta}\left(\theta^2 - r\right)^+ + \frac{\delta}{2}y^4 \tag{3.23}$$

$$\frac{\theta\sqrt{r}}{\max\left(|\theta|,\sqrt{r}\right)} \leq \frac{r\theta^2}{c\max\left(\theta^2, r\right)} + \frac{c}{4} \leq c^{-1}r + \frac{c}{4} \tag{3.24}$$

$$-(\theta+z) \leq c^{-1}(\theta+z)^2 + \frac{c}{4} \tag{3.25}$$

we obtain the following estimate for all $x \in \mathbb{R}^2, d \in \mathbb{R}$:

$$\dot{W} \leq -\frac{c}{2}W(x) + \frac{1}{2\delta}\left(\theta^2 - r\right)^+ + \frac{1}{c}d^2 \tag{3.26}$$

Integrating directly estimate (3.26) and assuming that $d \in L^\infty(\mathbb{R}_+;\mathbb{R})$ we obtain the following estimate for every solution of (3.18) and for all $t \geq 0$:

$$|y(t)| \leq \exp\left(-\frac{c}{4}t\right)|y(0)| + \frac{2}{c}\|d\|_\infty + \sqrt{\frac{2}{c\delta}\left(\theta^2 - r\right)^+} \tag{3.27}$$

Estimate (3.27) shows that the disturbance-free version of system (3.18) satisfies the p-UGAOS property for all $\theta \in \mathbb{R}$ with residual constant $\alpha = \sqrt{\frac{2}{c\delta}(\theta^2 - r)^+}$. Moreover, estimate (3.27) shows that system (3.18) satisfies the p-IOS property for all $\theta \in \mathbb{R}$ with residual constant $\alpha = \sqrt{\frac{2}{c\delta}(\theta^2 - r)^+}$. Moreover, the gain function of the disturbance $\gamma(s) = \frac{2}{c}s$ for all $s \geq 0$ is assignable. Furthermore, convergence is exponential with assignable convergence rate (equal to $c/4$).

It is clear that the disturbance-free version of system (3.18) has many stability properties (Lagrange and Lyapunov stability, GAOS and p-UGAOS for all $\theta \in \mathbb{R}$, UGAOS for $\theta \leq c^{-1}r + \frac{3c}{4}$, exponential convergence for $\theta < c^{-1}r + \frac{3c}{4}$) which guarantee satisfactory performance. This is shown in Figure 3.3, where the plots are given for the closed-loop system (3.1) with the adaptive controller (3.16) of the state variables and the control input in the disturbance-free case.

Figure 3.3 shows that no identification is performed. This is expected and the parameter estimate $\hat{\theta}(t)$ works as a dynamic gain (as in many other adaptive controllers that we have already presented).

Moreover, the p-IOS property that holds for the closed-loop system (3.1), (3.16) is a robustness property that should not be neglected. However, there are severe robustness problems for the closed-loop system (3.1), (3.16): there is no guarantee that the state remains bounded for an arbitrary persistent disturbance $d \in L^\infty(\mathbb{R}_+; \mathbb{R})$ with $d \notin L^2(\mathbb{R}_+; \mathbb{R})$. Moreover, the residual constant in the p-IOS estimate depends on the unknown parameter $\theta \in \mathbb{R}$. The poor robustness properties of the closed-loop system (3.1), (3.16) are shown in Figure 3.4, where the phenomenon of parameter drift is presented for a persistent disturbance.

3.4 ▪ Robustness + Performance + Identification: Feedback of Delayed State and Input

Delays were used in BaLSI aiming at the exact identification of the unknown parameter in finite time. However, the discontinuous nature of BaLSI made very difficult the theoretical analysis of the robustness properties of the closed-loop system. Therefore, one can think of a "continuous version" of BaLSI leading to an infinite-dimensional closed-loop system, for which a Lyapunov robustness analysis is feasible.

Indeed, this task was performed in [74]. To describe the feedback design procedure, we start by noticing that it is evident from (3.1) that for every $r > 0$ the following identity is valid for $t \geq r$:

$$y^2(t) - y^2(t-r) - 2\int_{t-r}^{t} y(s)u(s)ds = 2\theta\int_{t-r}^{t} y^2(s)ds + 2\int_{t-r}^{t} y(s)d(s)ds \tag{3.28}$$

It is tempting to design a feedback law for (3.1) of the form $u(t) = -\left(2c + \hat{\theta}(t)\right)y(t)$, where $c > 0$ is a constant and the parameter estimate $\hat{\theta}(t)$ is based on the following formula for the unknown parameter $\theta \in \mathbb{R}$ when $\int_{t-r}^{t} y^2(s)ds > 0$:

$$\hat{\theta}(t) = \frac{y^2(t) - y^2(t-r) - 2\int_{t-r}^{t} y(s)u(s)ds}{2\int_{t-r}^{t} y^2(s)ds} \tag{3.29}$$

There are two major issues for the implementation of this idea. First, the feedback law will give the current value of $u(t)$ based on the estimate $\hat{\theta}(t)$ which also depends on the history of $u(t)$. Therefore, the closed-loop system will not be a conventional delay system. Second, the mapping that gives the estimate $\hat{\theta}(t)$ is not continuous (it is not even locally bounded).

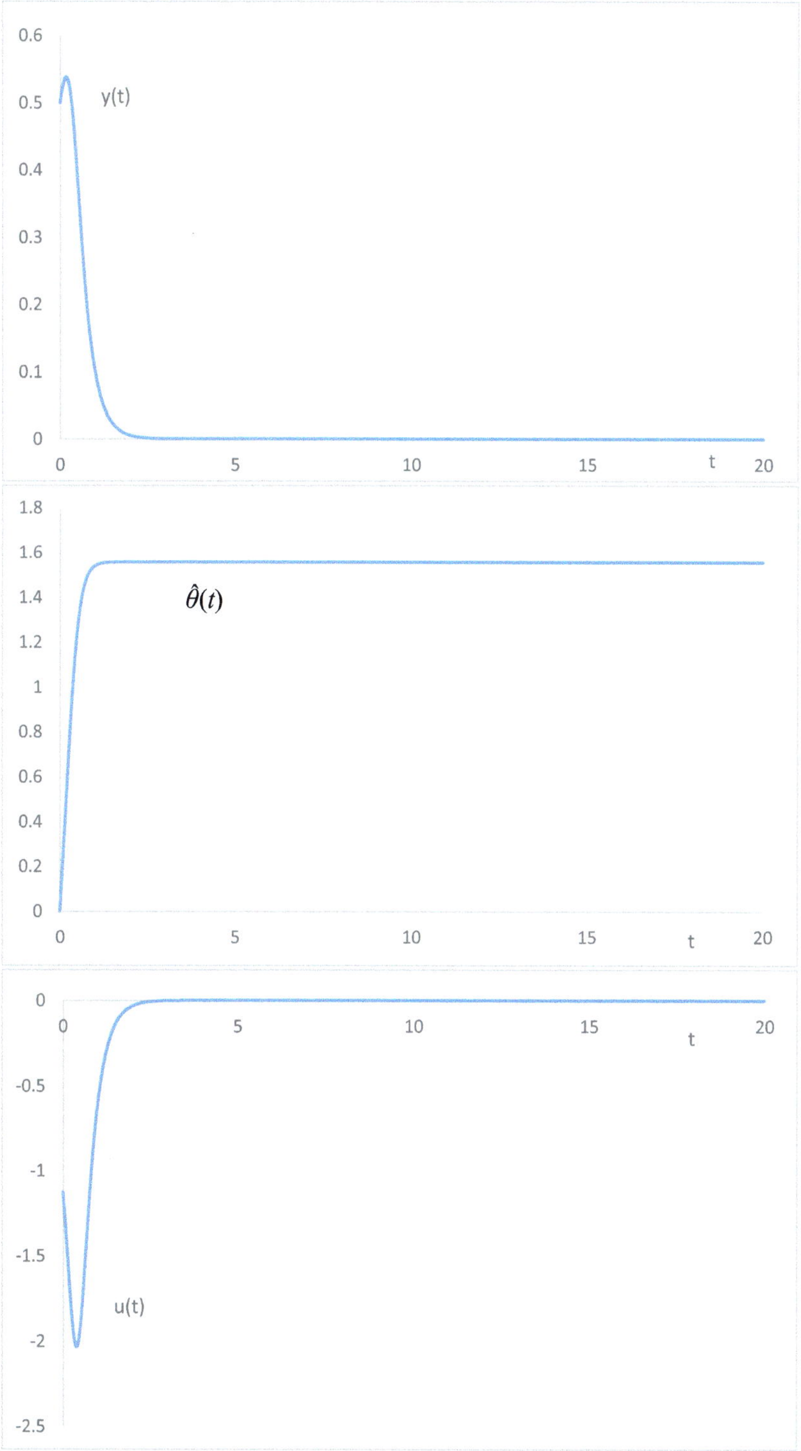

Figure 3.3. *The evolution of the state variables and the control input for the closed-loop system* (3.1), (3.16), *with* $d \equiv 0$, $\delta = 2$, $\theta = 3$, $c = 1, \Gamma = 10, r = 1$ *and* $y(0) = 0.5$, $\hat{\theta}(0) = 0$.

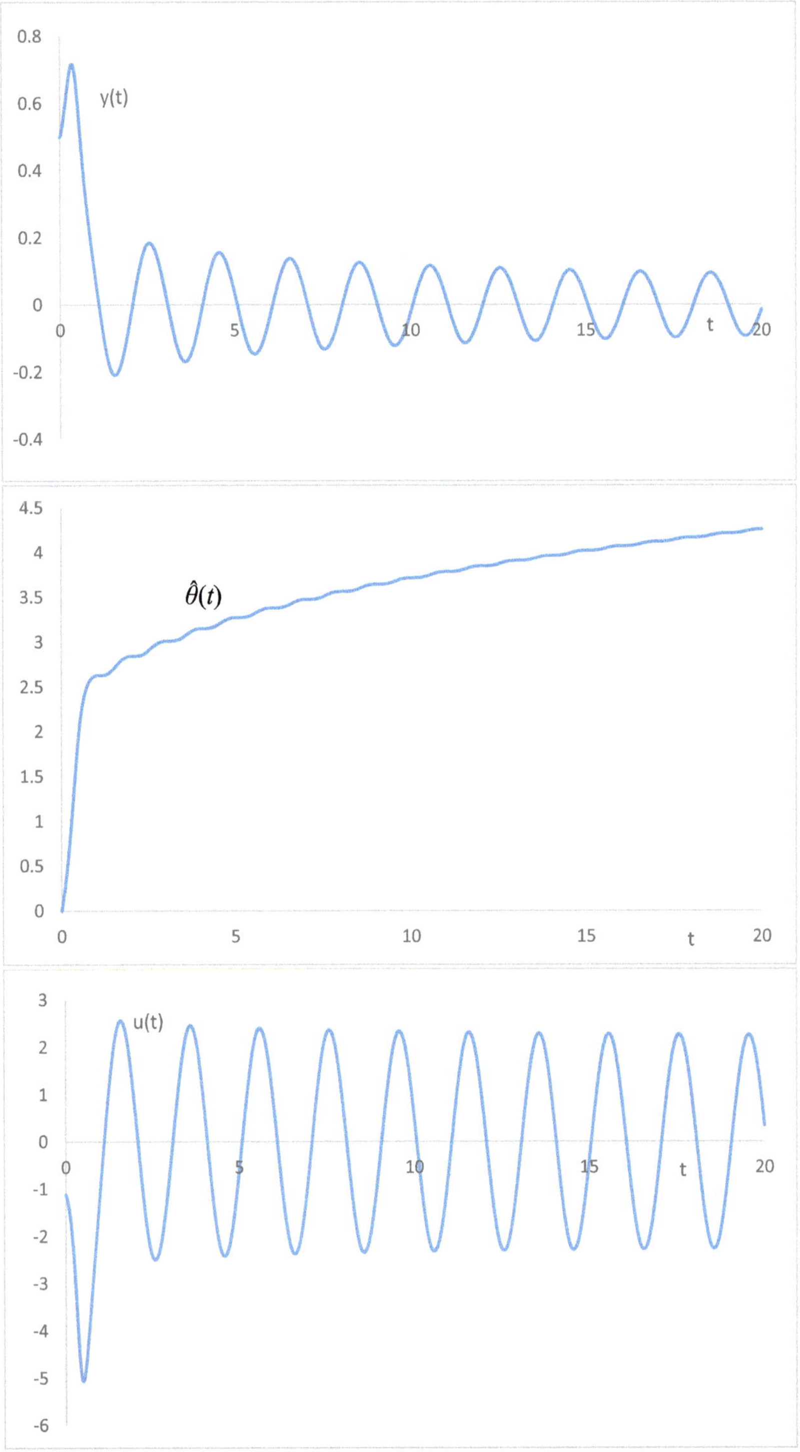

Figure 3.4. *The evolution of the state variables and the control input for the closed-loop system* (3.1), (3.16), *with* $d(t) = 2\sin(\pi t)$, $\delta = 2$, $\theta = 3$, $c = 1, \Gamma = 10, r = 1$ *and* $y(0) = 0.5$, $\hat{\theta}(0) = 0$.

Consequently, there will be severe problems for establishing the well-posedness (and robustness) of the closed-loop system.

A modification of formula (3.29) gives the adaptive controller

$$u(t) = -\left(2c + \frac{\left(y^2(t) - y^2(t-r) - 2\int_{t-r}^{t} y(s)u(s)ds\right)^+ + cry^2(t)}{2\int_{t-r}^{t} y^2(s)ds + 2\left(\varepsilon^2 - y^2(t)\right)^+}\right) y(t) \tag{3.30}$$

where $\varepsilon, c, r > 0$ are constants that can guarantee the well-posedness of the closed-loop system as well as a set of properties that cannot be guaranteed by other adaptive controllers. More specifically, we get the following result.

Theorem 2 in [74]: *Let* $\omega \in (0, c]$ *be an (arbitrary) constant with* $\exp(\omega r) < 2$. *For every* $\theta \in \mathbb{R}$ *there exist a constant* $M > 0$, *a function* $a \in K_\infty$ *and a non-decreasing function* $\gamma : \mathbb{R}_+ \to \mathbb{R}_+$ *such that for every* $(y_0, u_0) \in X$, $d \in L^\infty(\mathbb{R}_+)$ *where*

$$\begin{aligned} X &:= \left\{ (y,u) \in S \,:\, u(0) = -\left(\frac{\left(y^2(0) - y^2(-r) - 2\int_{-r}^{0} y(s)u(s)ds\right)^+ + cry^2(0)}{2\int_{-r}^{0} y^2(s)ds + 2\left(\varepsilon^2 - y^2(0)\right)^+} + 2c \right) y(0) \right\} \\ S &:= W^{1,\infty}((-r,0)) \times C^0([-r,0]) \end{aligned} \tag{3.31}$$

there exists a unique pair of continuous mappings $y, u : \mathbb{R}_+ \to \mathbb{R}$ *with* $y \in W^{1,\infty}(\mathbb{R}_+)$, $y(s) = y_0(s)$, $u(s) = u_0(s)$ *for all* $s \in [-r, 0]$ *for which (3.1) holds for* $t \geq 0$ *a.e.,* (3.30) *holds for all* $t \geq 0$, *and the following estimates hold for all* $t \geq 0$:

$$|y(t)| \leq M \exp\left(-c(t-r)^+\right) |y_0(0)| + \varepsilon + \left(1 + \sqrt{r} M \exp\left(-c(t-r)^+\right)\right) \frac{\|d\|_\infty}{2^{1/4} c} \tag{3.32}$$

$$\begin{aligned} \max_{t-r \leq s \leq t}(|u(s)|) &\leq \exp(-\omega t) a\left(\max_{-r \leq s \leq 0}(|y_0(s)|) + \sup_{-r<s<0}(|\dot{y}_0(s)|) + \max_{-r \leq s \leq 0}(|u_0(s)|) \right) \\ &+ \gamma\left(\max_{-r \leq s \leq 0}(|y_0(s)|) + \sup_{-r<s<0}(|\dot{y}_0(s)|) + \max_{-r \leq s \leq 0}(|u_0(s)|) + \|d\|_\infty \right) \|d\|_\infty + \varepsilon\rho \end{aligned} \tag{3.33}$$

where $\rho := 2|\theta| + \frac{1}{2\sqrt{2r}} + c(r+3)$.

The stability estimate (3.32) is the infinite-dimensional analogue of the p-IOS property with respect to the disturbance d with assignable asymptotic gain (equal to $2^{-1/4}c^{-1}$), assignable exponential convergence rate (equal to c) and assignable residual constant (equal to ε). Moreover, the stability estimate (3.33) combined with (3.32) provides the infinite-dimensional analogue of the p-UBIBS property for the closed-loop system (3.1) with (3.30).

It should be noticed that the feedback law (3.30) does not use a parameter estimate $\hat{\theta}$; the feedback law (3.30) is a direct adaptive controller. However, an identifier can be proposed for system (3.1) by exploiting the ideas described above. The identifier can work in parallel with any direct adaptive controller for (3.1) and it is given by the following equations:

$$\frac{d\hat{\theta}}{dt}(t) = 0 \text{ for all } t \in [ir, (i+1)r),\ i \in \mathbb{N} \tag{3.34}$$

$$\hat{\theta}((i+1)r) = \hat{\theta}(ir) \text{ if } \max_{ir \leq l \leq (i+1)r}\left(\int_{l-r}^{l} y^2(s)ds\right) < \sigma^2,\ i \in \mathbb{N} \tag{3.35}$$

$$\begin{aligned} \hat{\theta}((i+1)r) &= \frac{y^2(\tau) - y^2(\tau-r) - 2\int_{\tau-r}^{\tau} y(s)u(s)ds}{2\int_{\tau-r}^{\tau} y^2(s)ds} \text{ if } \max_{ir \leq l \leq (i+1)r}\left(\int_{l-r}^{l} y^2(s)ds\right) \geq \sigma^2,\ i \in \mathbb{N} \\ \text{with } \tau &= \max\left\{ t \in [ir, (i+1)r] \,:\, \int_{t-r}^{t} y^2(s)ds = \max_{ir \leq l \leq (i+1)r}\left(\int_{l-r}^{l} y^2(s)ds\right) \right\} \end{aligned} \tag{3.36}$$

where $\sigma > 0$ is a constant. It is clear that the identifier (3.34), (3.35), (3.36) uses estimate (3.29) at the most recent time in the interval $[ir, (i+1)r]$ for which the $L^2(-r, 0)$ norm of y is greater or equal to σ. When there is no time in the interval $[ir, (i+1)r]$ for which the $L^2(-r, 0)$ norm of y is greater or equal to σ, then the identifier (3.34), (3.35), (3.36) does not update the parameter estimate $\hat{\theta}$.

Generalized form of Theorem 3 in [74]: *For every $\theta \in \mathbb{R}$, $d, u \in L^\infty(-r, +\infty)$ and for every $y \in C^0([-r, +\infty))$ that satisfies* (3.1) *for $t \geq 0$ a.e. there exists $\hat{\theta} \in L^\infty(\mathbb{R}_+)$ that satisfies (3.24), (3.25), (3.26) for all $t \geq 0$. Moreover, the following property holds: if there exists $\tau \geq r$ with $\int_{\tau-r}^{\tau} y^2(s)ds \geq \sigma^2$ then there exists (a finite time) $T \in [\tau, \tau + r]$ for which the following estimate holds:*

$$\left|\hat{\theta}(t) - \theta\right| \leq \frac{\sqrt{r}}{\sigma} \|d\|_\infty \; for \; all \; t \geq T \tag{3.37}$$

Estimate (3.37) guarantees that the parameter estimate $\hat{\theta}(t)$ is always bounded (even when disturbances are present) and exact estimation in finite time of the parameter $\theta \in \mathbb{R}$ when there exists sufficient excitation (i.e., when there exists $\tau \geq r$ with $\int_{\tau-r}^{\tau} y^2(s)ds \geq \sigma^2$) and the disturbance is absent. Moreover, estimate (3.37) guarantees an ultimate estimation error which is proportional to the magnitude of the disturbance when there exists sufficient excitation of the state; this feature cannot be provided by other adaptive schemes in the literature.

The gain of the disturbance for the estimation error (equal to $\sqrt{r}/\sigma$) indicates a tradeoff between "learning" and "robustness": when $\sigma > 0$ is large then the gain of the disturbance is small (indicating strong robustness) but this implies that it is less likely the "learning" process to be complete (since it is less likely that there exists $\tau \geq r$ with $\int_{\tau-r}^{\tau} y^2(s)ds \geq \sigma^2$). On the other hand, when $\sigma > 0$ is small then the gain of the disturbance is large (indicating lack of robustness) but this implies that it is more possible the "learning" process to be complete.

The hybrid identifier (3.34), (3.35), (3.36) can be implemented without much difficulty, because the integrals $\int_{t-r}^{t} y(s)u(s)ds$, $\int_{t-r}^{t} y^2(s)ds$ needed in formulas (3.35), (3.36) can be evaluated numerically with great accuracy (particularly in the case where u is piecewise smooth). Another important feature of the identifier (3.34), (3.35), (3.36) is the fact that we have a criterion that shows whether or not identification is achieved: identification is achieved if there exists $\tau \geq r$ with $\int_{\tau-r}^{\tau} y^2(s)ds \geq \sigma^2$. When this criterion is not satisfied then we cannot have any trust in the parameter estimate $\hat{\theta}(t)$.

It should be noticed that estimate (3.37) guarantees that the gain of the disturbance for the estimation error (equal to $\sqrt{r}/\sigma$) is assignable and not the overall estimation error (less or equal to $\sqrt{r}\,\|d\|_\infty/\sigma$). The overall estimation error depends on the magnitude of the disturbance $d \in \mathbb{R}$ which is not a priori known.

The advantages of the delayed scheme (3.30), (3.34), (3.35), (3.36) are shown in Figure 3.5 and Figure 3.6. No parameter drift is present, all states are bounded even in the case of a persistent disturbance, and the plant state converges exponentially in the disturbance-free case to an ε-neighborhood of the origin. Moreover, for many initial conditions it is possible to achieve exact identification of the unknown parameter in the disturbance-free case.

The disadvantage of the delayed scheme (3.30), (3.34), (3.35), (3.36) is its complexity. First of all, the closed-loop system is infinite-dimensional with both y and u as states. Second, the extension of this scheme to systems that satisfy the matching condition or strict-feedback systems is very hard. Exactly, as in the case of BaLSI, one needs a kind of an observability property for the unknown parameters in order to be able to extend the scheme. However, an additional difficulty arises here: showing existence/uniqueness as well as the stability estimates for the closed-loop system is not a simple task (see [74] for the case (3.1)).

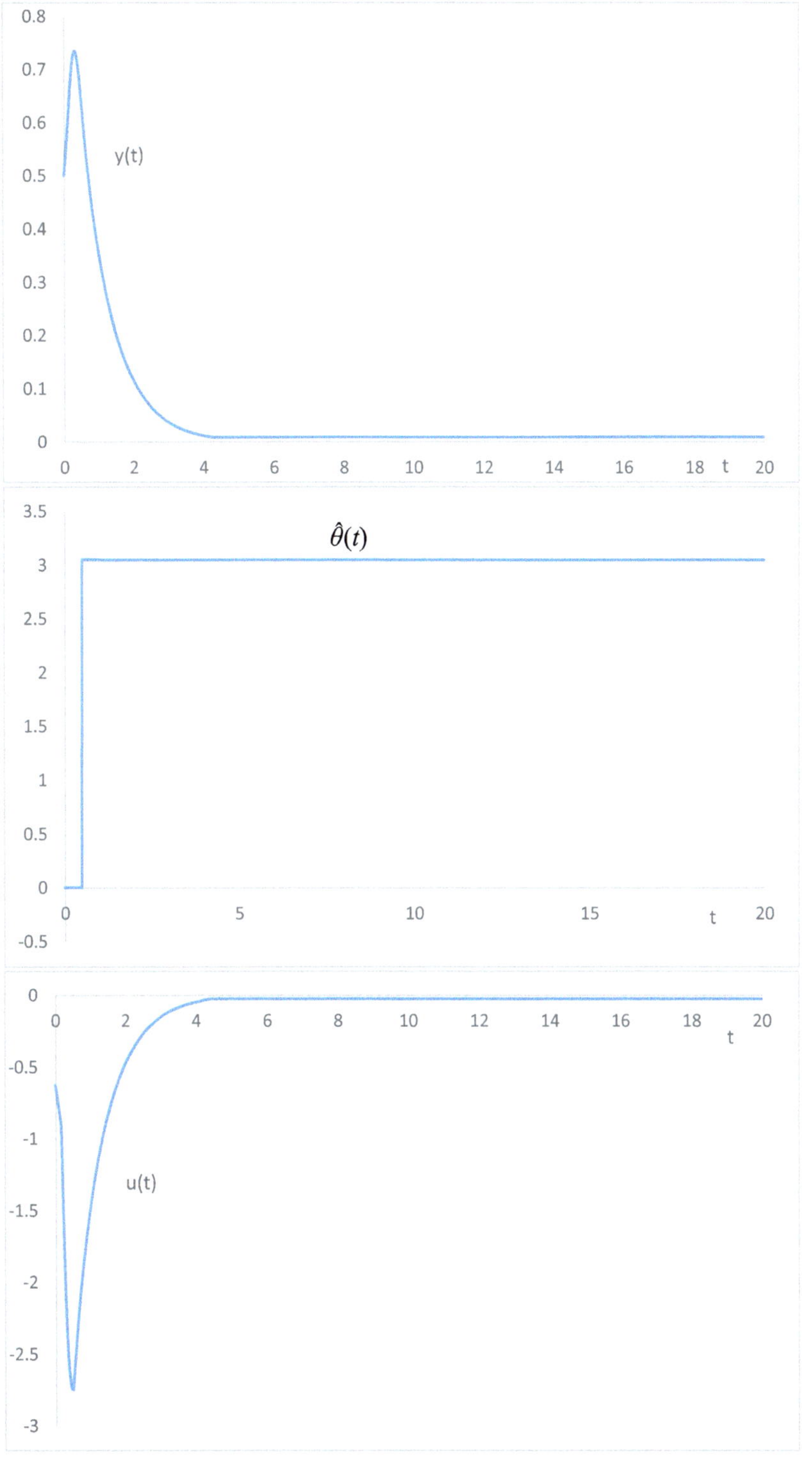

Figure 3.5. *The evolution of the state variables for the closed-loop system* (3.1), (3.30), (3.34), (3.35), (3.36), *with* $d(t) \equiv 0$, $\theta = 3$, $c = r = 0.5$, $\varepsilon = 0.01$, $\sigma = 0.447$ *and* $y(s) = 0.5$, $u(s) = 1$ *for* $s \in [-0.5, 0]$, $\hat{\theta}(0) = 0$.

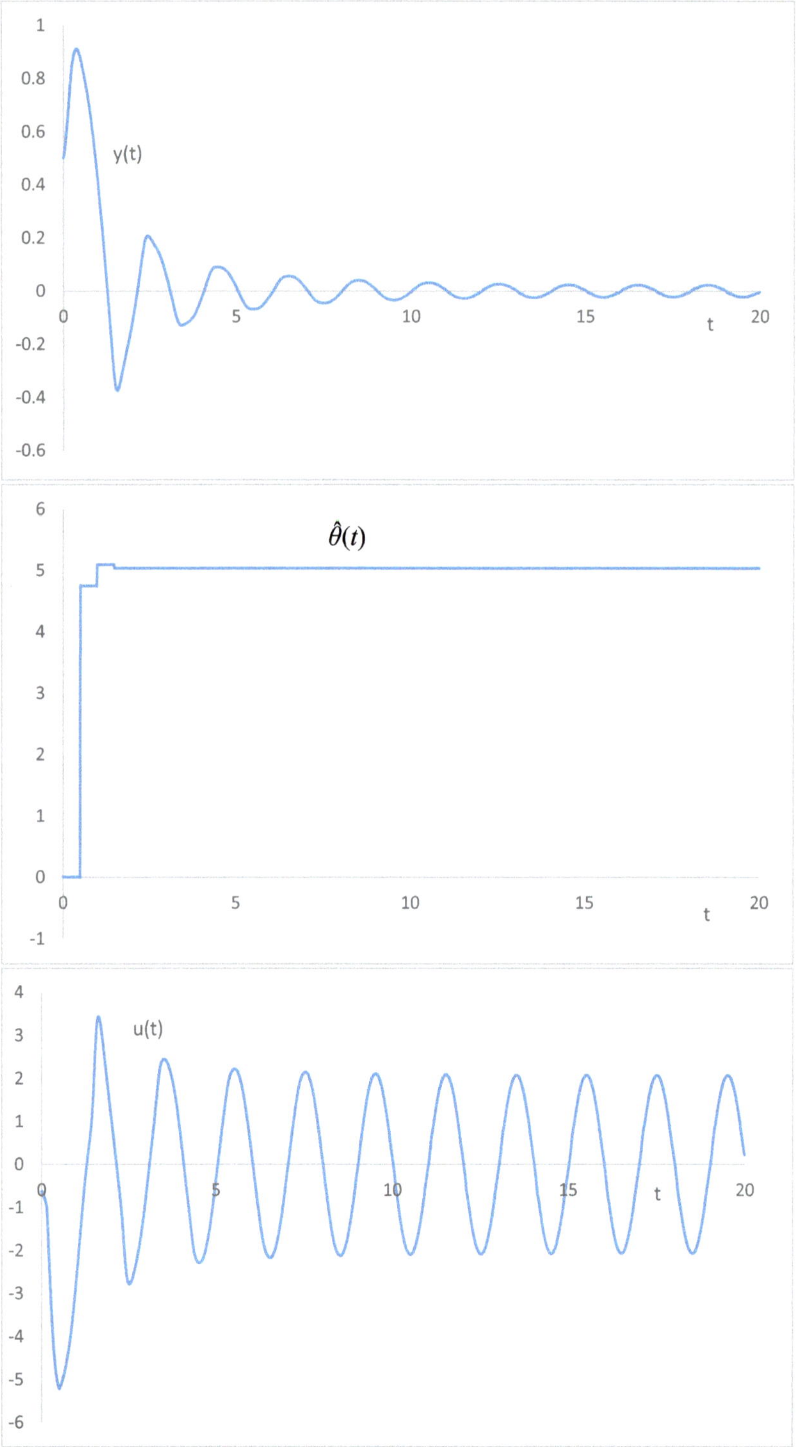

Figure 3.6. *The evolution of the state variables for the closed-loop system* (3.1), (3.30), (3.34), (3.35), (3.36), *with* $d(t) = 2\sin(\pi t)$, $\theta = 3$, $c = r = 0.5$, $\varepsilon = 0.01$, $\sigma = 0.447$ *and* $y(s) = 0.5$, $u(s) = 1$ *for* $s \in [-0.5, 0]$, $\hat{\theta}(0) = 0$.

Nevertheless, the idea that the identifier can work in parallel with any direct adaptive controller and, consequently, the identifier can focus on the task of identification only and not work as a dynamic gain, can be exploited as the following chapters show. Finally, the delayed adaptive scheme (3.30) showed that it is possible to assign the asymptotic residual constant for an uncertain system. It is, therefore, natural to search for simple adaptive schemes that can assign the asymptotic residual constant and at the same they can guarantee no gain drift.

3.5 ▪ Notes and comments

1. The use of delays in feedback control of finite-dimensional, continuous-time control systems is an idea that was proposed many years ago in control theory. The idea appears in the book [117] and has been exploited by many researchers: for instance, in [33], [64], [65] it is shown that the intentional use of delays to a delay-free system can guarantee properties that cannot be guaranteed by delay-free designs.

2. The difficulty of achieving robustness in nonlinear closed loops under adaptive control and the inherent limitations hidden are well known in the control literature: see the relevant discussions in [58], [59], [87], [115], [130] as well as the limitations for robustness that appear in discrete-time systems under adaptive control in [40] (but see also the review paper [41] and references therein). In the supervisory framework of direct adaptive control, practical stability in the presence of external disturbances for plants belonging to known bounded sets has been studied in [45], [46], [106].

3. Least-squares identification is attractive because of the ability of the Riccati equation to adjust the adaptation gain in real time to the actual signal content of the regressor matrix so that all the channels of the parameter estimation vector evolve at comparable speed, rather than at vastly different time scales (with the slowest one being dominant). Recent advances and applications in least-squares-based adaptive control are in [23], [84], [116], [151], [156]. However, in most of the least-squares adaptive control methods, the parameter convergence is not guaranteed without persistency of excitation (PE), which is seldom verifiable a priori. A different least-squares estimator, which uses a hybrid dead-beat observer with delays, was proposed in [68] and [69]. The estimator in [68] does not require PE but only the weaker assumption of strong observability and was shown to be robust with respect to measurement noise.

4. Dead-beat identifiers were first proposed in [1], [2], [43]. They require a PE assumption and are not of a least-squares type.

Chapter 4

Deadzone-Adapted Disturbance Suppression (DADS): The Scalar Case

4.1 ▪ Introduction

In this chapter we focus on a slightly more general version of the scalar example studied in Chapter 3, namely,

$$\dot{y} = \theta y \varphi(y) + u + d \tag{4.1}$$

where $y \in \mathbb{R}$ is the plant state, $u \in \mathbb{R}$ is the control input, $\theta \in \mathbb{R}$ is an unknown constant parameter, $d \in \mathbb{R}$ is an unknown possibly time-varying disturbance, and φ is a smooth function. For the scalar system (4.1) we provide a simple, novel adaptive control scheme that combines two elements: (a) nonlinear damping with dynamic gains, and (b) a Lyapunov-based deadzone in the update law. The proposed adaptive scheme is direct; here no identification is performed (which also explains the simplicity of the proposed scheme). We call the proposed controller a Deadzone-Adapted Disturbance Suppression (DADS) controller.

The DADS controller is explained in detail in the following chapters. This chapter provides the intuition behind the construction of the DADS controller: the update law "pumps up" the controller dynamic gain to a sufficient size, adapting the gain to the unknown disturbance, and shuts off further pumping up of the gain once the state is in the prespecified residual set, for the disturbance signal thus far. The only way the gain can grow without bound is if the disturbance grows without bound. Moreover, we show many properties that are guaranteed by the DADS controller: attenuation of the plant state to an *assignable* small level, despite the presence of disturbances and unknown parameters of arbitrary and unknown bounds, i.e., we achieve the zero p-OAG property with an asymptotic residual constant which is independent of the constant parameters and is assignable. The DADS controller also prevents gain and state drift regardless of the size of the disturbance and unknown parameter. The latter property is a consequence of the p-UBIBS property and the p-IOS property that both hold for the closed-loop system.

4.2 ▪ Goals

In our approach to robust adaptive control, namely, to control in the presence of unmeasured disturbance d of unlimited size and an unknown parameter θ which is also unrestricted in size, our goal is to drive the plant state y to a residual set of an arbitrarily small, desired size. This can be expressed, using a Lyapunov function $V(y)$, as the goal that

$$\limsup_{t \to +\infty} \left(V(y(t)) \right) \leq \varepsilon \tag{4.2}$$

where $\varepsilon > 0$ is an arbitrarily small, desired constant. We pursue this goal by using a feedback law that employs a positive gain that is increased ("pumped up") sufficiently high so that both the disturbing effect of d and the destabilizing effect of θ are overcome and $V(y)$ is ultimately brought below $\varepsilon > 0$. At the same time, the gain that is being pumped up should be prevented from drifting to infinity. One of the classical tools for preventing parameter drift, though never previously successful for large disturbances, is the deadzone. Hence, we want to pump up the controller's gain until $V(y)$ has entered the deadzone $V(y) \leq \varepsilon$. Once $V(y)$ is in the deadzone, the pumping stops.

4.3 ▪ Gain Pumping Design

As we indicated above, the controller gain being pumped is positive. Since such a gain is a dynamic variable, and is thus a state variable of the closed-loop system, the gain being positive, rather than, more generally, real, presents some mathematical inconvenience. The state space in that case becomes only "half" of a Euclidean space. For instance, with a dynamic gain $\kappa(t)$, used to control the state $y(t) \in \mathbb{R}$ of the scalar example (4.1), the state space is only the half plane $(y, \kappa) \in \mathbb{R} \times (0, +\infty)$.

To avoid this mathematical inconvenience, we represent κ as $\kappa = \exp(z)$, where $z \in \mathbb{R}$, so that the state (y, z) of the overall system is $\mathbb{R}^2$. This way, the global results that we obtain are free of mathematical awkwardness and imprecision.

Let us now return to designing a gain-pumping update with a drift-preventing deadzone. Let $\Gamma > 0$ denote the adaptation gain. As a reminder, the deadzone of deadzone band $\varepsilon > 0$, acting on $V(y) \geq 0$, is denoted by $(V(y) - \varepsilon)^+$, where, following the notation given in the notation section, the $(\cdot)^+$ denotes

$$s^+ = (0, s) = \mathrm{ReLU}(s) \tag{4.3}$$

where ReLU is the notation common in neural networks, representing the "rectifier linear unit."

If we were pumping a positive gain κ, using a deadzone, the update would be

$$\dot{\kappa} = \Gamma \left(V(y) - \varepsilon\right)^+ \tag{4.4}$$

If, instead, we represent κ as $\kappa = \exp(z)$, the update

$$\frac{d}{dt}(\exp(z)) = \Gamma \left(V(y) - \varepsilon\right)^+ \tag{4.5}$$

is more appropriately rewritten, using the chain rule, as

$$\dot{z} = \Gamma \exp(-z) \left(V(y) - \varepsilon\right)^+ \tag{4.6}$$

For the scalar example (4.1) we will take the simplest Lyapunov function $V(y) = y^2/2$, and the update law for z will be

$$\boxed{\dot{z} = \Gamma \exp(-z) \left(\frac{1}{2}y^2 - \varepsilon\right)^+} \tag{4.7}$$

4.4 ▪ Control Law Design

We have now designed our gain update, and explained its purpose, but we have not even begun to discuss the control law and how the gain $\exp(z)$ will be employed in the control law. Recalling that the scalar system is $\dot{y} = \theta y \varphi(y) + u + d$ and Lyapunov function $V(y) = y^2/2$, the derivative of the Lyapunov function is

$$\dot{V} = \theta y^2 \varphi(y) + yd + yu \tag{4.8}$$

The first two of the three terms in this expression are problematic, whereas the third term, with the control u, is to be designed to counteract the problematic terms.

Let us first examine the simpler of the problematic terms, yd, using Young's inequality:

$$yd \leq \frac{1+\exp(z)}{4a}y^2 + \frac{a}{1+\exp(z)}d^2 \tag{4.9}$$

where $a > 0$ is a gain that provides a degree of ISS gain assignment flexibility. Why are we using a "Young" coefficient $1 + \exp(z) > 0$, rather than simply $\exp(z) > 0$? The margin of 1 will prove to be useful and guarantees that the coefficient of d^2 is always less than a no matter what the value of z is.

Let us now turn our attention to the term $\theta y^2\varphi(y)$ in $\dot{V}$, which is more complicated to handle when we want not only $\limsup_{t\to+\infty} (V(y(t))) \leq \varepsilon$ but also $V(y(t)) \to 0$ when $d(t) \to 0$, namely in the disturbance-vanishing case. We proceed through two applications of Young's inequality and use of a deadzone:

$$\begin{aligned}
&\theta y^2\varphi(y) \leq |\theta|\,|\varphi(y)|\,y^2 \\
&= (|\theta| - 1 - \exp(z))\,|\varphi(y)|\,y^2 + (1+\exp(z))\,|\varphi(y)|\,y^2 \\
&\leq (|\theta| - 1 - \exp(z))^+\,|\varphi(y)|\,y^2 + (1+\exp(z))\,|\varphi(y)|\,y^2 \\
&\leq a\tfrac{\left((|\theta|-1-\exp(z))^+\right)^2}{1+\exp(z)} + \tfrac{1+\exp(z)}{4a}\varphi^2(y)y^4 + (1+\exp(z))\,|\varphi(y)|\,y^2 \\
&\leq a\tfrac{\left((|\theta|-1-\exp(z))^+\right)^2}{1+\exp(z)} + \tfrac{1+\exp(z)}{4a}\varphi^2(y)y^4 + \tfrac{c}{2}(1+\exp(z))\,y^2 + \tfrac{(1+\exp(z))}{2c}\varphi^2(y)y^2
\end{aligned} \tag{4.10}$$

In summary, with the bounds on yd and $\theta y^2\varphi(y)$, one obtains

$$\begin{aligned}
\dot{V} \leq\ & y\left(u + \tfrac{1+\exp(z)}{2}\left(\tfrac{\varphi^2(y)y^2+1}{2a} + c + \tfrac{\varphi^2(y)}{c}\right)y\right) \\
&+a\tfrac{d^2+\left((|\theta|-1-\exp(z))^+\right)^2}{1+\exp(z)}
\end{aligned} \tag{4.11}$$

Taking the control law

$$\boxed{u = -\frac{c}{2}y - \frac{1+\exp(z)}{2}\left(\frac{1+\varphi^2(y)y^2}{2a} + c + \frac{\varphi^2(y)}{c}\right)y} \tag{4.12}$$

written alternatively as

$$u = -py - (1+\exp(z))\left(p + q + \frac{1}{4p}\varphi^2(y) + qy^2\varphi^2(y)\right)y \tag{4.13}$$

with $p, q > 0$, which employs nonlinear damping and dynamic gain, we obtain

$$\dot{V} \leq -cV(y) + a\frac{d^2 + \left((|\theta| - 1 - \exp(z))^+\right)^2}{1+\exp(z)}. \tag{4.14}$$

The same inequality (4.14) is also obtained with a somewhat different completion of squares, leading to a somewhat different feedback law

$$u = -\left((1+\exp(z))\frac{1+\varphi^2(y)y^2}{4a} + c + \frac{\varphi^2(y)}{2c}(1+\exp(z))^2\right)y \tag{4.15}$$

It is this feedback law that we use in the simulations at the end of this section and elsewhere. But we highlight (4.12) as the simplest choice, linear in the dynamic gain $1+\exp(z)$.

Let us now discuss why (4.14) is relevant for achieving (4.2) and even $V(y(t)) \to 0$ when $d(t) \to 0$. We provide only intuition. The proofs are to follow. Recall that the state z is being pumped up until $V(y)$ falls below ε and the deadzone turns off the pumping. Let us also note that $1+\exp(z)$ can be pumped to as large a value as it takes to force $V(y)$ under ε by dominating $a\left(d^2+\left((|\theta|-1-\exp(z))^+\right)^2\right)$ in $\dot{V}$. Moreover, if $1+\exp(z)$ is pumped above $|\theta|$, in the disturbance-vanishing case, y is regulated to zero.

Next we prove four major properties guaranteed by the adaptive controller (4.7), (4.12), in four subsections, colloquially titled for the four properties: practical IOS, no parameter drift, practical regulation for arbitrary disturbance, and asymptotic gain to disturbance.

4.5 ▪ Practical IOS

It follows from (4.14) that every solution of the closed-loop system (4.1) with (4.7), (4.12) satisfies

$$\begin{aligned} V(y(t)) &\le \exp(-ct)V(y(0)) + a\int_0^t \exp(-c(t-s))\frac{d^2(s)+\left((|\theta|-1-\exp(z(s)))^+\right)^2}{1+\exp(z(s))}ds \\ &\le \exp(-ct)V(y(0)) + a\int_0^t \exp(-c(t-s))\frac{\|d\|_\infty^2+\left((|\theta|-1-\exp(z(s)))^+\right)^2}{1+\exp(z(s))}ds \end{aligned} \tag{4.16}$$

Since $z(t)$ is non-decreasing, $z(t) \ge z(0)$, we get

$$\begin{aligned} V(y(t)) &\le \exp(-ct)V(y(0)) + a\int_0^t \exp(-c(t-s))ds\frac{\|d\|_\infty^2+\left((|\theta|-1-\exp(z(0)))^+\right)^2}{1+\exp(z(0))} \\ &\le \exp(-ct)V(y(0)) + a\frac{\|d\|_\infty^2+\left((|\theta|-1-\exp(z(0)))^+\right)^2}{c(1+\exp(z(0)))} \end{aligned} \tag{4.17}$$

Thus, the y-component of the state satisfies

$$\boxed{V(y(t)) \le \exp(-ct)V(y(0)) + a\frac{\|d\|_\infty^2+\left((|\theta|-1-\exp(z(0)))^+\right)^2}{c\left(1+\exp(z(0))\right)}} \tag{4.18}$$

From estimate (4.18) we can infer that

$$|y(t)| \le \exp(-ct/2)\,|y(0)| + \sqrt{\frac{2a}{c}}\,\|d\|_\infty + \sqrt{\frac{2a}{c}}\,(|\theta|-1)^+ \tag{4.19}$$

Estimate (4.19) holds as long as the solution of the closed-loop system (4.1) with (4.7), (4.12) exists. Once, we establish forward completeness, we can say that this is a property of practical Input-to-Output Stability (p-IOS), and is desirable, but it does not represent, qualitatively, an improvement over the σ-modification (with nonlinear damping) since the residual constant $\sqrt{\frac{2a}{c}}\,(|\theta|-1)^+$ depends on the unknown θ. A property superior to σ-modification is established shortly. What follows is the longest and the most novel part of our proof.

4.6 ▪ No Parameter Drift

Next, we derive a bound which $z(t)$ does not exceed. The bound is complicated and depends on the initial states $y(0)$, $z(0)$, the unknown θ, and the unmeasured bounded d.

Lemma 4.1: *Let $T \in (0, +\infty]$ be given. Consider two absolutely continuous functions $z, V : [0,T) \to \mathbb{R}$ for which the following differential inequalities hold for $t \in [0,T)$ a.e.:*

$$0 \leq \dot{z}(t) \leq \Gamma \exp(-z(t))\,(V(t) - \varepsilon)^{+} \tag{4.20}$$

$$\dot{V}(t) \leq -cV(t) + \frac{\mu}{1 + \exp(z(t))} \tag{4.21}$$

for certain constants $c, \varepsilon, \Gamma > 0$ and $\mu \geq 0$. Then the following estimate holds for all $t \in [0,T)$:

$$z(0) \leq z(t) \leq \ln\left(\max\left(\frac{\mu}{c\varepsilon} - 1, \exp(z(0))\right) + \frac{\Gamma}{c}\left(V(0) + \frac{\mu}{c(1 + \exp(z(0)))} - \varepsilon\right)^{+}\right) \tag{4.22}$$

Before proving Lemma 4.1, we discuss its meaning. The lemma deals with feedback dynamics, where V is an input to the z-subsystem and z is an input to the V-subsystem. The V-subsystem is already bounded since (4.21) can be majorized as $\dot{V}(t) \leq -cV(t) + \mu$ and bounded as $V(t) \leq V(0) + \mu/c$. Hence, the z-subsystem is the subject of study of this feedback loop. The z-subsystem is nothing more than an integrator of $(V - \varepsilon)^{+}$ with an output $\exp(z)$. Such a system's output is not necessarily bounded, depending on the input signal $V(t)$. But, if V is governed by, to abuse terminology, the "reciprocal ISS" dynamics (4.21) relative to the input z, where a larger $\exp(z)$ makes V smaller (hence the term "reciprocal"), Lemma 4.1 establishes that the boundedness of $z(t)$ is maintained. Since an integrator is the simplest form of a passive system, it is good to recall the standard passivity theorem, which establishes that a *negative* feedback loop of passive systems is stable (and hence, if globally, with bounded states). While the z-subsystem is passive, the V-subsystem $\dot{V} \leq -cV + \frac{\mu}{1+\exp(z)}$ is neither passive nor negatively fed into the z-subsystem. However, the reciprocal appearance of $\exp(z)$ in the V-subsystem acts as a "repressing" function on V. And, since the relative degree from z to V is unity, Lemma 4.1 aligns with the intuition from the passivity theorem. But the proof of Lemma 4.1 differs immensely from the simple "sum of storage functions" proof of the passivity theorem. The bound (4.22) is derived through a series of estimates derived for the z-subsystem (4.20) and the V-subsystem (4.21).

Proof of Lemma 4.1: Our first step is to show that, when $z(t)$ grows dominant over μ, the difference $V(t) - \varepsilon$ decays *exponentially*, based on the assumed inequality (4.21). Let $\delta > 0$ be a given positive constant (arbitrary). We distinguish two cases:

Case 1: $\exp(z(t)) < \max\left(\frac{\mu}{c\varepsilon} - 1, \exp(z(0)) + \delta\right)$ for all $t \in [0,T)$.

Case 2: There exists $\tau \in [0,T)$ with $\exp(z(\tau)) \geq \max\left(\frac{\mu}{c\varepsilon} - 1, \exp(z(0)) + \delta\right)$. Since $\exp(z(0)) < \max\left(\frac{\mu}{c\varepsilon} - 1, \exp(z(0)) + \delta\right)$, by continuity there exists $\tilde{T} \in [0,\tau]$ with

$$\exp(z(\tilde{T})) = \max\left(\frac{\mu}{c\varepsilon} - 1, \exp(z(0)) + \delta\right). \tag{4.23}$$

Since $z(t)$ is non-decreasing (recall (4.10)), we get $\exp(z(t)) \geq \max\left(\frac{\mu}{c\varepsilon} - 1, \exp(z(0)) + \delta\right)$ and thus, $\exp(z(t)) \geq \frac{\mu}{c\varepsilon} - 1$, for all $t \geq \tilde{T}$. Hence, we get from (4.21) the estimate for $t \in \left[\tilde{T}, T\right)$ a.e.

$$\frac{d}{dt}(V - \varepsilon) \leq -c(V - \varepsilon) \tag{4.24}$$

Differential inequality (4.24) implies the estimate

$$(V(t)-\varepsilon) \leq \exp(-c(t-\tilde{T}))\left(V(\tilde{T})-\varepsilon\right) \tag{4.25}$$

and the estimate

$$(V(t)-\varepsilon)^{+} \leq \exp(-c(t-\tilde{T}))\left(V(\tilde{T})-\varepsilon\right)^{+} \tag{4.26}$$

for all $t \in \left[\tilde{T}, T\right)$.

Next, we show that the exponential decay of $(V(t)-\varepsilon)^{+}$ bounds the rate $\frac{d}{dt}(\exp(z(t)))$ to be *exponentially* decaying as well. Indeed, from (4.20), we get

$$\frac{d}{dt}(\exp(z(t))) \leq \Gamma \exp(-c(t-\tilde{T}))\left(V(\tilde{T})-\varepsilon\right)^{+} \tag{4.27}$$

for $t \in \left[\tilde{T}, T\right)$ a.e.. Finally, as a result of the exponential decay in the rate of $z(t)$ in (4.27), the growth of $z(t)$ settles exponentially to a certain bound on $\exp(z(t))$, which depends on everything: $\mu, c, \Gamma, z(0), V(0)$, and ε. We proceed to prove this. Differential inequality (4.21) and the fact that $z(t)$ is non-decreasing (recall (4.10)), implies the following estimate for all $t \in [0, T)$:

$$\begin{aligned} & V(t) \leq \exp(-ct)V(0) + \textstyle\int_0^t \exp(-c(t-s))\frac{\mu}{1+\exp(z(s))}ds \\ & \leq \exp(-ct)V(0) + \frac{\mu}{1+\exp(z(0))}\textstyle\int_0^t \exp(-c(t-s))ds \\ & \leq V(0) + \frac{\mu}{c(1+\exp(z(0)))} \end{aligned} \tag{4.28}$$

Integrating and using (4.27) we obtain the estimate for all $t \in \left[\tilde{T}, T\right)$

$$\begin{aligned} & \exp(z(t)) = \exp(z(\tilde{T})) + \textstyle\int_{\tilde{T}}^t \frac{d}{ds}(\exp(z(s)))\, ds \\ & \leq \exp(z(\tilde{T})) + \textstyle\int_{\tilde{T}}^t \Gamma \exp(-c(s-\tilde{T}))\left(V(\tilde{T})-\varepsilon\right)^{+} ds \\ & \leq \exp(z(\tilde{T})) + \frac{\Gamma}{c}\left(V(\tilde{T})-\varepsilon\right)^{+} \end{aligned} \tag{4.29}$$

To complete the proof, the above bound on $\exp(z(t))$ expressed in terms of $z(\tilde{T})$, $V(\tilde{T})$, is expressed in terms of $z(0)$, $V(0)$:

$$\begin{aligned} & \exp(z(t)) \leq \max\left(\frac{\mu}{c\varepsilon}-1, \exp(z(0))+\delta\right) + \frac{\Gamma}{c}\left(V(\tilde{T})-\varepsilon\right)^{+} \\ & \leq \max\left(\frac{\mu}{c\varepsilon}-1, \exp(z(0))+\delta\right) + \frac{\Gamma}{c}\left(V(0)+\frac{\mu}{c(1+\exp(z(0)))}-\varepsilon\right)^{+} \end{aligned} \tag{4.30}$$

where (4.28) gas been used in the last step. Since $z(t)$ is non-decreasing, we conclude that estimate (4.30) holds for all $t \in [0, T)$.

Taking into account both cases, we conclude that (4.30) holds for both cases for all $t \in [0, T)$, and, since $\delta > 0$ is arbitrary, by continuity it also holds for $\delta = 0$ and for all for all $t \in [0, T)$:

$$\exp(z(t)) \leq \max\left(\frac{\mu}{c\varepsilon}-1, \exp(z(0))\right) + \frac{\Gamma}{c}\left(V(0)+\frac{\mu}{c\,(1+\exp(z(0)))}-\varepsilon\right)^{+} \tag{4.31}$$

Estimate (4.22) is a direct consequence of the above inequality and the fact that $z(t)$ is non-decreasing. The proof is complete. ◁

Inequality (4.14) allows to show that the z-component of the state is bounded. Indeed, inequality (4.14) and the fact that $z(t)$ is non-decreasing implies that (4.21) holds with

$$\mu = a\left(\|d\|_\infty^2 + \left((|\theta| - 1 - \exp(z(0)))^+\right)^2\right) \tag{4.32}$$

Consequently, Lemma 4.1 implies that the following estimate holds as long as the solution of the closed-loop system (4.1) with (4.7), (4.12) exists

$$\boxed{z(0) \le z(t) \le \Omega\left(\|d\|_\infty, |\theta| - 1, |y(0)|, \exp(z(0))\right)} \tag{4.33}$$

where $\Omega : \mathbb{R}_+ \times \mathbb{R} \times \mathbb{R}_+ \times (0, +\infty) \to \mathbb{R}$ is the continuous function

$$\begin{aligned} &\Omega(s_1, s_2, s_3, s_4) \\ &:= \ln\left(\max\left(\tfrac{a}{c\varepsilon}\left(s_1^2 + \left((s_2 - s_4)^+\right)^2\right) - 1, s_4\right) + \tfrac{\Gamma}{c}\left(\tfrac{1}{2}s_3^2 + a\tfrac{s_1^2 + \left((s_2-s_4)^+\right)^2}{c(1+s_4)} - \varepsilon\right)^+\right) \end{aligned} \tag{4.34}$$

which is non-decreasing in s_1, s_2, s_3 but non-monotone with respect to s_4.

Therefore, the z-component of the state remains bounded as long as the solution exists. Based on all the above a priori estimates, we conclude that the solution exists for all $t \ge 0$ and satisfies estimates (4.18), (4.33) for all $t \ge 0$.

The simplest observation from (4.33) is the absence of an unbounded parameter drift, achieved thanks to the deadzone in the update (4.7), as well as the use of the adapted gain z in the nonlinear damping terms in the controller (4.12). But parameter drift is also stopped with the σ-modification. We show next a practical regulation property, independent of the unknown parameter size, and heretofore unachieved in conventional robust adaptive control, where the σ-modification is prevalent.

4.7 ▪ Practical Regulation for Arbitrary Disturbance

Next, we show the key property of practical regulation (4.2). Practical regulation is shown with Barbălat's Lemma, whose conditions are satisfied by virtue of the no-drift property (4.33) and the p-IOS property (4.19). Estimate (4.33) and $z(t)$ being non-decreasing guarantees that $z(t)$ has a finite limit as $t \to +\infty$, as does $\exp(z(t))$. From (4.1), (4.12) and since d is bounded, so is $\frac{d}{dt}(V(y(t)))$. It follows that $\frac{d}{dt}(\exp(z(t))) = \Gamma\,(V(y(t)) - \varepsilon)^+$ is uniformly continuous, in addition to $\int_0^{+\infty} \frac{d}{dt}(\exp(z(t)))\,dt = \lim_{t\to+\infty}(\exp(z(t))) - \exp(z(0)) < +\infty$. From Barbălat's Lemma (see [73])

$$\lim_{t\to+\infty}\left(\frac{d}{dt}(\exp(z(t)))\right) = \Gamma \lim_{t\to+\infty}\left((V(y(t)) - \varepsilon)^+\right) = 0 \tag{4.35}$$

Therefore, we obtain (4.2), namely,

$$\boxed{\limsup_{t\to+\infty}(V(y(t))) \le \varepsilon} \tag{4.36}$$

Property (4.2) is the "marquee" property of DADS, the one that is most unequivocally unachievable with previous robust adaptive schemes. As a reminder, we call this property zero p-OAG. Before possibly jumping to the conclusion that the zero p-OAG property is merely the result of a regulation (Lyapunov) based update and the use of a deadzone, we remind the reader that such an update would not be drift-free under a large disturbance if it weren't for the update-based nonlinear damping in the controller (4.12). This can be discerned from the proof of Lemma 4.1 (the calculation leading to (4.16)).

4.8 ▪ Asymptotic Gain to Disturbance

Using (4.18) and the semigroup property for the solution of the closed-loop system (4.1), (4.7), (4.12), we get

$$\boxed{\limsup_{t\to+\infty}\left(V(y(t))\right) \le a\frac{\left(\limsup\limits_{t\to+\infty}\left(|d(t)|\right)\right)^2+\left(\left(|\theta|-1-\exp\left(\lim\limits_{t\to+\infty}(z(t))\right)\right)^+\right)^2}{c\left(1+\exp\left(\lim\limits_{t\to+\infty}(z(t))\right)\right)}} \tag{4.37}$$

Combining the above estimate with (4.2) we obtain

$$\limsup_{t\to+\infty}\left(|y(t)|\right) \le \sqrt{2\min\left(\varepsilon, a\frac{\left(\limsup\limits_{t\to+\infty}(|d(t)|)\right)^2+\left(\left(|\theta|-1-\exp\left(\lim\limits_{t\to+\infty}(z(t))\right)\right)^+\right)^2}{c\left(1+\exp\left(\lim\limits_{t\to+\infty}(z(t))\right)\right)}\right)} \tag{4.38}$$

Estimates (4.18), (4.33) show that the DADS controller (4.7), (4.12) achieves boundedness of the state for all initial conditions and for all bounded disturbances. Estimate (4.38) shows that the DADS controller (4.7), (4.12) can "push" the plant state into the region $|y| \le \sqrt{2\varepsilon}$ for an arbitrarily small desired $\varepsilon > 0$, no matter what the bounded disturbance is. By examining Ω as defined in (4.34), one sees that the upper bound on the gain z increases as the desired regulation bound $\varepsilon > 0$ decreases toward zero, as expected, and also as the adaptation gain $\Gamma > 0$ increases.

The results in the following chapters show that all these properties are guaranteed even if $\theta \in \mathbb{R}$ is time-varying. More specifically, we show that all these properties are valid when $\theta \in L^\infty(\mathbb{R}_+;\mathbb{R})$. The results of the following chapters also show that there can be more degrees of freedom for the design of the DADS controller. However, the simulations that are shown below and the conclusions that can be deduced from the above estimates do not change qualitatively.

Estimate (4.19) shows that the convergence of the plant state to the region

$$|y| \le \sqrt{2a\frac{\|d\|_\infty^2+\left(\left(|\theta|-1-\exp(z(0))\right)^+\right)^2}{c\left(1+\exp(z(0))\right)}} \tag{4.39}$$

is exponential with assignable convergence rate equal to $c/2$. On the other hand, the convergence of the dynamic gain to its limit value and the convergence of the plant state to the region $|y| \le \sqrt{2\min\left(\varepsilon, a\frac{\left(\limsup\limits_{t\to+\infty}(|d(t)|)\right)^2+\left(\left(|\theta|-1-\exp\left(\lim\limits_{t\to+\infty}(z(t))\right)\right)^+\right)^2}{c\left(1+\exp\left(\lim\limits_{t\to+\infty}(z(t))\right)\right)}\right)}$, and in particular to the residual interval $\left[-\sqrt{2\varepsilon}, \sqrt{2\varepsilon}\right]$, is in general non-uniform and may be slow for certain initial conditions and certain disturbances.

4.9 ▪ Regulation under Asymptotically Vanishing Disturbance

It should also be noted that estimate (4.38) and the fact that $z(t)$ is non-decreasing guarantees that at least one of the following holds when $\lim\limits_{t\to+\infty}(d(t)) = 0$ (the disturbance-vanishing case):

1. $\lim_{t \to +\infty} (y(t)) = 0$,

2. $|\theta| > 1$ and $z(t) \leq \lim_{s \to +\infty} (z(s)) < \ln(|\theta| - 1)$ for all $t \geq 0$.

Consequently, convergence of the plant state to zero is not excluded even for disturbances with infinite energy (i.e., when $d \notin L^2(\mathbb{R}_+; \mathbb{R})$). In order to increase the possibility of having the plant state to converge to zero, the control practitioner can increase $\Gamma > 0$ (in order to increase the limit value $\lim_{s \to +\infty} (z(s))$).

4.10 ▪ Simulations

The performance of the closed-loop system (4.1) with (4.7), (4.15) with $\varphi(y) \equiv 1$ in the disturbance-free case is shown in Figure 4.1. It is clear that the performance is satisfactory. The output converges very rapidly to zero with a very small overshoot, while no excessive control effort is needed. Moreover, the dynamic gain $z(t)$ converges to a value slightly less than 1.054.

The performance of the closed-loop system (4.1) with (4.7), (4.15) with $\varphi(y) \equiv 1$ in the case where persistent disturbances are present is shown in Figure 4.2. This time the plant state and the dynamic gain $z(t)$ need more time to converge (the plant state to a periodic solution of amplitude $\sqrt{2\varepsilon} = 0.05$ and the dynamic gain to a value slightly less than 2.02), while no excessive control effort is needed (the control input tries to annihilate the effect of the disturbance).

Convergence to the set $\{(y, z) \in \mathbb{R}^2 : |y| \leq \sqrt{2\varepsilon}\}$ may become very slow when persistent disturbances are present and we pick a very small value of $\varepsilon > 0$ (very small compared to $\|d\|_\infty$). This is shown in Figure 4.3 where even after 100 time units the output oscillates with an amplitude of 0.03 (notice that $\sqrt{2\varepsilon} = 0.01$) and the dynamic gain has not converged yet. There is a way to accelerate convergence to the set $\{(y, z) \in \mathbb{R}^2 : |y| \leq \sqrt{2\varepsilon}\}$: by using larger values for $\Gamma > 0$. This is shown in Figure 4.4, where a much larger value of $\Gamma > 0$ was used.

4.11 ▪ Notes and comments

1. The idea of the DADS controller first appeared in [76].

2. A detailed stability analysis of the closed-loop system (4.1) with (4.7), (4.15) with $\varphi(y) \equiv 1$ was presented in Chapter 2 (Example 2.9).

3. The reader should compare the DADS controller (4.7), (4.15) for the simple system (4.1) with $\varphi(y) \equiv 1$, with the feedback adaptive controller of delayed state and input (3.30) that was given in the previous chapter (as well as in [74]). Although the controller (3.30) guaranteed similar properties as the DADS controller, it is clear that the DADS controller is far simpler and easier to implement.

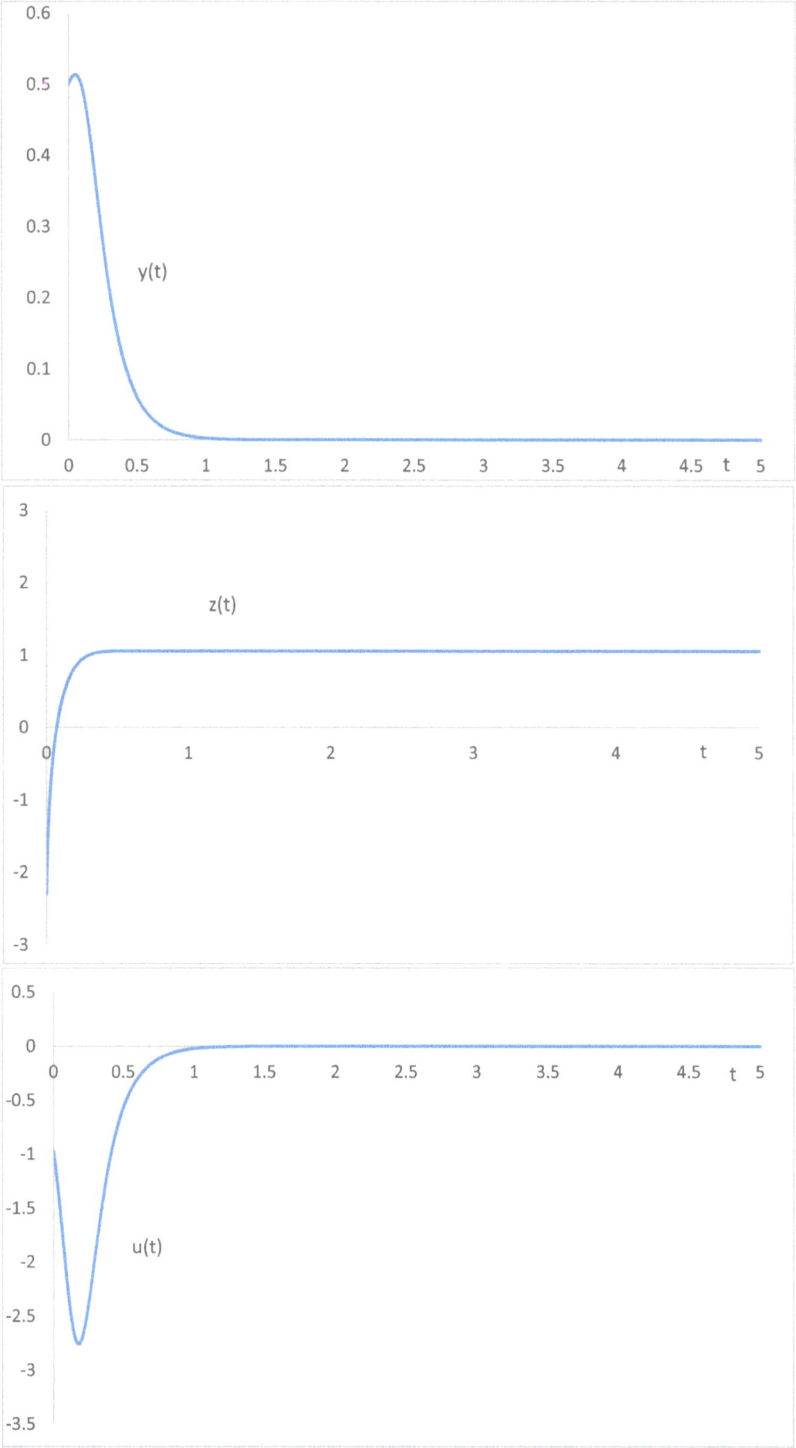

Figure 4.1. *The evolution of the state and the control input for the closed-loop system* (4.1), (4.7), (4.15) *with* $\varphi(y) \equiv 1$, $d \equiv 0$, $\theta = 3$, $c = a = 1, \Gamma = 100, \varepsilon = 0.00125$ *and* $y(0) = 0.5$, $z(0) = -\ln(10)$.

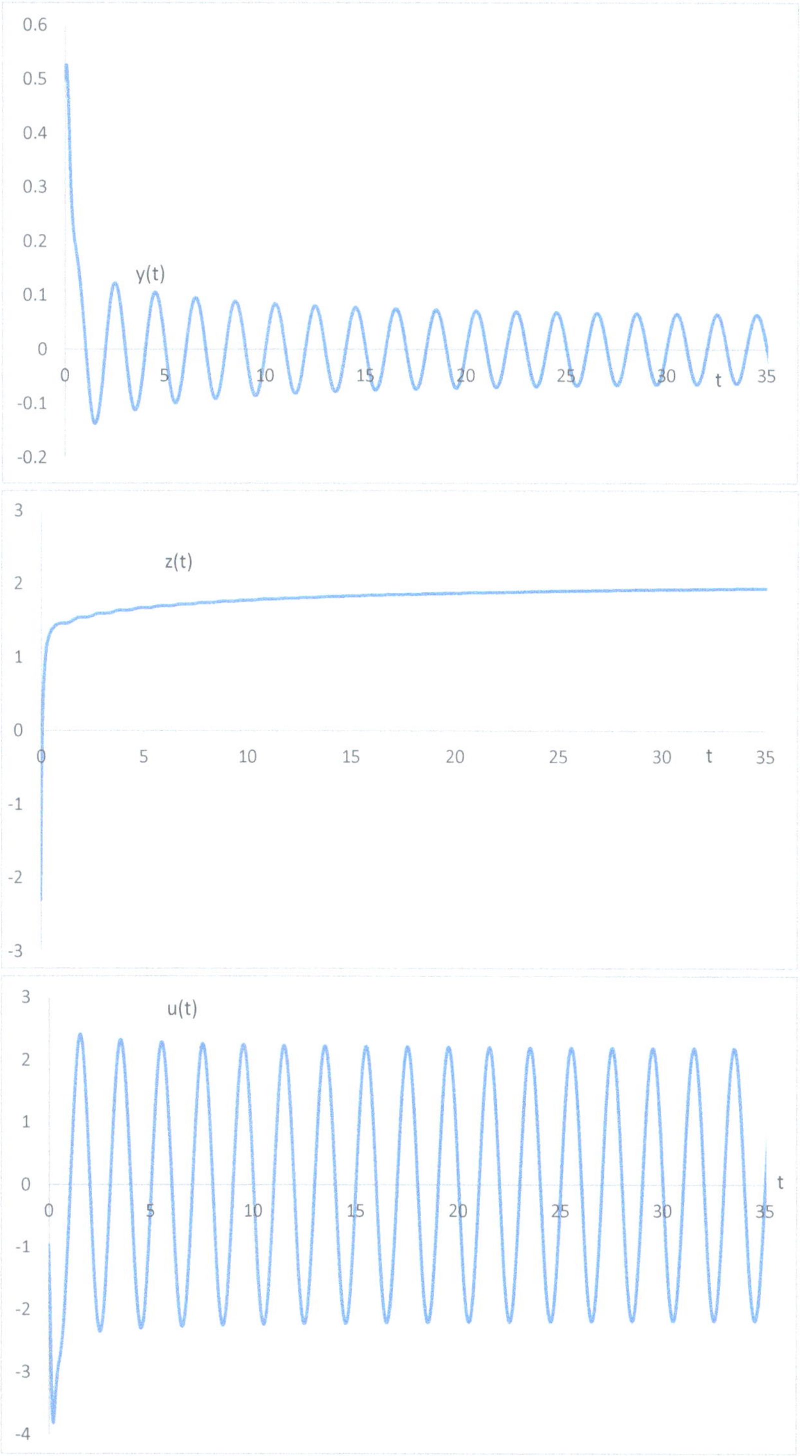

Figure 4.2. *The evolution of the state and the control input for the closed-loop system* (4.1), (4.7), (4.15) *with* $\varphi(y) \equiv 1$, $d = 2\sin(\pi t)$, $\theta = 3$, $c = a = 1, \Gamma = 100, \varepsilon = 0.00125$ *and* $y(0) = 0.5$, $z(0) = -\ln(10)$.

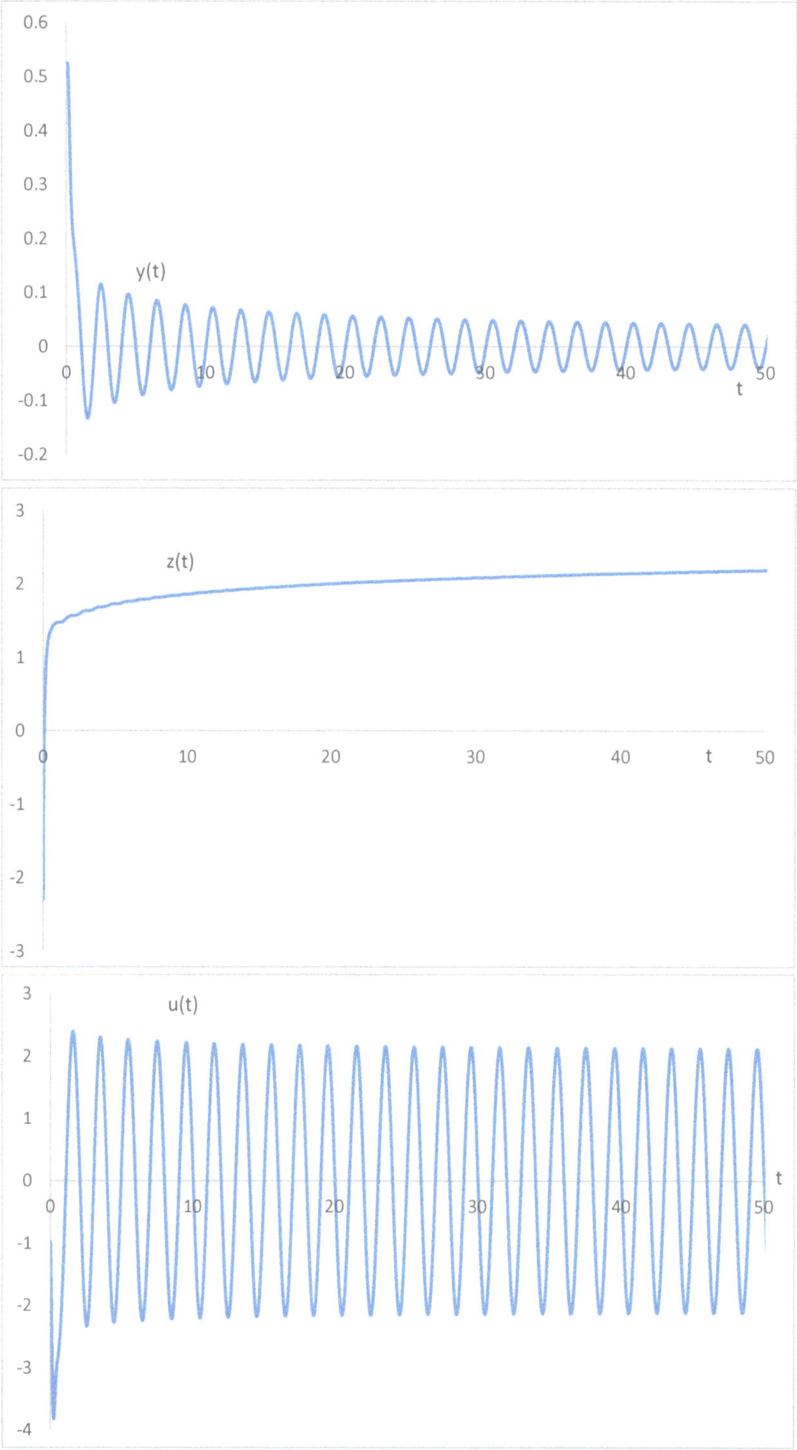

Figure 4.3. *The evolution of the state and the control input for the closed-loop system* (4.1), (4.7), (4.15) *with* $\varphi(y) \equiv 1$, $d = 2\sin(\pi t)$, $\theta = 3$, $c = a = 1, \Gamma = 100, \varepsilon = 0.00005$ *and* $y(0) = 0.5$, $z(0) = -\ln(10)$.

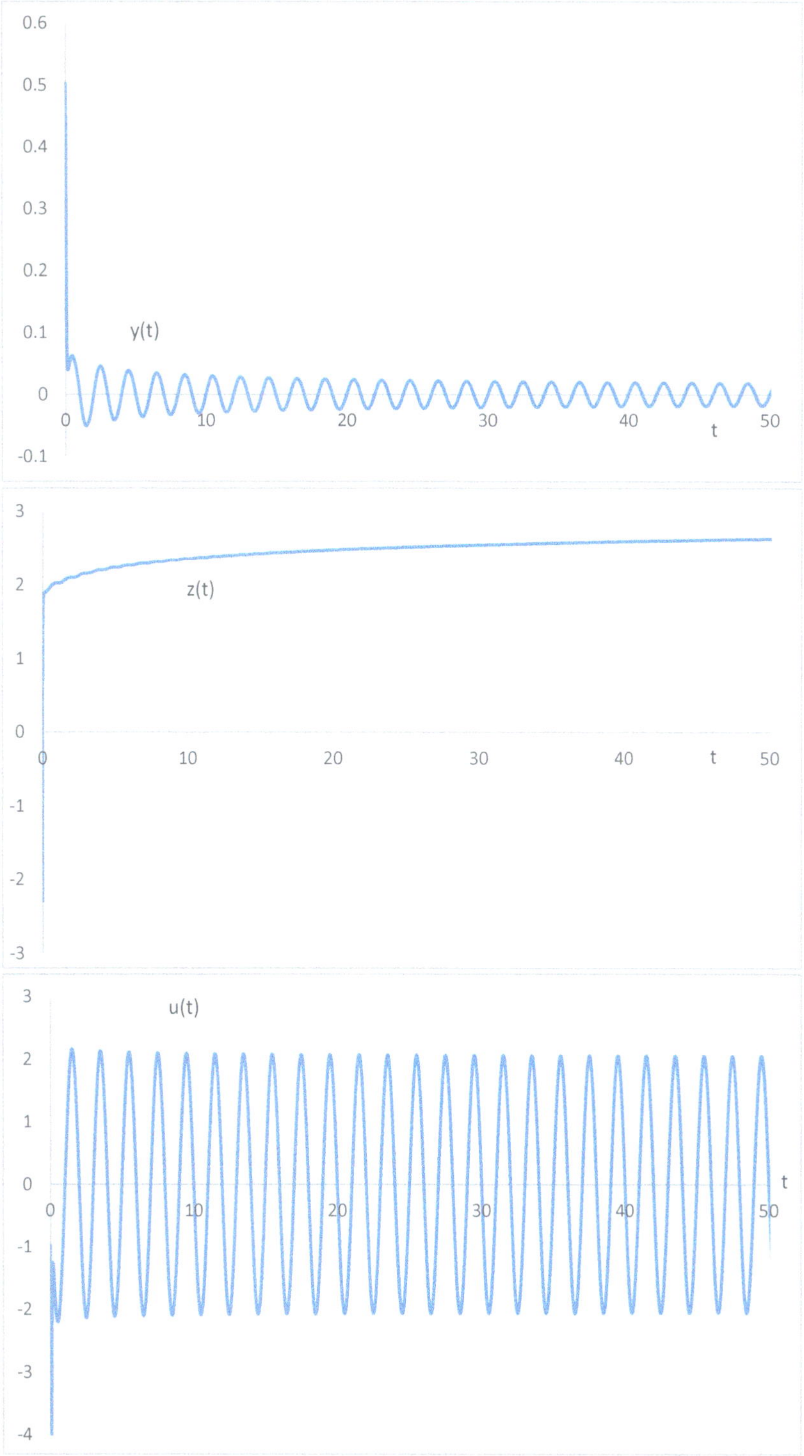

Figure 4.4. *The evolution of the state and the control input for system* (4.1), (4.7), (4.15) *with* $\varphi(y) \equiv 1$, $d = 2\sin(\pi t)$, $\theta = 3$, $c = a = 1, \Gamma = 1000, \varepsilon = 0.00005$ *and* $y(0) = 0.5$, $z(0) = -\ln(10)$.

Chapter 5

Matched Uncertainties: Deadzone-Adapted Disturbance Suppression Control

5.1 ▪ Introduction

In this chapter we study nonlinear control systems of the form

$$\begin{aligned} &\dot{y} = f(y) + g(y)\left(u + \varphi'(y)\theta + A'(x)d\right) \\ &y \in \mathbb{R}^n, u \in \mathbb{R}, d \in \mathbb{R}^m, \theta \in \mathbb{R}^p \end{aligned} \tag{5.1}$$

where $f, g : \mathbb{R}^n \to \mathbb{R}^n$, $\varphi : \mathbb{R}^n \to \mathbb{R}^p$, $A : \mathbb{R}^n \to \mathbb{R}^m$ are smooth mappings with $f(0) = 0$, $\varphi(0) = 0$, $y \in \mathbb{R}^n$ is the plant state, $u \in \mathbb{R}$ is the control input, and $\theta \in \mathbb{R}^p$, $d \in \mathbb{R}^m$ are unknown disturbances. The reader should notice the difference between θ and d when both are considered to be perturbations: while $\theta \in \mathbb{R}^p$ is a vanishing perturbation (due to the fact that $\varphi(0) = 0$), $d \in \mathbb{R}^m$ is—in general—a non-vanishing perturbation. Systems of the form (5.1) are systems that satisfy the so-called matching condition, i.e., the effect of both $\theta \in \mathbb{R}^p$ and $d \in \mathbb{R}^m$ can be canceled by the control input $u \in \mathbb{R}$ if they are known. We assume next that $d \in L^\infty\left(\mathbb{R}_+; \mathbb{R}^m\right)$, $\theta \in L^\infty\left(\mathbb{R}_+; \mathbb{R}^p\right)$ but we assume no bounds for $\theta \in \mathbb{R}^p$ and $d \in \mathbb{R}^m$.

For systems of the form (5.1) we extend the ideas of the previous chapter and we provide a simple, direct, adaptive control scheme that combines three elements: (a) nonlinear damping with dynamic gains, (b) single-gain adjustment (the dynamic feedback has only one state), and (c) deadzone in the update law. The proposed adaptive scheme is direct; here no identification is performed (which also explains the simplicity of the proposed scheme). We call the proposed controller a Deadzone-Adapted Disturbance Suppression (DADS) controller.

The advantages of the DADS controller (besides its simplicity) are many. The proposed adaptive control scheme achieves an attenuation of the plant state to an *assignable* small level, despite the presence of disturbances and unknown parameters of arbitrary and unknown bounds, i.e., we achieve the zero p-OAG property with an asymptotic residual constant which is independent of the constant parameters and is assignable. Moreover, the DADS controller prevents gain and state drift regardless of the size of the disturbance and unknown parameter. The latter property is a consequence of the p-UBIBS property and the p-IOS property that both hold for the closed-loop system. The important (and so rarely achieved) combination of robustness properties that are guaranteed for the closed-loop system in conjunction with the simplicity of the controller justify our focus on this special class of systems (systems with matched uncertainties).

There are hundreds of robust adaptive control papers. Yet, the attenuation of the plant state to an assignable small level, despite unknown bounds on the parameters and disturbance, without a drift of the gain, state, and input is previously unachieved. The reason for that is the following. Among the robustification tools for the update law, only the σ-modification (see [58], [59], [115],

[130]) is free of a requirement of a known bound on the parameter. Projection requires a known bound and deadzone alone is known to provide only *local* robustness (local in the initial condition and the disturbance size).

5.2 ▪ Feedback Design for Systems with Matched Uncertainties

We perform our Lyapunov-based feedback design for system (5.1) under some assumptions. Our first assumption provides a Lyapunov function.

Assumption (A): *There exist smooth mappings $V, Q : \mathbb{R}^n \to \mathbb{R}_+$, $k : \mathbb{R}^n \to \mathbb{R}$ with $k(0) = 0$, V, Q being positive definite and radially unbounded such that the following inequality holds for all $y \in \mathbb{R}^n$:*

$$\nabla V(y)\left(f(y) + g(y)k(y)\right) \leq -Q(y) \tag{5.2}$$

Assumption (A) guarantees the existence of global feedback stabilizer when $\theta = 0$ and $d = 0$. However, due to the presence of $\theta \in \mathbb{R}^p$ and $d \in \mathbb{R}^m$ we cannot apply the feedback law $u = k(y)$ without some modification. To this purpose we need an additional assumption.

Assumption (B): *There exists a smooth mapping $\mu : \mathbb{R}^n \to (0, +\infty)$ such that the following inequality holds for all $y \in \mathbb{R}^n$:*

$$|\varphi(y)|^2 \leq \mu(y) Q(y) \tag{5.3}$$

Assumption (B) holds automatically when there is a constant $\bar{Q} > 0$ such that the inequality $Q(y) \geq \bar{Q}\,|y|^2$ holds for all $y \in \mathbb{R}^n$. Proposition 4.4 in [67] guarantees that both assumptions (A), (B) hold automatically if there exists a smooth mapping $k : \mathbb{R}^n \to \mathbb{R}$ with $k(0) = 0$, such that the feedback law $u = k(y)$ guarantees UGAS and local exponential stability for system $\dot{y} = f(y) + g(y)k(y)$.

Supposing that assumptions (A) and (B) hold, we can have the following modifications of the feedback law $u = k(y)$.

1) Robust control modification. Assuming that $|\theta| \leq \rho$ where $\rho > 0$ is a constant, the feedback law

$$u = k(y) - \frac{1}{4a}\left(2a\rho^2\mu(y) + |A(y)|^2\right)\nabla V(y) g(y) \tag{5.4}$$

where $a > 0$ is a parameter of the controller, guarantees the following inequality for all $y \in \mathbb{R}^n$ and $d \in \mathbb{R}^m$:

$$\nabla V(y)\left(f(y) + g(y)\left(u + \varphi'(y)\theta + A'(y)d\right)\right) \leq -\frac{1}{2}Q(y) + a\,|d|^2 \tag{5.5}$$

Inequality (5.5) is shown by using (5.2), (5.3), (5.4), the fact that $|\theta| \leq \rho$, and the inequalities

$$\nabla V(y) g(y) A'(y) d \leq \frac{1}{4a}\,|A(y)|^2 \left(\nabla V(y) g(y)\right)^2 + a\,|d|^2 \tag{5.6}$$

$$\begin{aligned} \nabla V(y) g(y) \varphi'(y)\theta &\leq |\nabla V(y) g(y)|\,|\varphi(y)|\,|\theta| \\ &\leq \tfrac{|\varphi(y)|^2}{2\mu(y)} + \tfrac{\mu(y)}{2}\,|\theta|^2 \left(\nabla V(y) g(y)\right)^2 \end{aligned} \tag{5.7}$$

Inequality (5.5) shows that V is an ISS-Lyapunov function for the closed-loop system (5.1) with (5.4) and the ISS property from the disturbance $d \in L^\infty(\mathbb{R}_+; \mathbb{R}^m)$ holds. The state is eventually led to a ball around the origin and the radius of the ball is an increasing function of $a\,\|d\|_\infty^2$.

The disadvantage of the feedback law (5.4) is the fact that it works only when $|\theta| \leq \rho$ and (since $\theta \in L^\infty(\mathbb{R}_+; \mathbb{R}^p)$ is unknown) we cannot guarantee the validity of the inequality $|\theta| \leq \rho$.

2) Adaptive control modification. In this case, by assuming that $\theta \in \mathbb{R}^p$ is constant and using the so-called σ-modification, the dynamic feedback law

$$\begin{aligned} u &= k(y) - \varphi'(y)\hat{\theta} - c\,|A(y)|^2\, \nabla V(y) g(y) \\ \tfrac{d\hat{\theta}}{dt} &= \nabla V(y) g(y) \Gamma \varphi(y) - \sigma\,\hat{\theta} \end{aligned} \tag{5.8}$$

where $c, \sigma > 0$ are parameters of the controller and $\Gamma \in \mathbb{R}^{p\times p}$ is a symmetric positive definite matrix guarantees the p-ISS property from the disturbance $d \in L^\infty(\mathbb{R}_+; \mathbb{R}^m)$ for the closed-loop system (5.1) with (5.8) (with state $x = (y,\, \hat{\theta} - \theta) \in \mathbb{R}^n \times \mathbb{R}^p$). Indeed, this is shown by using the ISS-Lyapunov function

$$W(x) = V(y) + \frac{1}{2}\left(\hat{\theta} - \theta\right)' \Gamma^{-1} \left(\hat{\theta} - \theta\right) \tag{5.9}$$

which satisfies the following inequality for all $y \in \mathbb{R}^n$, $d \in \mathbb{R}^m$, $\hat{\theta}, \theta \in \mathbb{R}^p$, and $\lambda \in (0,1)$:

$$\dot{W} \le -Q(y) - \sigma(1-\lambda)\left(\hat{\theta} - \theta\right)' \Gamma^{-1} \left(\hat{\theta} - \theta\right) + \frac{\sigma}{4\lambda}\theta'\Gamma^{-1}\theta + \frac{|d|^2}{4c} \tag{5.10}$$

where $\dot{W}$ is the derivative of $W(x)$ along the trajectories of the closed-loop system (5.1) with (5.8). Inequality (5.10) shows that the residual constant in the p-ISS property is an increasing function of $\frac{\sigma}{4\lambda}\theta'\Gamma^{-1}\theta$. As explained in Chapter 2, this is the disadvantage of the approach: the state is eventually led (even in the disturbance-free case) to a ball around the origin of unknown radius.

In order to present a third option, we need an additional assumption.

Assumption (C): *There exist constants $\delta, \eta > 0$ such that, for Q introduced in* (5.2) *and* (5.3), *the following inequality holds for all $y \in \mathbb{R}^n$ with $|y| \le \delta$:*

$$Q(y) \ge 2\eta V(y) \tag{5.11}$$

Assumption (C) allows us to suggest a different option for the robust stabilization of (5.1). Consider the dynamic feedback law

$$u = k(y) - \frac{1+\kappa(\exp(z))}{4a}\left(|\varphi(y)|^2 + |A(y)|^2 + 2a\mu(y)\frac{(b+\lambda(\exp(z)))^2}{1+\kappa(\exp(z))}\right)\nabla V(y) g(y) \tag{5.12}$$

$$\dot{z} = \Gamma \exp(-z)\,(V(y) - \varepsilon)^+, \quad z \in \mathbb{R} \tag{5.13}$$

where $\varepsilon, \Gamma, a > 0$, $b \ge 0$ are parameters of the controller (constants) and $\kappa, \lambda \in K_\infty$ are smooth functions. We call the controller (5.12), (5.13) a Deadzone-Adapted Disturbance Suppression (DADS) controller. The controller (5.12), (5.13) combines the use of deadzone (in (5.11)) and dynamic nonlinear damping (in (5.10)).

The controller (5.12), (5.13) is clearly an extension of the controller (5.4) with z being adapted by means of the update law (5.13) and by treating both $\theta \in \mathbb{R}^p$ and $d \in \mathbb{R}^m$ as disturbances. The controller (5.12), (5.13) combines both the adaptation idea and the robust control methodology that leads to controller (5.4). However, notice that $\dot{z}$ becomes zero, i.e., the adaptation stops, when the plant state y enters the region defined by $V(y) \le \varepsilon$. This is the effect of the deadzone and it is a characteristic that does not appear in the update law of (5.8). The deadzone prevents the dynamic gain z from growing without bound in the case where a bounded disturbance is present.

The controller (5.12), (5.13) is simple: only one integrator is being used. The dynamic controller gain increases in order to overcome the effect of both $\theta \in \mathbb{R}^p$ and $d \in \mathbb{R}^m$.

The following theorem clarifies the performance characteristics that the DADS controller (5.12), (5.13) can guarantee for the closed-loop system.

Theorem 5.1: *Suppose that Assumptions (A), (B), and (C) hold. Let* $\varepsilon, \Gamma, a > 0$, $b \geq 0$ *be given constants and let* $\kappa, \lambda \in K_\infty$ *be given smooth functions. Then there exist functions* $\beta \in KL$, $\gamma \in K_\infty$ *and* $B \in C^0(\mathbb{R}^n \times \mathbb{R} \times \mathbb{R}_+ \times \mathbb{R}_+)$, *such that for every* $(y_0, z_0) \in \mathbb{R}^n \times \mathbb{R}$, $d \in L^\infty(\mathbb{R}_+; \mathbb{R}^m)$, $\theta \in L^\infty(\mathbb{R}_+; \mathbb{R}^p)$ *the unique solution of the initial-value problem* (5.1), (5.12), (5.13) *with initial condition* $(y(0), z(0)) = (y_0, z_0)$ *satisfies the following estimates for all* $t \geq 0$*:*

$$\limsup_{t \to +\infty} (V(y(t))) \leq \varepsilon \tag{5.14}$$

$$\limsup_{t \to +\infty} (V(y(t))) \leq \gamma\left(\chi\left(\limsup_{t \to +\infty}(|d(t)|), \limsup_{t \to +\infty}(|\theta(t)|), \exp\left(\lim_{t \to +\infty}(z(t))\right)\right)\right) \tag{5.15}$$

$$V(y(t)) \leq \beta(V(y_0), t) + \gamma(\chi(\|d\|_\infty, \|\theta\|_\infty, \exp(z_0))) \tag{5.16}$$

$$z_0 \leq z(t) \leq \lim_{s \to +\infty}(z(s)) \leq B(y_0, z_0, \|d\|_\infty, \|\theta\|_\infty) \tag{5.17}$$

where $\chi(s_1, s_2, s_3) := a\frac{s_1^2 + ((s_2 - b - \lambda(s_3))^+)^2}{1 + \kappa(s_3)}$ *for all* $s_1, s_2, s_3 \geq 0$. *Moreover, if inequality* (5.11) *holds for all* $y \in \mathbb{R}^n$ *then* $\gamma(s) = \eta^{-1}s$ *for all* $s \geq 0$ *and* $\beta(s,t) = \exp(-\eta t)s$ *for all* $t, s \geq 0$.

Remarks on Theorem 5.1: (a) Since V is positive definite and radially unbounded, it follows from (5.14) that the zero p-OAG property $\limsup_{t \to +\infty}(|y(t)|) \leq \alpha$ from the input d to the output y with a constant $\alpha > 0$ completely independent of the parameter θ holds. More specifically, the asymptotic residual constant $\alpha > 0$ depends only on the constant $\varepsilon > 0$.

(b) For constant $\theta \in \mathbb{R}^p$ inequality (5.16) shows that the p-IOS property from the disturbance $d \in L^\infty(\mathbb{R}_+; \mathbb{R}^m)$ holds for the closed-loop system (5.1), (5.12), (5.13). Indeed, the following estimate is a direct consequence of (5.16):

$$V(y(t)) \leq \beta(V(y(0)), t) + \gamma\left(2a\|d\|_\infty^2\right) + \gamma\left(2a\left((|\theta| - b)^+\right)^2\right) \tag{5.18}$$

The above estimate guarantees the p-IOS property from the disturbance $d \in L^\infty(\mathbb{R}_+; \mathbb{R}^m)$ for the closed-loop system (5.1), (5.12), (5.13). Estimates (5.16), (5.17) guarantee the p-UBIBS property while estimate (5.14) guarantees the zero p-OAG property with an *assignable* asymptotic residual constant.

(c) Taking into account the above remark, we can say that in the disturbance-free case (i.e., when $d \equiv 0$), the closed-loop system (5.1), (5.12), (5.13) for constant $\theta \in \mathbb{R}^p$
i) is Lagrange stable,
ii) satisfies the p-UGAOS property,
iii) satisfies the UGAOS property when $|\theta| \leq b$,
iv) the set $\{(0, z) \in \mathbb{R}^n \times \mathbb{R}\}$ is invariant,
v) is p-GOA with an *assignable* asymptotic residual constant,
vi) at least one of the following holds:

1. $\lim_{t \to +\infty}(y(t)) = 0$,

2. $|\theta| > b$ and $z(t) \leq \lim_{s \to +\infty}(z(s)) < \ln\left(\lambda^{-1}(|\theta| - b)\right)$ for all $t \geq 0$.

It is interesting to notice that in the disturbance-free case the closed-loop system (5.1), (5.8) for constant $\theta \in \mathbb{R}^p$ with $\sigma = 0$
i) is Lagrange stable,
ii) is Lyapunov stable,
iii) satisfies the GAOS property,
iv) the set $\left\{ (0, \hat{\theta}) \in \mathbb{R}^n \times \mathbb{R}^p \right\}$ is invariant.

However, it should also be noted that the closed-loop system (5.1), (5.8) with $\sigma = 0$ does not guarantee the BIBS property when disturbances are present and that is why σ-modification (i.e., (5.6) with $\sigma > 0$) is being used when disturbances may be present. The closed-loop system (5.1), (5.8) with $\sigma > 0$ satisfies the p-ISS property when disturbances are present and the p-UGAS property in the disturbance-free case with a residual constant that depends on $\theta \in \mathbb{R}^p$.

(d) Theorem 5.1 is also valid for systems of the form

$$\begin{aligned} &\dot{y} = f(y) + g(y)u + \Phi(y)\theta + G(y)d \\ &y \in \mathbb{R}^n, u \in \mathbb{R}, d \in \mathbb{R}^m, \theta \in \mathbb{R}^p \end{aligned} \tag{5.19}$$

where $f, g : \mathbb{R}^n \to \mathbb{R}^n$, $\Phi : \mathbb{R}^n \to \mathbb{R}^{n \times p}$, $G : \mathbb{R}^n \to \mathbb{R}^{n \times m}$ are smooth mappings with $f(0) = 0$, $\Phi(0) = 0$ for which there exist smooth mappings $\varphi : \mathbb{R}^n \to \mathbb{R}^p$, $A : \mathbb{R}^n \to \mathbb{R}^m$ such that $\nabla V(y)\Phi(y) = \nabla V(y)g(y)\varphi'(y)$, $\nabla V(y)G(y) = \nabla V(y)g(y)A'(y)$ for all $y \in \mathbb{R}^n$ and for which assumptions (A), (B), (C) hold. The class of systems of the form (5.19) includes systems that do not necessarily satisfy the matching condition.

(e) Proposition 4.4 in [67] guarantees that all assumptions (A), (B), (C) hold automatically if there exists a smooth mapping $k : \mathbb{R}^n \to \mathbb{R}$ with $k(0) = 0$, such that the feedback law $u = k(y)$ guarantees UGAS and local exponential stability for system $\dot{y} = f(y) + g(y)k(y)$. In other words, there exist smooth mappings $V, Q : \mathbb{R}^n \to \mathbb{R}_+$, $\mu : \mathbb{R}^n \to (0, +\infty)$ with V, Q being positive definite and radially unbounded for which assumptions (A), (B), (C) hold for system (5.1).

(f) Theorem 5.1 can be stated for less regular mappings $f, g : \mathbb{R}^n \to \mathbb{R}^n$, $\varphi : \mathbb{R}^n \to \mathbb{R}^p$, $V, Q : \mathbb{R}^n \to \mathbb{R}_+$, $k : \mathbb{R}^n \to \mathbb{R}$, and $A : \mathbb{R}^n \to \mathbb{R}^m$, but here for simplicity reasons we assume that all mappings are smooth.

Assumptions (A), (B), (C) are automatically satisfied for systems of the form

$$\begin{aligned} &\dot{y} = Fy + G\left(u + \varphi'(y)\theta + A'(y)d\right) \\ &y \in \mathbb{R}^n, u \in \mathbb{R}, \theta \in \mathbb{R}^p, d \in \mathbb{R}^m \end{aligned} \tag{5.20}$$

where $F \in \mathbb{R}^{n \times n}$, $G \in \mathbb{R}^n$ is a given stabilizable pair of matrices and $\varphi : \mathbb{R}^n \to \mathbb{R}^p$, $A : \mathbb{R}^n \to \mathbb{R}^m$ are given smooth mappings with $\varphi(0) = 0$. Using the methodology of the proof of Theorem 5.1, we obtain the following result.

Theorem 5.2: *Let $F \in \mathbb{R}^{n \times n}$, $G \in \mathbb{R}^n$ be a given stabilizable pair of matrices and let $\varphi : \mathbb{R}^n \to \mathbb{R}^p$, $A : \mathbb{R}^n \to \mathbb{R}^m$ be given smooth mappings with $\varphi(0) = 0$. Let $k \in \mathbb{R}^n$ be a vector for which $(F - Gk') \in \mathbb{R}^{n \times n}$ is Hurwitz and let $P \in \mathbb{R}^{n \times n}$ be a symmetric positive definite matrix for which $Q = -\left(F - Gk'\right)' P - P\left(F - Gk'\right)$ is a positive definite matrix. Let $\eta > 0$ be a constant and let $\mu : \mathbb{R}^n \to (0, +\infty)$ be a smooth function for which the following inequality holds for all $y \in \mathbb{R}^n$:*

$$y'Qy \geq \eta y'Py + \frac{|\varphi(y)|^2}{2\mu(y)} \tag{5.21}$$

Let $\varepsilon, \Gamma, a > 0$, $b \geq 0$ be given constants and let $\kappa, \lambda \in K_\infty$ be given smooth functions. Consider *the feedback law*

$$\begin{aligned} u &= -k'y - \tfrac{1+\kappa(\exp(z))}{2a}\left(|\varphi(y)|^2 + |A(y)|^2 + 2a\mu(y)\tfrac{(b+\lambda(\exp(z)))^2}{1+\kappa(\exp(z))}\right)G'Py \\ \dot{z} &= \Gamma\exp(-z)\left(y'Py-\varepsilon\right)^+, \quad z\in\mathbb{R} \end{aligned} \tag{5.22}$$

Then there exists a function $B \in C^0(\mathbb{R}^n \times \mathbb{R} \times \mathbb{R}_+ \times \mathbb{R}_+)$ *such that for every* $(y_0, z_0) \in \mathbb{R}^n \times \mathbb{R}$, $d \in L^\infty(\mathbb{R}_+;\mathbb{R}^m)$, $\theta \in L^\infty(\mathbb{R}_+;\mathbb{R}^p)$ *the unique solution of the initial-value problem* (5.20), (5.22) *with initial condition* $(y(0), z(0)) = (y_0, z_0)$ *satisfies estimate* (5.17) *for all* $t \geq 0$ *as well as the following estimates for all* $t \geq 0$*:*

$$\begin{aligned} &\limsup_{t\to+\infty}\left(y'(t)Py(t)\right) \\ &\leq \min\left(\varepsilon\,,\,\eta^{-1}\chi\left(\limsup_{t\to+\infty}(|d(t)|)\,,\limsup_{t\to+\infty}(|\theta(t)|)\,,\exp\left(\lim_{t\to+\infty}(z(t))\right)\right)\right) \end{aligned} \tag{5.23}$$

$$y'(t)Py(t) \leq \exp(-\eta t)\,y_0'Py_0 + \eta^{-1}\chi\left(\|d\|_\infty, \|\theta\|_\infty, \exp(z_0)\right) \tag{5.24}$$

where $\chi(s_1,s_2,s_3) := a\frac{s_1^2+\left((s_2-b-\lambda(s_3))^+\right)^2}{1+\kappa(s_3)}$ *for all* $s_1, s_2, s_3 \geq 0$.

Remarks on Theorem 5.2: It should be noticed that systems of the form (5.20) include the chain of integrators

$$\begin{aligned} \dot{y}_i &= y_{i+1}, \quad i=1,\dots,n-1 \\ \dot{y}_n &= u + \textstyle\sum_{i=1}^{p}\varphi_k(y)\theta_k + \sum_{i=1}^{m}A_k(y)d_k \\ y &= (y_1,\dots,y_n)' \in \mathbb{R}^n, \theta = (\theta_1,\dots,\theta_p)' \in \mathbb{R}^p, d = (d_1,\dots,d_m)' \in \mathbb{R}^m \end{aligned} \tag{5.25}$$

where $\varphi_k : \mathbb{R}^n \to \mathbb{R}$ $(k=1,\dots,p)$ and $A_k : \mathbb{R}^n \to \mathbb{R}$ $(k=1,\dots,m)$ are smooth functions with $\varphi_k(0) = 0$ $(k=1,\dots,p)$.

In particular for the system

$$\begin{aligned} \dot{y}_i &= y_{i+1}, \quad i=1,\dots,n-1 \\ \dot{y}_n &= u + \theta'\Phi y + d \\ y &= (y_1,\dots,y_n)' \in \mathbb{R}^n, u\in\mathbb{R}, d\in\mathbb{R}, \theta\in\mathbb{R}^p \end{aligned} \tag{5.26}$$

where $\Phi \in \mathbb{R}^{p\times n}$ is a constant matrix, we are able to study tracking problems where the $C^n(\mathbb{R}_+;\mathbb{R})$ signal $r(t)$ is to be tracked by $y_1(t)$. Defining $e_i(t) = y_i(t) - r^{(i-1)}(t)$ for $i = 1,\dots,n$ as well as the vector $\bar{r}(t) = \left(r(t), \dot{r}(t), \dots, r^{(n-1)}(t)\right)' \in \mathbb{R}^n$ we obtain the system

$$\begin{aligned} \dot{e}_i &= e_{i+1}, \quad i=1,\dots,n-1 \\ \dot{e}_n &= u + \theta'\Phi e + \tilde{d} \end{aligned} \tag{5.27}$$

where $\tilde{d} = d + \theta'\Phi\bar{r} - r^{(n)}$. Applying Theorem 5.2 with $\kappa(s) = \lambda(s) = s$ for $s \geq 0$, $b = 1$, we conclude that there exist a vector $k \in \mathbb{R}^n$, a symmetric positive definite matrix $P \in \mathbb{R}^{n\times n}$, and a constant $\mu > 0$ such that for every $\varepsilon, \Gamma, a > 0$ and for every $(y_0, z_0) \in \mathbb{R}^n \times \mathbb{R}$, $d \in L^\infty(\mathbb{R}_+;\mathbb{R}^m)$, $\theta \in L^\infty(\mathbb{R}_+;\mathbb{R}^p)$, $r \in C^n(\mathbb{R}_+;\mathbb{R})$ with $\sup_{t\geq 0}\left(\sum_{i=0}^{n}\left|r^{(i)}(t)\right|\right) < +\infty$ the solution of the closed-loop system (5.26) with the *nonlinear* adaptive feedback law

$$\begin{aligned} u &= -k'e - \tfrac{1+\exp(z)}{2a}\left(|\Phi e|^2 + 1 + 2a\mu\left(1+\exp(z)\right)\right)(0,\dots,0,1)\,Pe \\ \dot{z} &= \Gamma\exp(-z)\left(e'Pe-\varepsilon\right)^+, \quad z\in\mathbb{R} \end{aligned} \tag{5.28}$$

and $(y(0), z(0)) = (y_0, z_0)$ satisfies $\limsup_{t\to+\infty}\left(e'(t)Pe(t)\right) \leq \varepsilon$ and $y \in L^\infty(\mathbb{R}_+;\mathbb{R}^n)$, $z \in L^\infty(\mathbb{R}_+;\mathbb{R})$. Notice that the implementation of the adaptive feedback law (5.28) requires knowledge of the signal $r(t)$ and its first $n-1$ derivatives.

5.3 ▪ Illustrative Examples

The first example deals with a system of the form (5.20) and presents an additional comparison between the DADS controller and the adaptive controllers obtained by σ-modification.

Example 5.1: Consider the planar system

$$\begin{aligned}&\dot{y}_1 = y_2\\&\dot{y}_2 = \theta_1 y_1 + \theta_2 y_2 + \theta_3 y_1^2 + u + d\\&y = (y_1, y_2) \in \mathbb{R}^2, u \in \mathbb{R}, d \in \mathbb{R}, \theta = (\theta_1, \theta_2, \theta_3)' \in \mathbb{R}^3\end{aligned} \tag{5.29}$$

System (5.29) is a system of the form (5.20) for which we can apply Theorem 5.2 with

$$n = 2, p = 3, A(y) \equiv 1 \tag{5.30}$$

$$F = \begin{bmatrix} 0 & 1 \\ 0 & 0 \end{bmatrix}, G = \begin{bmatrix} 0 \\ 1 \end{bmatrix}, P = \frac{1}{2}\begin{bmatrix} 5 & 2 \\ 2 & 1 \end{bmatrix}, k = \begin{bmatrix} 5 \\ 4 \end{bmatrix}, Q = 4P \tag{5.31}$$

$$\varphi(y) = \begin{bmatrix} y_1 & y_2 & y_1^2 \end{bmatrix}' \tag{5.32}$$

Moreover, inequality (5.21) holds with $\eta = 3$ and $\mu(y) = 6 + y_1^2$. Indeed, inequality (5.21) is equivalent to the following inequality:

$$(4 - \eta)\left(y_1^2 + (y_2 + 2y_1)^2\right)\mu(y) \geq y_1^2 + y_2^2 + y_1^4 \tag{5.33}$$

The above inequality holds for all $y \in \mathbb{R}^2$ with $\eta = 3$ and $\mu(y) = 6 + y_1^2$ since the following inequalities hold for all $y \in \mathbb{R}^2$:

$$(1 + y_1^2)\left(y_1^2 + (y_2 + 2y_1)^2\right) \geq y_1^2 + y_1^4 \tag{5.34}$$

$$5\left(y_1^2 + (y_2 + 2y_1)^2\right) \geq y_2^2 \tag{5.35}$$

Using Theorem 5.2 with $\kappa(s) = \lambda(s) = s$, $b = 1$, and exploiting the inequality $y_1^2 + (y_2 + 2y_1)^2 \geq (3 - 2\sqrt{2})\,|y|^2$ that holds for all $y \in \mathbb{R}^2$, we establish that for every $\varepsilon, \Gamma, a > 0$ the dynamic feedback law

$$\begin{aligned}&\dot{z} = \Gamma \exp(-z)\left(\tfrac{1}{2}y_1^2 + \tfrac{1}{2}(y_2 + 2y_1)^2 - \varepsilon\right)^+\\&u = -5y_1 - 4y_2 - \tfrac{1+\exp(z)}{4a}\left(y_1^2 + y_2^2 + y_1^4 + 1\right)(y_2 + 2y_1)\\&\quad - \tfrac{(1+\exp(z))^2}{2}\left(6 + y_1^2\right)(y_2 + 2y_1)\end{aligned} \tag{5.36}$$

guarantees the existence of a function $B \in C^0\left(\mathbb{R}^2 \times \mathbb{R} \times \mathbb{R}_+ \times \mathbb{R}_+\right)$ with the property that for every $(y_0, z_0) \in \mathbb{R}^2 \times \mathbb{R}$, $d \in L^\infty(\mathbb{R}_+; \mathbb{R})$, $\theta = (\theta_1, \theta_2, \theta_3)' \in \mathbb{R}^3$ the unique solution of the initial-value problem (5.29), (5.36) with initial condition $(y(0), z(0)) = (y_0, z_0)$ satisfies estimate (5.17) as well as the following estimates for $t \geq 0$:

$$\begin{aligned}&y_1^2(t) + (y_2(t) + 2y_1(t))^2 \leq \exp(-3t)\left(y_1^2(0) + (y_2(0) + 2y_1(0))^2\right)\\&+ \tfrac{2a}{3}\tfrac{\|d\|_\infty^2 + \left((|\theta| - 1 - \exp(z_0))^+\right)^2}{1+\exp(z_0)}\end{aligned} \tag{5.37}$$

$$\begin{aligned}&(3 - 2\sqrt{2})\limsup_{t\to+\infty}\left(|y(t)|^2\right) \leq \limsup_{t\to+\infty}\left(y_1^2(t) + (y_2(t) + 2y_1(t))^2\right)\\&\leq \min\left(2\varepsilon\,,\ \tfrac{2a}{3}\frac{\left(\limsup\limits_{t\to+\infty}(|d(t)|)\right)^2 + \left(\left(|\theta| - 1 - \exp\left(\lim\limits_{t\to+\infty}(z(t))\right)\right)^+\right)^2}{1+\exp\left(\lim\limits_{t\to+\infty}(z(t))\right)}\right)\end{aligned} \tag{5.38}$$

We next compare the DADS controller (5.36) with the adaptive controller that we obtained by σ-modification using the Lyapunov function

$$V(y,\hat{\theta}) = \frac{1}{2}y_1^2 + \frac{1}{2}(y_2 + 2y_1)^2 + \frac{1}{2\Gamma_1}\left(\hat{\theta}_1 - \theta_1\right)^2 + \frac{1}{2\Gamma_2}\left(\hat{\theta}_2 - \theta_2\right)^2 + \frac{1}{2\Gamma_3}\left(\hat{\theta}_3 - \theta_3\right)^2 \tag{5.39}$$

with constants $\Gamma_i > 0$ $(i = 1,2,3)$ and the nominal feedback law $k(y) = -k'y = -5y_1 - 4y_2$:

$$\begin{aligned} u &= -5y_1 - 4y_2 - c(y_2 + 2y_1) - \hat{\theta}_1 y_1 - \hat{\theta}_2 y_2 - \hat{\theta}_3 y_1^2 \\ \tfrac{d\hat{\theta}_1}{dt} &= \Gamma_1 y_1 (y_2 + 2y_1) - \sigma\hat{\theta}_1 \\ \tfrac{d\hat{\theta}_2}{dt} &= \Gamma_2 y_2 (y_2 + 2y_1) - \sigma\hat{\theta}_2 \\ \tfrac{d\hat{\theta}_3}{dt} &= \Gamma_3 y_1^2 (y_2 + 2y_1) - \sigma\hat{\theta}_3 \end{aligned} \tag{5.40}$$

where $c, \sigma > 0$ are constants.

We apply the controllers (5.36) and (5.40) to system (5.29) with $\theta_1 = 10$, $\theta_2 = \theta_3 = 1$. The parameters for the controllers were

$$a = 1, \sigma = \Gamma_1 = \Gamma_2 = \Gamma_3 = 1, \Gamma = 20, \varepsilon = 0.004 \tag{5.41}$$

The time evolution of the Euclidean norm of the plant state for the solutions of the closed-loop systems (5.29), (5.36) and (5.29), (5.40) is shown in Fig. 5.1 (disturbance-free case) and in Fig. 5.2 for the disturbance $d(t) = 3\cos(10t)$. It is clear that in both cases the controller (5.36) achieves bringing the plant state very close to 0 while the controller (5.40) fails to achieve this objective.

The time evolution of $z(t)$ for the solution of the closed-loop system (5.29), (5.36) is shown in Fig. 5.3. We found that in both cases (the disturbance-free case and the case where $d(t) = 3\cos(10t)$) the evolution is similar: $z(t)$ increases and approaches quickly a limit value (1.368 in the disturbance-free case and 1.335 in the case where $d(t) = 3\cos(10t)$). For $t \geq 2$, $z(t)$ remains almost constant with values approximately equal to its limit value.

It should also be noticed that Fig. 5.2 shows that the DADS controller (5.36) achieves almost perfect disturbance rejection (as expected). This is a consequence of the zero p-OAG property from the input d to the plant state y, i.e., inequality (5.38). $\triangleleft$

The second example shows that the use of deadzone in (5.13) is absolutely essential for achieving the zero p-OAG property $\limsup_{t\to+\infty}(|y(t)|) \leq \alpha$ from the input d to the output y with a constant $\alpha > 0$ independent of the constant parameter θ, i.e., property (5.14).

Example 5.2: As explained above, the mechanism that guarantees the zero p-OAG property $\limsup_{t\to+\infty}(|y(t)|) \leq \alpha$ from the input d to the output y with a constant $\alpha > 0$ independent of the constant parameter θ, i.e., property (5.14), in conjunction with the boundedness of the whole state vector (y, z) even in the presence of disturbances, is the combination of the use of deadzone in (5.13) and dynamic nonlinear damping in (5.12). The use of deadzone in (5.13) is absolutely essential. To understand this point, we compare the controller (5.12), (5.13) with a controller for which (5.13) has been replaced by an update with leakage

$$\frac{d}{dt}(\exp(z)) = \Gamma V(x) - \sigma\exp(z) \tag{5.42}$$

where $\sigma > 0$ is a constant. Such a controller may achieve various objectives such as the p-IOS property and the boundedness of the whole state vector (y, z) even in the presence of disturbances. However, such a controller cannot achieve the zero p-OAG property $\limsup_{t\to+\infty}(|y(t)|) \leq \alpha$ with a constant $\alpha > 0$ independent of the constant parameter θ, i.e., property (5.14).

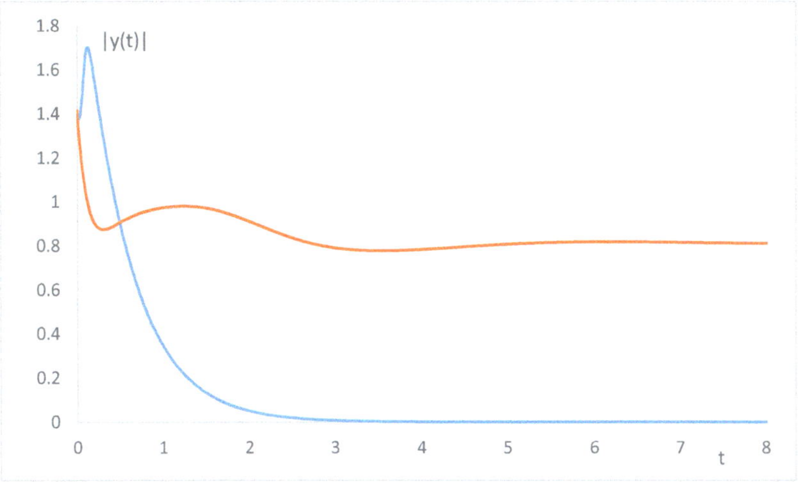

Figure 5.1. *The evolution of the Euclidean norm of the plant state for the solutions of the closed-loop systems* (5.29), (5.36) *(blue line—DADS) and* (5.29), (5.40) *(red line—leakage) with* $d \equiv 0$. *Initial condition* $x_1(0) = -1$, $x_2(0) = 1$, $z(0) = -\ln(10)$, $\hat{\theta}_1(0) = \hat{\theta}_2(0) = \hat{\theta}_3(0) = 0$.

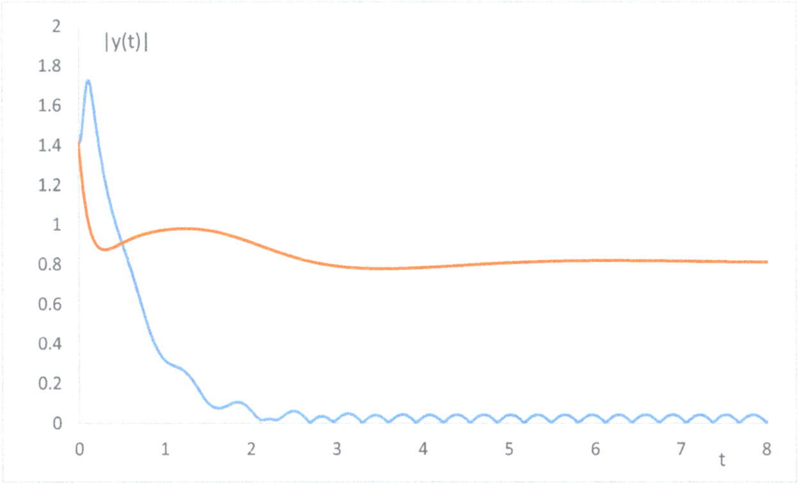

Figure 5.2. *The evolution of the Euclidean norm of the plant state for the solutions of the closed-loop systems* (5.29), (5.36) *(blue line—DADS) and* (5.29), (5.40) *(red line—leakage) with* $d(t) = 3\cos(10t)$. *Initial condition* $x_1(0) = -1$, $x_2(0) = 1$, $z(0) = -\ln(10)$, $\hat{\theta}_1(0) = \hat{\theta}_2(0) = \hat{\theta}_3(0) = 0$.

To see this, let us study again the simple example (4.1) with $\varphi(y) \equiv 1$ that was studied in the previous chapter, namely, the scalar system $\dot{y} = \theta y + u + d$. A controller based on (5.12), (5.42) and $V(y) = y^2/2$, $k(y) = -cy$, $\kappa(s) = \lambda(s) = s$ for $s \geq 0$, $b = 1$ would give a controller of the form

$$\begin{aligned} u &= -cy - \tfrac{\rho}{4a}\left(y^2 + 1 + 2c^{-1}a\rho\right)y \\ \dot{\rho} &= My^2 + \sigma - \sigma\rho \end{aligned} \tag{5.43}$$

where c, a, M, $\sigma > 0$ are constants and $\rho = 1 + \exp(z)$. The equilibrium points for the

Figure 5.3. *The evolution of $z(t)$ for the solution of the closed-loop system* (5.29), (5.36) *(DADS) with $d \equiv 0$. Initial condition $x_1(0) = -1$, $x_2(0) = 1$, $z(0) = -\ln(10)$, $\hat{\theta}_1(0) = \hat{\theta}_2(0) = \hat{\theta}_3(0) = 0$.*

closed-loop system (5.2), (5.43) with $d \equiv 0$ are given by the equations

$$\begin{aligned} &\left(\theta - c - \tfrac{\rho}{4a}\left(y^2+1\right) - \tfrac{1}{2c}\rho^2\right) y = 0 \\ &My^2 + \sigma = \sigma\rho \end{aligned} \tag{5.44}$$

It is clear that when $\theta - c > \frac{2a+c}{4ac}$ there are non-zero equilibrium points of the form $(y,\rho) = \left(\pm\left(\frac{\sigma(\rho-1)}{M}\right)^{1/2}, \rho\right)$ with $\rho > 1$ being the unique positive solution of the equation $\frac{c\sigma+2aM}{4acM}\rho^2 - \frac{\sigma-M}{4aM}\rho = \theta - c$. Notice that the unique positive solution satisfies the inequality $\rho \geq \left(M\frac{4ac(\theta-c)-c-2a}{c\sigma+2aM} + 1\right)^{1/2}$. Consequently, when $\theta - c > \frac{2a+c}{4ac}$ there are solutions of the closed-loop system (5.2), (5.43) with $d \equiv 0$ for which

$$\limsup_{t\to+\infty}\left(|y(t)|^2\right) \geq M^{-1}\sigma\left(\left(M\frac{4ac(\theta-c)-c-2a}{c\sigma+2aM}+1\right)^{1/2} - 1\right) \tag{5.45}$$

Therefore, it is clear that—no matter how the constants $c, a, M, \sigma > 0$ are selected—property (5.14) cannot hold with a constant $\varepsilon > 0$ independent of θ.

We conclude that the use of deadzone in (5.13) is essential for achieving the zero p-OAG property $\limsup_{t\to+\infty}(|y(t)|) \leq \alpha$ with a constant $\alpha > 0$ independent of the parameter θ. $\triangleleft$

The third example shows that the zero p-OAG property $\limsup_{t\to+\infty}(|y(t)|) \leq \alpha$ with a constant $\alpha > 0$ independent of the constant parameter θ, i.e., property (5.14), imposes severe structural constraints on the system. In other words, this property cannot be achieved for arbitrary stabilizable systems.

Example 5.3 (A disturbed chain of two integrators): Consider the planar system

$$\begin{aligned} &\dot{y}_1 = y_2 + d, \quad \dot{y}_2 = u \\ &y = (y_1, y_2)' \in \mathbb{R}^2, u \in \mathbb{R}, d \in \mathbb{R} \end{aligned} \tag{5.46}$$

We claim that the zero p-OAG property $\limsup_{t\to+\infty}(|y(t)|) \leq \alpha$ with an arbitrary constant $\alpha > 0$ cannot be achieved for system (5.46) (the constant parameter θ is irrelevant for this example). The proof of this fact is made by contradiction. Suppose that there exists a feedback controller for which the corresponding closed-loop system satisfies the zero p-OAG property $\limsup_{t\to+\infty}(|y(t)|) \leq \alpha$ with a constant $\alpha > 0$. Consider the solution of (5.46) corresponding to the constant disturbance $d(t) \equiv \alpha + 2$. By assumption, there exists $T > 0$ such that $|y(t)| \leq \alpha + 1$ for $t \geq T$. Using (5.46), we then have for $t \geq T$ a.e.

$$\dot{y}_1(t) = y_2(t) + d(t) \geq \alpha + 2 - |y_2(t)| \geq \alpha + 2 - |y(t)| \geq 1 \tag{5.47}$$

Consequently, we obtain for $t \geq T$ the inequality $\alpha + 1 \geq |y(t)| \geq |y_1(t)| \geq y_1(t) \geq y_1(T) + t - T$, which obviously leads to a contradiction.

It should be noted that the matching condition does not hold for system (5.46). ◁

The final example shows that even if we relax our requirements and try to achieve the p-OAG property $\limsup_{t\to+\infty}(|y(t)|) \leq \alpha\left(\|d\|_{\infty}\right)$ from the input d to the plant state y with a non-decreasing function $\alpha : \mathbb{R}_+ \to (0, +\infty)$ independent of the constant parameter θ, then we still need to impose severe structural constraints on the system.

Example 5.4 (A linear stabilizable uncertain system): Consider the system

$$\begin{array}{l} \dot{y}_1 = -y_1 + d, \quad \dot{y}_2 = \theta y_1 + y_3, \quad \dot{y}_3 = u \\ y = (y_1, y_2, y_3)' \in \mathbb{R}^3, u \in \mathbb{R}, \theta \in \mathbb{R}, d \in \mathbb{R} \end{array} \tag{5.48}$$

We claim that the p-OAG property $\limsup_{t\to+\infty}(|y(t)|) \leq \alpha\left(\|d\|_{\infty}\right)$ from the input d to the entire plant state $y = (y_1, y_2, y_3)$, as opposed to the portion (y_1, y_2) (that will be studied in the following chapter) with a non-decreasing function $\alpha : \mathbb{R}_+ \to (0, +\infty)$ independent of the constant parameter θ cannot be achieved for system (5.48). The proof of this fact is made by contradiction. Suppose that there exists a feedback controller for which the corresponding closed-loop system satisfies the p-OAG property $\limsup_{t\to+\infty}(|y(t)|) \leq \alpha\left(\|d\|_{\infty}\right)$ with a non-decreasing function $\alpha : \mathbb{R}_+ \to (0, +\infty)$ independent of the constant parameter θ. Consider the solution of (5.48) corresponding to the constant disturbance $d(t) \equiv 1$, constant parameter $\theta = 2 + \alpha(1)$, and initial condition $y_1(0) = 1$. By assumption, there exists $T > 0$ such that $|y(t)| \leq \alpha(1) + 1$ for $t \geq T$. Moreover, $y_1(t) = 1$ for all $t \geq 0$. Using (5.48), we then have for $t \geq T$ a.e.

$$\begin{array}{l} \dot{y}_2(t) = \theta y_1(t) + y_3(t) = \theta + y_3(t) \\ \geq \theta - |y_3(t)| \geq \theta - |y(t)| \geq \theta - \alpha(1) - 1 \end{array} \tag{5.49}$$

Consequently, since $\theta = 2 + \alpha(1)$ we obtain for $t \geq T$ the inequality $\alpha(1) + 1 \geq |y(t)| \geq |y_2(t)| \geq y_2(t) \geq y_2(T) + t - T$, which obviously leads to a contradiction.
It should be noted again that the matching condition does not hold for system (5.48). ◁

The last two examples show that the implications of the matching condition are far reaching and should not be neglected by the control theorist or the control practitioner.

5.4 ▪ Proofs

We start by providing the proof of Theorem 5.1.

Proof of Theorem 5.1: Due to (5.1), (5.12) we have for all $(y, z) \in \mathbb{R}^n \times \mathbb{R}$, $d \in \mathbb{R}^m, \theta \in \mathbb{R}^p$

$$\begin{array}{l} \dot{V}(y, z, d, \theta) = \nabla V(y)\left(f(y) + g(y)k(y)\right) \\ +\nabla V(y)g(y)\varphi'(y)\theta + \nabla V(y)g(y)A'(y)d \\ -\frac{1+\kappa(\exp(z))}{4a}\left(|\varphi(y)|^2 + |A(y)|^2 + 2a\mu(y)\frac{(b+\lambda(\exp(z)))^2}{1+\kappa(\exp(z))}\right)\left(\nabla V(y)g(y)\right)^2 \end{array} \tag{5.50}$$

where $\dot{V}(y,z)$ is the derivative of $V(y)$ along the trajectories of the closed-loop system (5.1) with (5.12), (5.13). Using (5.2) and the inequality

$$\nabla V(y)g(y)A'(y)d \leq \frac{1}{4a}\left(1+\kappa\left(\exp(z)\right)\right)|A(y)|^2\left(\nabla V(y)g(y)\right)^2 + \frac{a\,|d|^2}{1+\kappa\left(\exp(z)\right)} \tag{5.51}$$

we get from (5.50) for all $(y,z) \in \mathbb{R}^n \times \mathbb{R}$, $d \in \mathbb{R}^m, \theta \in \mathbb{R}^p$

$$\begin{aligned} &\dot{V}(y,z,d,\theta) \leq -Q(y) + |\nabla V(y)g(y)|\,|\varphi(y)|\,|\theta| + \tfrac{a|d|^2}{1+\kappa(\exp(z))} \\ &-\tfrac{1+\kappa(\exp(z))}{4a}\left(|\varphi(y)|^2 + 2a\mu(y)\tfrac{(b+\lambda(\exp(z)))^2}{1+\kappa(\exp(z))}\right)\left(\nabla V(y)g(y)\right)^2 \end{aligned} \tag{5.52}$$

Inequality (5.52) and the fact that $|\theta| \leq \left(|\theta| - b - \lambda\left(\exp(z)\right)\right)^+ + b + \lambda\left(\exp(z)\right)$ gives for all $(y,z) \in \mathbb{R}^n \times \mathbb{R}$, $d \in \mathbb{R}^m, \theta \in \mathbb{R}^p$

$$\begin{aligned} &\dot{V}(y,z,d,\theta) \leq -Q(y) + |\nabla V(y)g(y)|\,|\varphi(y)|\left(|\theta| - b - \lambda\left(\exp(z)\right)\right)^+ \\ &+|\nabla V(y)g(y)|\,|\varphi(y)|\left(b+\lambda\left(\exp(z)\right)\right) + \tfrac{a|d|^2}{1+\kappa(\exp(z))} \\ &-\tfrac{1+\kappa(\exp(z))}{4a}\left(|\varphi(y)|^2 + 2a\mu(y)\tfrac{(b+\lambda(\exp(z)))^2}{1+\kappa(\exp(z))}\right)\left(\nabla V(y)g(y)\right)^2 \end{aligned} \tag{5.53}$$

Using the inequalities

$$\begin{aligned} &\left(b+\lambda\left(\exp(z)\right)\right)|\nabla V(y)g(y)|\,|\varphi(y)| \\ &\leq \tfrac{\mu(y)}{2}\left(b+\lambda\left(\exp(z)\right)\right)^2\left(\nabla V(y)g(y)\right)^2 + \tfrac{1}{2\mu(y)}|\varphi(y)|^2 \end{aligned} \tag{5.54}$$

$$\begin{aligned} &|\nabla V(y)g(y)|\,|\varphi(y)|\left(|\theta| - b - \lambda\left(\exp(z)\right)\right)^+ \\ &\leq \tfrac{1}{4a}\left(1+\kappa\left(\exp(z)\right)\right)|\varphi(y)|^2\left(\nabla V(y)g(y)\right)^2 + a\tfrac{\left((|\theta|-b-\lambda(\exp(z)))^+\right)^2}{1+\kappa(\exp(z))} \end{aligned} \tag{5.55}$$

we get from (5.53) for all $(y,z) \in \mathbb{R}^n \times \mathbb{R}$, $d \in \mathbb{R}^m, \theta \in \mathbb{R}^p$

$$\dot{V}(y,z,d,\theta) \leq -Q(y) + \frac{|\varphi(y)|^2}{2\mu(y)} + a\frac{|d|^2 + \left(\left(|\theta| - b - \lambda\left(\exp(z)\right)\right)^+\right)^2}{1+\kappa\left(\exp(z)\right)} \tag{5.56}$$

Inequality (5.56) in conjunction with (5.3) gives for all $(x,z) \in \mathbb{R}^n \times \mathbb{R}$, $d \in \mathbb{R}^m, \theta \in \mathbb{R}^p$

$$\dot{V}(y,z,d,\theta) \leq -\frac{1}{2}Q(y) + a\frac{|d|^2 + \left(\left(|\theta| - b - \lambda\left(\exp(z)\right)\right)^+\right)^2}{1+\kappa\left(\exp(z)\right)} \tag{5.57}$$

Since V, Q are smooth, positive definite, and radially unbounded functions, Proposition 2.2 on page 107 in [66] guarantees the existence of a function $\tilde{\rho} \in K_\infty$ such that the following inequality holds for all $y \in \mathbb{R}^n$:

$$\frac{1}{2}Q(y) \geq \tilde{\rho}\left(V(y)\right) \tag{5.58}$$

Since V is a continuous, positive definite function, there exists $\tilde{\delta} > 0$ such that the following implication holds:

$$V(y) \leq \tilde{\delta} \Rightarrow |y| \leq \delta \tag{5.59}$$

where $\delta > 0$ is the constant appearing in assumption (C). Implication (5.59) and (5.11) implies that the following implication holds:

$$V(y) \leq \tilde{\delta} \Rightarrow \frac{1}{2}Q(y) \geq \eta V(y) \tag{5.60}$$

where $\eta > 0$ is the constant appearing in assumption (C).

We next claim that there exists$\rho \in K_\infty$ and $c > 0$ such that the following implications hold:

$$s \in \left[0, \tilde{\delta}\right] \Rightarrow \max\left(\eta\, s, \tilde{\rho}\,(s)\right) \geq \rho\,(s) \geq cs \tag{5.61}$$

$$s \geq \tilde{\delta} \Rightarrow \rho\,(s) = \tilde{\rho}\,(s) \tag{5.62}$$

Indeed, we can give a formula for the function $\rho \in K_\infty$ that satisfies (5.61) and (5.62), namely, the formula for $s \geq 0$:

$$\rho\,(s) = \max\left(\frac{\tilde{\rho}\left(\tilde{\delta}\right)\eta\, s}{\tilde{\rho}\left(\tilde{\delta}\right) + \eta s}, \tilde{\rho}\,(s)\right) \tag{5.63}$$

Formula (5.63) implies (5.61) with $c = \frac{\tilde{\rho}(\tilde{\delta})\eta}{\tilde{\rho}(\tilde{\delta})+\eta\,\tilde{\delta}}$. Moreover, the fact that $\frac{\tilde{\rho}(\tilde{\delta})\eta\, s}{\tilde{\rho}(\tilde{\delta})+\eta s} \leq \tilde{\rho}\left(\tilde{\delta}\right)$ for all $s \geq 0$ gives implication (5.62).

Inequality (5.58) and implications (5.60), (5.61), (5.62) guarantee that the following inequality holds for all $y \in \mathbb{R}^n$:

$$\frac{1}{2}Q(y) \geq \rho\,(V(y)) \tag{5.64}$$

If inequality (5.11) holds for all $y \in \mathbb{R}^n$ then $\rho\,(s) = \eta s$ for all $s \geq 0$.

Combining (5.57) and (5.64) we get for all $(y, z) \in \mathbb{R}^n \times \mathbb{R}$, $d \in \mathbb{R}^m, \theta \in \mathbb{R}^p$

$$\dot{V}(y, z, d, \theta) \leq -\rho\,(V(y)) + a\frac{|d|^2 + \left(\left(|\theta| - b - \lambda\left(\exp(z)\right)\right)^+\right)^2}{1 + \kappa\left(\exp(z)\right)} \tag{5.65}$$

The proof becomes complete by using (5.65) and Theorem 2.8 in Chapter 2. ◁

We finish this section by providing the proof of Theorem 5.2.

Proof of Theorem 5.2: Using the function $V(y) = y'Py$ and equations (5.20), (5.22) we get for all $(y, z) \in \mathbb{R}^n \times \mathbb{R}$, $d \in \mathbb{R}^m, \theta \in \mathbb{R}^p$

$$\begin{aligned}\dot{V}(y, z, d, \theta) &= y'\left(P\,(F - Gk') + (F - Gk')'\,P\right)y + 2y'PG\,(\varphi'(y)\theta + A'(y)d) \\ &-\tfrac{1+\kappa(\exp(z))}{a}\left(|\varphi(y)|^2 + |A(y)|^2 + 2a\mu(y)\tfrac{(b+\lambda(\exp(z)))^2}{1+\kappa(\exp(z))}\right)|G'Py|^2\end{aligned} \tag{5.66}$$

Using the fact that $Q = -\,(F - Gk')'\,P - P\,(F - Gk')$, we obtain from (5.66) for all $(y, z) \in \mathbb{R}^n \times \mathbb{R}$, $d \in \mathbb{R}^m, \theta \in \mathbb{R}^p$

$$\begin{aligned}\dot{V}(y, z, d, \theta) &\leq -y'Qy + 2\,|G'Py|\,|\varphi(y)|\,|\theta| + 2\,|G'Py|\,|A(y)|\,|d| \\ &-\tfrac{1+\kappa(\exp(z))}{a}\left(|\varphi(y)|^2 + |A(y)|^2 + 2a\mu(y)\tfrac{(b+\lambda(\exp(z)))^2}{1+\kappa(\exp(z))}\right)|G'Py|^2\end{aligned} \tag{5.67}$$

The inequality $2\,|G'Py|\,|A(y)|\,|d| \leq \frac{1+\kappa(\exp(z))}{a}\,|A(y)|^2\,|G'Py|^2 + \frac{a|d|^2}{1+\kappa(\exp(z))}$ in conjunction with (5.67) gives for all $(y, z) \in \mathbb{R}^n \times \mathbb{R}$, $d \in \mathbb{R}^m, \theta \in \mathbb{R}^p$

$$\begin{aligned}\dot{V}(y, z, d, \theta) &\leq -y'Qy + 2\,|G'Py|\,|\varphi(y)|\,|\theta| + \tfrac{a|d|^2}{1+\kappa(\exp(z))} \\ &-\tfrac{1+\kappa(\exp(z))}{a}\left(|\varphi(y)|^2 + 2a\mu(y)\tfrac{(b+\lambda(\exp(z)))^2}{1+\kappa(\exp(z))}\right)|G'Py|^2\end{aligned} \tag{5.68}$$

Inequality (5.68) and the fact that $|\theta| \leq (|\theta| - b - \lambda(\exp(z)))^+ + b + \lambda(\exp(z))$ gives for all $(y,z) \in \mathbb{R}^n \times \mathbb{R}$, $d \in \mathbb{R}^m, \theta \in \mathbb{R}^p$

$$\begin{aligned} \dot{V}(y,z,d,\theta) &\leq -y'Qy + 2|G'Py|\,|\varphi(y)|\,(|\theta| - b - \lambda(\exp(z)))^+ \\ &+ 2|G'Py|\,|\varphi(y)|\,(b + \lambda(\exp(z))) + \tfrac{a|d|^2}{1+\kappa(\exp(z))} \\ &- \tfrac{1+\kappa(\exp(z))}{a}\left(|\varphi(y)|^2 + 2a\mu(y)\tfrac{(b+\lambda(\exp(z)))^2}{1+\kappa(\exp(z))}\right)|G'Py|^2 \end{aligned} \tag{5.69}$$

Exploiting the fact that

$$\begin{aligned} &2|G'Py|\,|\varphi(y)|\,(b + \lambda(\exp(z))) \\ &\leq 2\mu(y)\,(b + \lambda(\exp(z)))^2\,|G'Py|^2 + \tfrac{1}{2\mu(y)}\,|\varphi(y)|^2 \end{aligned} \tag{5.70}$$

we get from (5.69) for all $(y,z) \in \mathbb{R}^n \times \mathbb{R}$, $d \in \mathbb{R}^m, \theta \in \mathbb{R}^p$

$$\begin{aligned} \dot{V}(y,z,d,\theta) &\leq -y'Qy + 2|G'Py|\,|\varphi(y)|\,(|\theta| - b - \lambda(\exp(z)))^+ \\ &+ \tfrac{|\varphi(y)|^2}{2\mu(y)} + \tfrac{a|d|^2}{1+\kappa(\exp(z))} - \tfrac{1+\kappa(\exp(z))}{a}\,|\varphi(y)|^2\,|G'Py|^2 \end{aligned} \tag{5.71}$$

Inequality (5.21) in conjunction with (5.71) gives for all $(y,z) \in \mathbb{R}^n \times \mathbb{R}$, $d \in \mathbb{R}^m, \theta \in \mathbb{R}^p$

$$\begin{aligned} \dot{V}(y,z,d,\theta) &\leq -\eta\, y'Py + 2|G'Py|\,|\varphi(y)|\,(|\theta| - b - \lambda(\exp(z)))^+ \\ &+ \tfrac{a|d|^2}{1+\kappa(\exp(z))} - \tfrac{1+\kappa(\exp(z))}{a}\,|\varphi(y)|^2\,|G'Py|^2 \end{aligned} \tag{5.72}$$

The fact that $V(y) = y'Py$ and inequality

$$\begin{aligned} &2|G'Py|\,|\varphi(y)|\,(|\theta| - b - \lambda(\exp(z)))^+ \\ &\leq \tfrac{1+\kappa(\exp(z))}{a}\,|\varphi(y)|^2\,|G'Py|^2 + \tfrac{a\left((|\theta|-b-\lambda(\exp(z)))^+\right)^2}{1+\kappa(\exp(z))} \end{aligned} \tag{5.73}$$

combined with (5.72) implies for all $(y,z) \in \mathbb{R}^n \times \mathbb{R}$, $d \in \mathbb{R}^m, \theta \in \mathbb{R}^p$

$$\dot{V}(y,z,d,\theta) \leq -\eta\, V(y) + a\frac{|d|^2 + \left((|\theta| - b - \lambda(\exp(z)))^+\right)^2}{1 + \kappa(\exp(z))} \tag{5.74}$$

The proof becomes complete by using (5.65) and Theorem 2.8 in Chapter 2. $\triangleleft$

5.5 ▪ Notes and comments

1. The results of the chapter first appeared in [76]. More precisely, the controllers in [76] are similar to but not exactly the same as the controllers presented in this chapter. Here, we have modified slightly the formulas in order to illustrate the fact that the same ideas are used for systems with matched uncertainties as well as other classes of systems (see the following chapter). Moreover, the proofs here are different from the proofs in [76]; here we exploit the general result of Theorem 2.8 in Chapter 2.

2. It is clear that the DADS controller for systems with matched uncertainties covers the three words "Robustness-Performance-Simplicity." Of course, the word "Identification" is not covered, since the DADS controller is a direct adaptive control scheme. In Chapter 7, we explain how to perform identification with arbitrary direct adaptive controllers.

3. Robustness in adaptive control is a major issue that has attracted the attention of many researchers in control theory. Various approaches have been proposed in the literature for time-invariant nonlinear control systems for which no persistence of excitation condition is assumed: nonlinear damping (see [73], [75], [87], [90]), leakage (see [58], [59], [115], [130]), projection methodologies (see [58] and Appendix E in [87]), supervision for direct adaptive schemes (see [45]), dynamic (high) gains or gain adjustment (see [62], [82], [101], [127]), and deadzone in the update law (introduced in the paper [123] and well explained in the book [58] in the context of indirect adaptive control; see also [57] and references therein for the "λ-tracking" approach in direct adaptive control). Every approach has its own advantages and disadvantages and some of the approaches require special assumptions (e.g., knowledge of bounds for the disturbances and/or the unknown parameters). However, the DADS controller is a novel control scheme that combines for the first time three elements: (a) nonlinear damping, (b) single-gain adjustment (the dynamic feedback has only one state), and (c) deadzone in the update law. That is why the DADS controller provides robustness/performance features that are not provided by other adaptive schemes: attenuation of the plant state to an assignable small level, despite the presence of disturbances and unknown parameters of arbitrary and unknown bounds, and absence of gain and state drift regardless of the size of the disturbance and unknown parameter.

Chapter 6

Strict-Feedback Systems: Deadzone-Adapted Disturbance Suppression Control

6.1 ▪ Introduction

In this chapter we study nonlinear control systems of the form

$$\dot{\xi}_i = \xi_{i+1} \text{ for } i = 1, ..., n-1 \text{ and } \dot{\xi}_n = w_1 \tag{6.1}$$

$$\begin{aligned} \dot{w}_j &= h_j(\xi, \bar{w}_j) + g_j(\xi, \bar{w}_j, \theta) w_{j+1} + \varphi_j'(\xi, \bar{w}_j)\theta + \alpha_j'(\xi, \bar{w}_j) d \text{ for } j = 1, ..., m-1 \\ \dot{w}_m &= h_j(\xi, w) + g_m(\xi, w, \theta) u + \varphi_m'(\xi, w)\theta + \alpha_m'(\xi, w) d \end{aligned} \tag{6.2}$$

where $\bar{w}_j = (w_1, ..., w_j)$, $(\xi, w) \in \mathbb{R}^n \times \mathbb{R}^m$ is the state with $\xi = (\xi_1, ..., \xi_n)$, $w = (w_1, ..., w_m)$, $u \in \mathbb{R}$ is the control input, $d \in \mathbb{R}^l$, $\theta \in \Theta \subseteq \mathbb{R}^p$ are disturbances and $\Theta \subseteq \mathbb{R}^p$ is a closed set, $h_j : \mathbb{R}^n \times \mathbb{R}^j \to \mathbb{R}$, $g_j : \mathbb{R}^n \times \mathbb{R}^j \times \Theta \to \mathbb{R}$, $\varphi_j : \mathbb{R}^n \times \mathbb{R}^j \to \mathbb{R}^p$, $\alpha_j : \mathbb{R}^n \times \mathbb{R}^j \to \mathbb{R}^l$ for $j = 1, ..., m$ are smooth mappings with $h_j(0,0) = 0$ and $\varphi_j(0,0) = 0$ for $j = 1, ..., m$. Systems of the form (6.2) are termed in the literature as triangular systems or systems in parametric strict-feedback form. It should be noted that $\theta \in \Theta \subseteq \mathbb{R}^p$ is a vanishing perturbation while $d \in \mathbb{R}^l$ is, in general, a non-vanishing perturbation. The subsystem (6.1) is a plain chain of integrators without a parametric uncertainty or disturbance. The reason why we include an integrator chain on top of strict-feedback system is that, in mechanical systems, integrators, relating positions with velocities, and angles with angular velocities, arise in the kinematics of such mechanical systems and are free of unknown parameters and disturbances.

If one hasn't made extensive attempts to expand adaptive control designs beyond the class of systems above, particularly the triangular class (6.2), one may regard this class as restrictive or special. Nothing is farther from fact. Except for routine generalizations to including zero dynamics (the so-called block-backstepping in Chapters 3 and 4 in [87]), comparably routine extensions to multi-input parametric-strict-feedback systems, or the very routine extensions where the nonlinearities, rather than being "triangular" are "triangularly bounded," the class (6.2) is in fact the most general class of nonlinear systems for which adaptive control (with perfect asymptotic tracking in the absence of disturbances) is possible *systematically*. A similar statement holds for robust nonlinear control of systems affine in disturbances. The consideration of (6.2) as being the most general systematically tractable class of systems for adaptive robust nonlinear control is not only based on the experience of decades of unsuccessful attempts since [87] to achieve greater generality. It is also based on the fact that necessary and sufficient coordinate-free differential geometric conditions for the transformation of completely general nonlinear systems affine in control and the parameter using a parameter-independent diffeomorphic transformation are given in Appendix G in [87]. It is thus for pedagogical reasons that we limit our attention to

(6.2), with (6.1), rather than pursuing the somewhat more general block-strict-feedback, multi-input, or triangularly bounded systems, for which the designs would get more complex (in a routine sense) but harder to understand on a first exposure. The class of systems (6.1), (6.2) is not only mathematically general. It is also of great physical generality. Fully actuated systems in robotics and vehicles, with the addition of actuator dynamics, do have the structure of kinematics $\to$ mechanical dynamics $\to$ actuator dynamics. Hence (6.1), (6.2) is vastly more general than this broad class of practical systems.

The examples of the previous chapter may have caused a disappointment to the reader by giving a false impression that the DADS control scheme cannot provide its guarantees for systems with unmatched uncertainties (such as system (6.1), (6.2)). This is true, technically, only if the output map coincides with the whole plant state, i.e., $y = (\xi, w)$. If different outputs are considered then the DADS scheme furnishes its guarantees. Indeed, we show in this chapter that the DADS control scheme is successful when the output map contains only the first component of the vector $w = (w_1, ..., w_m)$, i.e., when $y = (\xi, w_1)$. This restriction is easily understood: in order to keep w_1 in a prespecified neighborhood of zero, one has to manipulate u so that w_2 almost "cancels" the effect of the disturbances on w_1. Therefore, w_2 and all other components $w_3, ..., w_m$ must necessarily depend on the magnitude of the disturbance vector and cannot be kept in a prespecified neighborhood of zero.

Due to the generality of (6.1), (6.2), the algorithm for its robust adaptive control may appear complex. The complexity is not due to the DADS methodology. The complexity comes from the class of systems. The class of systems contains both unmatched unknown parameters and unmatched disturbances, and not only are they unmatched, but they allow an arbitrarily high "uncertainty relative degree" $m - 1$, which is the number of integrators separating the control input from the uncertainties. This is why, as m grows, the complexity necessarily grows. The attempts to reduce the complexity using "dynamic surface control" are not only unsuccessful but also misguided (and hence not cited here) since they trade the algebraic complexity for (a) an explosion in the dynamic order of the compensator through filters, (b) severe shrinking of the region of attraction, unless extreme gains are employed, and (c) the loss of asymptotic regulation in the disturbance-free case.

The main results of the present chapter first appeared in [78].

6.2 ▪ Main Results

In this section we simply state the main results of the chapter. We start our study by making the following assumption for (6.1), (6.2).

(A) *There exist smooth mappings $\eta_j : \mathbb{R}^n \times \mathbb{R}^j \to (0, +\infty)$ for $j = 1, ..., m$, $\mu_j : \mathbb{R}^n \times \mathbb{R}^j \to (0, +\infty)$ for $j = 1, ..., m - 1$ such that the following inequalities hold:*

$$\begin{aligned} &\eta_j(\xi, w_1, ..., w_j) \le g_j(\xi, w_1, ..., w_j, \theta) \\ &\textit{for all } (\xi, w_1, ..., w_j) \in \mathbb{R}^n \times \mathbb{R}^j,\ \theta \in \Theta,\ j = 1, ..., m \end{aligned} \tag{6.3}$$

$$\begin{aligned} &g_j(\xi, w_1, ..., w_j, \theta) \le \mu_j(\xi, w_1, ..., w_j)\,(1 + |\theta|) \\ &\textit{for all } (\xi, w_1, ..., w_j) \in \mathbb{R}^n \times \mathbb{R}^j,\ \theta \in \Theta,\ j = 1, ..., m - 1 \end{aligned} \tag{6.4}$$

When $\Theta \subseteq \mathbb{R}^p$ is a bounded set, we can exploit (6.3) and (6.4) in order to design a smooth feedback law $u = k(\xi, w)$ that guarantees the ISS property with respect to $d \in \mathbb{R}^l$ (uniformly in θ); see [66]. On the other hand, when θ is constant (but unknown) and $d \equiv 0$ then we can apply the methodologies in [87] in order to design an adaptive controller that guarantees GAOS for the closed-loop system with output $y = (\xi, w)$ (notice that here we are talking about GAOS and not GAS since the closed-loop system has additional states: the states of the controller). When θ

is constant (and unknown) and d is not necessarily zero one can employ standard techniques in order to achieve the p-ISS property with respect to $d \in \mathbb{R}^l$ for the closed-loop system. One such technique that does not require the knowledge of bounds for θ is the so-called σ-modification, which guarantees the p-ISS property for the closed-loop system. However, for the σ-modification the residual constant is an increasing function of the unknown $|\theta|$, which means that the output $y = (\xi, w)$ (even in the case $d \equiv 0$) approaches a neighborhood of zero of unknown (and possibly large) radius.

As we saw in the previous chapter, an alternative approach to σ-modification is the Deadzone-Adapted Disturbance Suppression (DADS) controller. The DADS controller combines both the adaptation idea and the robust control methodology and uses only one integrator with deadzone. The DADS controller guarantees an attenuation of the plant output to an assignable small level, despite the presence of disturbances and unknown parameters of arbitrary and unknown bounds, while at the same time prevents gain and state drift regardless of the size of the disturbance and unknown parameter. In this chapter, we extend the DADS controller to the case (6.1), (6.2) where no matching condition holds. Our main result is stated next.

Theorem 6.1 (DADS controller for systems in parametric strict-feedback form): *Consider the control system* (6.1), (6.2) *under assumption (A). Then for every constant* $b, \Gamma, \varepsilon, c, a > 0$ *and for every* $\kappa, \lambda \in K_\infty \cap C^\infty(\mathbb{R}_+)$ *there exist functions* $k, V \in C^\infty(\mathbb{R}^n \times \mathbb{R}^m \times \mathbb{R})$ *with* $V(0,0,z) = k(0,0,z) = 0$ *for* $z \in \mathbb{R}$ *and* $V(\xi, w, z) > 0$ *for* $(\xi, w) \neq 0$, *a constant* $M \geq 1$ *and a function* $B \in C^0(\mathbb{R}^n \times \mathbb{R}^m \times \mathbb{R} \times \mathbb{R}_+ \times \mathbb{R}_+)$ *such that for every* $(\xi_0, w_0, z_0) \in \mathbb{R}^n \times \mathbb{R}^m \times \mathbb{R}$, $d \in L^\infty(\mathbb{R}_+; \mathbb{R}^l)$, $\theta \in L^\infty(\mathbb{R}_+; \Theta)$ *the unique solution of* (6.1), (6.2) *with*

$$u = k(\xi, w, z) \tag{6.5}$$

$$\dot{z} = \Gamma \exp(-z) \left(V(\xi, w, z) - \frac{\varepsilon^2}{2M} \right)^+ \tag{6.6}$$

and $(\xi(0), w(0), z(0)) = (\xi_0, w_0, z_0)$ *exists and satisfies the following estimates for all* $t \geq 0$:

$$\limsup_{t \to +\infty} (|\xi(t)| + |w_1(t)|) \leq \varepsilon, \limsup_{t \to +\infty} (V(\xi(t), w(t), z(t))) \leq \frac{\varepsilon^2}{2M} \tag{6.7}$$

$$\begin{aligned} &|\xi(t)|^2 + |w_1(t)|^2 \leq MV(\xi(t), w(t), z(t)) \\ &\leq M \exp(-ct) V(\xi_0, w_0, z_0) + a \frac{\|d\|_\infty^2 + \left(\left(\|\theta\|_\infty - b - \lambda(\exp(z_0)) \right)^+ \right)^2}{c(1 + \kappa(\exp(z_0)))} \end{aligned} \tag{6.8}$$

$$|\xi(t)| + |w(t)| \leq B(\xi_0, w_0, z_0, \|d\|_\infty, \|\theta\|_\infty) \tag{6.9}$$

$$z_0 \leq z(t) \leq \lim_{s \to +\infty} (z(s)) \leq B(\xi_0, w_0, z_0, \|d\|_\infty, \|\theta\|_\infty) \tag{6.10}$$

Moreover, if $\limsup_{t \to +\infty} (|\theta(t)|) \leq b + \lambda\left(\exp\left(\lim_{t \to +\infty} (z(t))\right)\right)$ *and* $\lim_{t \to +\infty} (d(t)) = 0$ *then* $\lim_{t \to +\infty} (|\xi(t)| + |w(t)|) = 0$.

Inequality (6.7) guarantees that the plant output $y = (\xi, w_1)$ is eventually led (even in the case where a disturbance is present) to a ball around the origin of *assignable radius*. Indeed, inequality (6.7) guarantees the zero p-OAG property $\limsup_{t \to +\infty} (|(\xi(t), w_1(t))|) \leq \alpha$ with a constant $\alpha > 0$ independent of θ. This is a feature that cannot be guaranteed by σ-modification when θ is constant. Moreover, the state, i.e., $(\xi(t), w(t), z(t))$, remains bounded for every initial condition and every disturbance $d \in L^\infty(\mathbb{R}_+; \mathbb{R}^l)$, $\theta \in L^\infty(\mathbb{R}_+; \Theta)$—see (6.10), (6.9); this is the p-UBIBS property. Inequality (6.8) shows that the p-IOS property from the disturbance $d \in L^\infty(\mathbb{R}_+; \mathbb{R}^l)$ holds when θ is constant for the closed-loop system (6.1), (6.2), (6.5), (6.6) with output $y=(\xi,w_1)$. When $d\equiv 0$ and θ is constant then the closed-loop system (6.1), (6.2), (6.5), (6.6) with output $y=(\xi,w_1)$: i) is Lagrange stable, ii) satisfies the p-UGAOS property, iii) satisfies the UGAOS property when $|\theta|\leq b$, and iv) has a globally attracting set, namely the set $\left\{(\xi,w,z)\in\mathbb{R}^n\times\mathbb{R}^m\times\mathbb{R} : V(\xi,w,z)\leq\frac{\varepsilon^2}{2M}\right\}$. Finally, it should also be noticed that convergence to $(\xi,w)=(0,0)$ is not excluded: it can be achieved when $\lim_{t\to+\infty}(d(t))=0$ and $\limsup_{t\to+\infty}(|\theta(t)|)\leq b+\lambda\left(\exp\left(\lim_{t\to+\infty}(z(t))\right)\right)$. This is another feature that cannot be guaranteed by σ-modification when θ is constant. For the DADS controller in the disturbance-free case with constant θ we have the following conclusion: *if the plant state is not led asymptotically to zero then* $|\theta|>b$ *and the estimate* $z(t)\leq\lim_{t\to+\infty}(z(t))<\ln\left(\lambda^{-1}(|\theta|-b)\right)$ *holds for all* $t\geq 0$.

Using the same tools, we can also study nonlinear strict-feedback systems of the form

$$\begin{aligned}\dot{w}_i &= h_i(\bar{w}_i)+g_i(\bar{w}_i,\theta)w_{i+1}+\varphi_i'(\bar{w}_i)\theta+\alpha_i'(\bar{w}_i)d \text{ for } i=1,\dots,n-1\\ \dot{w}_n &= h_n(w)+g_n(w,\theta)u+\varphi_n'(w)\theta+\alpha_n'(w)d\end{aligned} \tag{6.11}$$

where $\bar{w}_i=(w_1,\dots,w_i)$, $w=(w_1,\dots,w_n)\in\mathbb{R}^n$ is the state, $u\in\mathbb{R}$ is the control input, $d\in\mathbb{R}^l$, $\theta\in\Theta\subseteq\mathbb{R}^p$ are disturbances, $h_i:\mathbb{R}^i\to\mathbb{R}$, $g_i:\mathbb{R}^i\times\Theta\to\mathbb{R}$, $\varphi_i:\mathbb{R}^i\to\mathbb{R}^p$, $\alpha_i:\mathbb{R}^i\to\mathbb{R}^l$ for $i=1,\dots,n$ are smooth mappings with $h_i(0)=0$ and $\varphi_i(0)=0$ for $i=1,\dots,n$. In this case we obtain the following result.

Theorem 6.2 (DADS controller for systems in parametric strict-feedback form with no integrators): *Consider the control system* (6.11) *and assume that there exist smooth positive mappings* $\eta_i:\mathbb{R}^i\to(0,+\infty)$ *for* $i=1,\dots,n$, $\mu_i:\mathbb{R}^i\to(0,+\infty)$ *for* $i=1,\dots,n-1$ *such that the following inequalities hold:*

$$\begin{aligned}&\eta_i(w_1,\dots,w_i)\leq g_i(w_1,\dots,w_i,\theta)\\ &\textit{for all } (w_1,\dots,w_i)\in\mathbb{R}^i,\ \theta\in\Theta,\ i=1,\dots,n\end{aligned} \tag{6.12}$$

$$\begin{aligned}&g_i(w_1,\dots,w_i,\theta)\leq\mu_i(w_1,\dots,w_i)\,(1+|\theta|)\\ &\textit{for all } (w_1,\dots,w_i)\in\mathbb{R}^i,\ \theta\in\Theta,\ i=1,\dots,n-1\end{aligned} \tag{6.13}$$

Then for every constant $b,\Gamma,\varepsilon,a,c>0$ *and for every* $\kappa,\lambda\in K_\infty\cap C^\infty(\mathbb{R}_+)$ *there exist functions* $k,V\in C^\infty(\mathbb{R}^n\times\mathbb{R})$ *with* $V(0,z)=k(0,z)=0$ *for* $z\in\mathbb{R}$ *and* $V(w,z)>0$ *for* $w\neq 0$ *and a function* $B\in C^0(\mathbb{R}^n\times\mathbb{R}\times\mathbb{R}_+\times\mathbb{R}_+)$ *such that for every* $(w_0,z_0)\in\mathbb{R}^n\times\mathbb{R}$, $d\in L^\infty(\mathbb{R}_+;\mathbb{R}^l)$, $\theta\in L^\infty(\mathbb{R}_+;\Theta)$ *the unique solution of* (6.11) *with*

$$u=k(w,z) \tag{6.14}$$

$$\dot{z}=\Gamma\exp(-z)\left(V(w,z)-\varepsilon^2/2\right)^+ \tag{6.15}$$

and $(w(0),z(0))=(w_0,z_0)$ *exists and satisfies the following estimates for all* $t\geq 0$*:*

$$\limsup_{t\to+\infty}(|w_1(t)|)\leq\varepsilon \textit{ and } \limsup_{t\to+\infty}(V(w(t),z(t)))\leq\varepsilon^2/2 \tag{6.16}$$

$$\begin{aligned}&|w_1(t)|^2\leq 2V(w(t),z(t))\\ &\leq 2\exp(-ct)V(w_0,z_0)+2a\frac{\|d\|_\infty^2+\left(\left(\|\theta\|_\infty-b-\lambda(\exp(z_0))\right)^+\right)^2}{c(1+\kappa(\exp(z_0)))}\end{aligned} \tag{6.17}$$

$$|w(t)|\leq B\left(w_0,z_0,\|d\|_\infty,\|\theta\|_\infty\right) \tag{6.18}$$

$$z_0\leq z(t)\leq\lim_{s\to+\infty}(z(s))\leq B\left(w_0,z_0,\|d\|_\infty,\|\theta\|_\infty\right) \tag{6.19}$$

Moreover, if $\limsup_{t\to+\infty}(|\theta(t)|)\leq b+\lambda\left(\exp\left(\lim_{t\to+\infty}(z(t))\right)\right)$ *and* $\lim_{t\to+\infty}(d(t))=0$ *then* $\lim_{t\to+\infty}(|w(t)|)=0$.

6.3 ▪ Backstepping

The proofs of Theorem 6.1 and Theorem 6.2 are based on the following lemma, which allows a step-by-step construction of the DADS controller. The lemma deals with the control system

$$\begin{aligned}\dot{\xi} &= f(\xi) + g(\xi,\theta)w + \Phi(\xi)\theta + G(\xi)d \\ \dot{w} &= h(\xi,w) + \tilde{g}(\xi,w,\theta)u + \varphi'(\xi,w)\theta + \alpha'(\xi,w)d\end{aligned} \tag{6.20}$$

where $(\xi, w) \in \mathbb{R}^n \times \mathbb{R}$ is the state, $u \in \mathbb{R}$ is the control input, $d \in \mathbb{R}^l$, $\theta \in \Theta \subseteq \mathbb{R}^p$ are disturbances, $f : \mathbb{R}^n \to \mathbb{R}^n$, $g : \mathbb{R}^n \times \Theta \to \mathbb{R}^n$, $\Phi : \mathbb{R}^n \to \mathbb{R}^{n\times p}$, $G : \mathbb{R}^n \to \mathbb{R}^{n\times l}$, $h : \mathbb{R}^n \times \mathbb{R} \to \mathbb{R}$, $\varphi : \mathbb{R}^n \times \mathbb{R} \to \mathbb{R}^p$, $\alpha : \mathbb{R}^n \times \mathbb{R} \to \mathbb{R}^l$, $\tilde{g} : \mathbb{R}^n \times \mathbb{R} \times \Theta \to \mathbb{R}$ are smooth mappings with $f(0) = 0$, $\Phi(0) = 0$, $h(0,0) = 0$, and $\varphi(0,0) = 0$.

Lemma 6.1 (A backstepping lemma): *Consider system (6.20) and assume that there exist constants $a, b, c, \Gamma, \varepsilon, \eta > 0$, smooth functions $\kappa, \lambda \in K_\infty \cap C^\infty(\mathbb{R}_+)$, $\sigma, k, V \in C^\infty(\mathbb{R}^n \times \mathbb{R})$, $\mu \in C^\infty(\mathbb{R}^n)$, $\eta \in C^\infty(\mathbb{R}^n \times \mathbb{R}; (0, +\infty))$ with $V(0, z) = k(0, z) = 0$ for $z \in \mathbb{R}$ and $V(\xi, z) > 0$ for $\xi \neq 0$, $z \in \mathbb{R}$, such that the following inequalities hold for all $\theta \in \Theta$, $d \in \mathbb{R}^l$, $(\xi, w, z) \in \mathbb{R}^n \times \mathbb{R} \times \mathbb{R}$:*

$$|\xi|^2 \leq \sigma(\xi,z)\,V(\xi,z) \tag{6.21}$$

$$\begin{aligned}&\tfrac{\partial V}{\partial \xi}(\xi,z)\left(f(\xi) + g(\xi,\theta)k(\xi,z) + \Phi(\xi)\theta + G(\xi)d\right) \\ &+\tfrac{\partial V}{\partial z}(\xi,z)\Gamma\exp(-z)\left(V(\xi,z)-\varepsilon\right)^+ \\ &\leq -cV(\xi,z) + a\tfrac{|d|^2 + \left((|\theta| - b - \lambda(\exp(z)))^+\right)^2}{1+\kappa(\exp(z))}\end{aligned} \tag{6.22}$$

$$\tilde{g}(\xi,w,\theta) \geq \eta(\xi,w) \tag{6.23}$$

$$|g(\xi,\theta)| \leq \mu(\xi)\left(1 + |\theta|\right) \tag{6.24}$$

Define the following function for all $(\xi, w, z) \in \mathbb{R}^n \times \mathbb{R} \times \mathbb{R}$:

$$\bar{V}(\xi,w,z) = V(\xi,z) + \frac{1}{2}\left(w - k(\xi,z)\right)^2 \tag{6.25}$$

Then there exist functions $\bar{\sigma} \in C^\infty(\mathbb{R}^n \times \mathbb{R} \times \mathbb{R})$, $M \in C^\infty(\mathbb{R}^n \times \mathbb{R} \times \mathbb{R}; (0, +\infty))$ such that, with

$$\bar{k}(\xi,w,z) := -\frac{M(\xi,w,z)}{\eta(\xi,w)}\left(w - k(\xi,z)\right) \;\; \textit{for all} \;\; (\xi,w,z) \in \mathbb{R}^n \times \mathbb{R} \times \mathbb{R} \tag{6.26}$$

the following inequalities hold for all $d \in \mathbb{R}^l$, $\theta \in \Theta$ and $(\xi, w, z) \in \mathbb{R}^n \times \mathbb{R} \times \mathbb{R}$:

$$|\xi|^2 + |w|^2 \leq \bar{\sigma}(\xi,w,z)\,\bar{V}(\xi,w,z) \tag{6.27}$$

$$\begin{aligned}&\tfrac{\partial \bar{V}}{\partial \xi}(\xi,w,z)\left(f(\xi) + g(\xi,\theta)w + \Phi(\xi)\theta + G(\xi)d\right) \\ &+\tfrac{\partial \bar{V}}{\partial z}(\xi,w,z)\Gamma\exp(-z)\left(\bar{V}(\xi,w,z)-\varepsilon\right)^+ \\ &+\tfrac{\partial \bar{V}}{\partial w}(\xi,w,z)\left(h(\xi,w) + \tilde{g}(\xi,w,\theta)\bar{k}(\xi,w,z) + \varphi'(\xi,w)\theta + \alpha'(\xi,w)d\right) \\ &\leq -\tfrac{c}{2}\bar{V}(\xi,w,z) + 2a\tfrac{|d|^2 + \left((|\theta| - b - \lambda(\exp(z)))^+\right)^2}{1+\kappa(\exp(z))}\end{aligned} \tag{6.28}$$

Remarks: (a) The reader is encouraged to notice the analogy of (6.28) with inequality (3.4) in [89]. However, it is to be noted that here (6.28) is obtained by means of a "domination approach" (with nonlinear damping terms as in [95]) while inequality (3.4) in [89] is obtained by means of a "cancelation approach". Moreover, the reader is also encouraged to notice the analogy of (6.26) with equation (78) in [95].

(b) The functions $\bar{V}, \bar{k} \in C^{\infty}(\mathbb{R}^n \times \mathbb{R} \times \mathbb{R})$ defined by (6.25) and (6.26) satisfy $\bar{V}(0,0,z) = \bar{k}(0,0,z) = 0$ for $z \in \mathbb{R}$ and $\bar{V}(\xi,w,z) > 0$ for $(\xi,w) \neq 0$, $z \in \mathbb{R}$. Moreover, if the set $\left\{ (\xi,z) \in \mathbb{R}^n \times \left[-\tilde{M},\tilde{M}\right] \,:\, V(\xi,z) \leq \tilde{M} \right\}$ is bounded for each $\tilde{M} > 0$ then the set

$$\left\{ (\xi,w,z) \in \mathbb{R}^n \times \mathbb{R} \times \left[-\tilde{M},\tilde{M}\right] \,:\, \bar{V}(\xi,w,z) \leq \tilde{M} \right\} \tag{6.29}$$

is bounded for each $\tilde{M} > 0$. This is exactly the case where Theorem 2.8 can be applied in order to deduce useful estimates for the solution of the closed-loop system (6.20) with $u = \bar{k}(\xi,w,z)$ and $\dot{z} = \Gamma \exp(-z)\left(\bar{V}(\xi,w,z) - \varepsilon\right)^{+}$.

(c) When $\tilde{g}(\xi,w,\theta) \equiv 1$ and $g(\xi,\theta)$ is independent of θ, i.e., $g(\xi,\theta) = g(\xi)$, then we can provide a simple formula for $\bar{k}(\xi,w,z)$. We have

$$\bar{k}(\xi,w,z) = -\frac{\partial V}{\partial \xi}(\xi,z)g(\xi) + \frac{\partial k}{\partial \xi}(\xi,z)\left(f(\xi) + g(\xi)w\right) - M(\xi,w,z)\left(w - k(\xi,z)\right) \tag{6.30}$$

where $M : \mathbb{R}^n \times \mathbb{R} \times \mathbb{R} \to (0,+\infty)$ is a C^{∞} function that satisfies

$$\begin{aligned} &M(\xi,w,z) \geq \tfrac{\Gamma}{2}\exp(-z)\left(\left|\tfrac{\partial V}{\partial z}(\xi,z)\right| + |w - k(\xi,z)|\left|\tfrac{\partial k}{\partial z}(\xi,z)\right|\right) \\ &+\tfrac{r_1^2(\xi,w,z)}{c}\left(b + \lambda\left(\exp(z)\right)\right)^2 \\ &+\tfrac{c}{4} + \tfrac{\Gamma^2\exp(-2z)}{c}V(\xi,z)\left(\tfrac{\partial k}{\partial z}(\xi,z)\right)^2 + r_2(\xi,w,z)\left(b + \lambda\left(\exp(z)\right)\right) \\ &+\tfrac{1+\kappa(\exp(z))}{4a}\left(\left|\alpha'(\xi,w) - \tfrac{\partial k}{\partial \xi}(\xi,z)G(\xi)\right|^2 + \left|\varphi'(\xi,w) - \tfrac{\partial k}{\partial \xi}(\xi,z)\Phi(\xi)\right|^2\right) \end{aligned} \tag{6.31}$$

and $r_i(\xi,w,z)$, $i = 1,2$, are C^{∞} non-negative functions that satisfy the inequality for all $(\xi,w,z) \in \mathbb{R}^n \times \mathbb{R} \times \mathbb{R}$

$$\left|\varphi'(\xi,w) - \frac{\partial k}{\partial \xi}(\xi,z)\Phi(\xi)\right| \leq r_1(\xi,w,z)\sqrt{V(\xi,z)} + r_2(\xi,w,z)\left|w - k(\xi,z)\right| \tag{6.32}$$

If in addition $V(\xi,z)$ is independent of z, i.e.,

$$V(\xi,z) = V(\xi) \tag{6.33}$$

then a locally Lipschitz function $\bar{k}(\xi,w,z)$ that satisfies (6.28) is given by the formula

$$\begin{aligned} &\bar{k}(\xi,w,z) = -h(\xi,w) - \tfrac{\partial V}{\partial \xi}(\xi)g(\xi) + \tfrac{\partial k}{\partial \xi}(\xi,z)\left(f(\xi) + g(\xi)w\right) \\ &+\tfrac{\partial k}{\partial z}(\xi,z)\Gamma\exp(-z)\left(V(\xi) + \tfrac{1}{2}\left(w - k(\xi,z)\right)^2 - \varepsilon\right)^{+} - \tfrac{c}{4}\left(w - k(\xi,z)\right) \\ &-\left(b + \lambda\left(\exp(z)\right)\right)\left(\tfrac{b+\lambda(\exp(z))}{2c}r_1^2(\xi,w,z) + r_2(\xi,w,z)\right)\left(w - k(\xi,z)\right) \\ &-\tfrac{1+\kappa(\exp(z))}{4a}\left(\left|\alpha'(\xi,w) - \tfrac{\partial k}{\partial \xi}(\xi,z)G(\xi)\right|^2 + \left|\varphi'(\xi,w) - \tfrac{\partial k}{\partial \xi}(\xi,z)\Phi(\xi)\right|^2\right)\left(w - k(\xi,z)\right) \end{aligned} \tag{6.34}$$

Therefore, using Theorem 2.8 and following the proof of Lemma 6.1 step by step, we obtain the following corollary for the system:

$$\begin{aligned} \dot{\xi} &= f(\xi) + g(\xi)w + \Phi(\xi)\theta + G(\xi)d \\ \dot{w} &= h(\xi,w) + u + \varphi'(\xi,w)\theta + \alpha'(\xi,w)d \end{aligned} \tag{6.35}$$

where $(\xi,w) \in \mathbb{R}^n \times \mathbb{R}$ is the state, $u \in \mathbb{R}$ is the control input, $d \in \mathbb{R}^l$, $\theta \in \Theta \subseteq \mathbb{R}^p$ are disturbances, $f : \mathbb{R}^n \to \mathbb{R}^n$, $g : \mathbb{R}^n \to \mathbb{R}^n$, $\Phi : \mathbb{R}^n \to \mathbb{R}^{n \times p}$, $G : \mathbb{R}^n \to \mathbb{R}^{n \times l}$, $h : \mathbb{R}^n \times \mathbb{R} \to \mathbb{R}$,

$\varphi : \mathbb{R}^n \times \mathbb{R} \to \mathbb{R}^p$, $\alpha : \mathbb{R}^n \times \mathbb{R} \to \mathbb{R}^l$ are locally Lipschitz mappings with $f(0) = 0$, $\Phi(0) = 0$, $h(0,0) = 0$, and $\varphi(0,0) = 0$.

Corollary 6.1 (The addition of one integrator): *Consider system* (6.35) *and assume that there exist constants $a, b, c > 0$, locally Lipschitz functions $\kappa, \lambda \in K_\infty$, $k \in C^2(\mathbb{R}^n \times \mathbb{R})$, $V \in C^2(\mathbb{R}^n)$ with V being positive definite and radially unbounded and with $k(0,z) = 0$ for $z \in \mathbb{R}$, such that the following inequalities hold for all $\theta \in \Theta$, $d \in \mathbb{R}^l$, $(\xi, w, z) \in \mathbb{R}^n \times \mathbb{R} \times \mathbb{R}$:*

$$\left| \varphi'(\xi, w) - \frac{\partial k}{\partial \xi}(\xi, z)\Phi(\xi) \right| \le r_1(\xi, w, z)\sqrt{V(\xi)} + r_2(\xi, w, z)\left| w - k(\xi, z) \right| \tag{6.36}$$

$$\begin{aligned} &\tfrac{\partial V}{\partial \xi}(\xi)\left(f(\xi) + g(\xi)k(\xi, z) + \Phi(\xi)\theta + G(\xi)d\right) \\ &\le -cV(\xi) + a\tfrac{|d|^2 + \left((|\theta| - b - \lambda(\exp(z)))^+\right)^2}{1 + \kappa(\exp(z))} \end{aligned} \tag{6.37}$$

Let arbitrary $\Gamma, \varepsilon > 0$ be given and define $\bar{k}(\xi, w, z)$ for all $(\xi, w, z) \in \mathbb{R}^n \times \mathbb{R} \times \mathbb{R}$ by means of (6.34). *Then there exists a function $B \in C^0(\mathbb{R}^n \times \mathbb{R} \times \mathbb{R} \times \mathbb{R}_+ \times \mathbb{R}_+)$ such that for every $(\xi_0, w_0, z_0) \in \mathbb{R}^n \times \mathbb{R} \times \mathbb{R}$, $d \in L^\infty(\mathbb{R}_+; \mathbb{R}^l)$, $\theta \in L^\infty(\mathbb{R}_+; \Theta)$ the unique solution of* (6.35) *with*

$$u = \bar{k}(\xi, w, z) \tag{6.38}$$

$$\dot{z} = \Gamma \exp(-z)\left(V(\xi) + \frac{1}{2}\left(w - k(\xi, z)\right)^2 - \varepsilon\right)^+ \tag{6.39}$$

and $(\xi(0), w(0), z(0)) = (\xi_0, w_0, z_0)$ exists and satisfies the following estimates for all $t \ge 0$:

$$\limsup_{t \to +\infty}\left(V(\xi(t)) + \frac{1}{2}\left(w(t) - k(\xi(t), z(t))\right)^2\right) \le \varepsilon \tag{6.40}$$

$$\begin{aligned} &V(\xi(t)) + \tfrac{1}{2}\left(w(t) - k(\xi(t), z(t))\right)^2 \\ &\le \exp(-ct/2)\left(V(\xi_0) + \tfrac{1}{2}\left(w_0 - k(\xi_0, z_0)\right)^2\right) + 4a\tfrac{\|d\|_\infty^2 + \left((\|\theta\|_\infty - b - \lambda(\exp(z_0)))^+\right)^2}{c(1 + \kappa(\exp(z_0)))} \end{aligned} \tag{6.41}$$

$$|\xi(t)| + |w(t)| \le B\left(\xi_0, w_0, z_0, \|d\|_\infty, \|\theta\|_\infty\right) \tag{6.42}$$

$$z_0 \le z(t) \le \lim_{s \to +\infty}(z(s)) \le B\left(\xi_0, w_0, z_0, \|d\|_\infty, \|\theta\|_\infty\right) \tag{6.43}$$

Moreover, if $\limsup_{t \to +\infty}(|\theta(t)|) \le b + \lambda\left(\exp\left(\lim_{t \to +\infty}(z(t))\right)\right)$ *and* $\lim_{t \to +\infty}(d(t)) = 0$ *then* $\lim_{t \to +\infty}(|\xi(t)| + |w(t)|) = 0$.

Corollary 6.1 avoids all unnecessary regularity assumptions and can be used in the case where *only one* integrator is to be added.

We next provide the proof of Lemma 6.1.

Proof of Lemma 6.1: Without loss of generality, we may assume that $\sigma(\xi, z) > 0$ for all $\xi \in \mathbb{R}^n$, $z \in \mathbb{R}$.

We start by using the fact that for every smooth function $\psi : \mathbb{R}^n \times \mathbb{R} \to \mathbb{R}^m$ with $\psi(0, z) = 0$ for all $z \in \mathbb{R}$ there exists a smooth positive function $L_\psi : \mathbb{R}^n \times \mathbb{R} \to (0, +\infty)$ such that the inequality $|\psi(\xi, z)| \le L_\psi(\xi, z)\,|\xi|$ holds for all $(\xi, z) \in \mathbb{R}^n \times \mathbb{R}$.

Since $V(0, z) = 0$ for all $z \in \mathbb{R}$ and $V(\xi, z) > 0$ for all $\xi \ne 0$, $z \in \mathbb{R}$, it follows that for each fixed $z \in \mathbb{R}$ the mapping $\xi \mapsto V(\xi, z)$ has a global minimum at $\xi = 0$. Consequently, $\frac{\partial V}{\partial \xi}(0, z) = 0$ for all $z \in \mathbb{R}$. Moreover, since $k(0, z) = 0$ for all $z \in \mathbb{R}$ and since $f(0) = 0$, $\Phi(0) = 0$, $h(0, 0) = 0$, $\varphi(0, 0) = 0$, we conclude that there exist smooth positive functions

$R, \rho : \mathbb{R}^n \times \mathbb{R} \to (0, +\infty)$, $r : \mathbb{R}^n \to (0, +\infty)$ such that the following inequalities hold for all $(\xi, w, z) \in \mathbb{R}^n \times \mathbb{R} \times \mathbb{R}$:

$$\begin{aligned} &\left|\tfrac{\partial V}{\partial \xi}(\xi, z)\right| + |k(\xi, z)| + \left|\tfrac{\partial k}{\partial \xi}(\xi, z)\Phi(\xi)\right| \le R(\xi, z)\,|\xi| \\ &|f(\xi)| \le r(\xi)\,|\xi| \\ &|h(\xi, w)| + |\varphi(\xi, w)| \le \rho(\xi, w)\,|\xi| + \rho(\xi, w)\,|w| \end{aligned} \tag{6.44}$$

Define for all $(\xi, w, z) \in \mathbb{R}^n \times \mathbb{R} \times \mathbb{R}$

$$\bar{\sigma}(\xi, w, z) := \left(1 + 2R^2(\xi, z)\right)\sigma(\xi, z) + 4 \tag{6.45}$$

$$\begin{aligned} M(\xi, w, z) :=\ & \tfrac{c}{4} + \tfrac{\Gamma^2 \exp(-2z)}{4c}\left(1 + \left(\tfrac{\partial k}{\partial z}(\xi, z)\right)^2\right)^2 V(\xi, z) \\ &+ \tfrac{1}{2}\left(\tfrac{\Gamma \exp(-z)}{4}\left(1 + (w - k(\xi, z))^2\right) + \mu(\xi)\right)\left(1 + \left(\tfrac{\partial k}{\partial z}(\xi, z)\right)^2\right) \\ &+ \tfrac{1+\kappa(\exp(z))}{4a}\left|\alpha'(\xi, y) - \tfrac{\partial k}{\partial \xi}(\xi, z)G(\xi)\right|^2 + P(\xi, w, z)\left(b + \lambda\left(\exp(z)\right)\right) \\ &+ \tfrac{1+\kappa(\exp(z))}{2a}P^2(\xi, w, z)\left((w - k(\xi, z))^2 + |\xi|^2\right) + \rho(\xi, w) \\ &+ \tfrac{\Gamma \exp(-z)}{4}\left(1 + \left(\tfrac{\partial V}{\partial z}(\xi, z)\right)^2\right) + \tfrac{\sigma(\xi, z)}{c}P^2(\xi, w, z)\left(b + 1 + \lambda\left(\exp(z)\right)\right)^2 \end{aligned} \tag{6.46}$$

$$\begin{aligned} P(\xi, w, z) :=\ & \tfrac{r(\xi)+\mu(\xi)}{2}\left(1 + \left|\tfrac{\partial k}{\partial \xi}(\xi, z)\right|^2\right) + \rho(\xi, w) \\ &+ \left(1 + \tfrac{\mu(\xi)}{2}\left(3 + \left|\tfrac{\partial k}{\partial \xi}(\xi, z)\right|^2\right) + \rho(\xi, w)\right)R(\xi, z) \end{aligned} \tag{6.47}$$

where η is the function appearing in (6.23). In what follows, we denote by Λ the quantity appearing in the left-hand side of (6.28) and $d \in \mathbb{R}^l$, $\theta \in \Theta$, $(\xi, w, z) \in \mathbb{R}^n \times \mathbb{R} \times \mathbb{R}$ are arbitrary given vectors. For brevity, we also use the following notation:

$$s = w - k(\xi, z) \tag{6.48}$$

$$W = \varphi'(\xi, w) - \frac{\partial k}{\partial \xi}(\xi, z)\Phi(\xi) \tag{6.49}$$

$$Q = \alpha'(\xi, w) - \frac{\partial k}{\partial \xi}(\xi, z)G(\xi) \tag{6.50}$$

Using definition (6.25) we get

$$\begin{aligned} \Lambda =\ & \left(\tfrac{\partial V}{\partial z}(\xi, z) - s\tfrac{\partial k}{\partial z}(\xi, z)\right)\Gamma \exp(-z)\left(\bar{V}(\xi, w, z) - \varepsilon\right)^+ \\ &+ \left(\tfrac{\partial V}{\partial \xi}(\xi, z) - s\tfrac{\partial k}{\partial \xi}(\xi, z)\right)\left(f(\xi) + g(\xi, \theta)w + \Phi(\xi)\theta + G(\xi)d\right) \\ &+ s\left(h(\xi, w) + \tilde{g}(\xi, w, \theta)\bar{k}(\xi, w, z) + \varphi'(\xi, w)\theta + \alpha'(\xi, w)d\right) \end{aligned} \tag{6.51}$$

Inequality (6.22) in conjunction with (6.51) gives

$$\begin{aligned} \Lambda \le\ & -cV(\xi, z) - s\tfrac{\partial k}{\partial z}(\xi, z)\Gamma \exp(-z)\left(\bar{V}(\xi, w, z) - \varepsilon\right)^+ \\ &+ \tfrac{\partial V}{\partial z}(\xi, z)\Gamma \exp(-z)\left(\left(\bar{V}(\xi, w, z) - \varepsilon\right)^+ - \left(V(\xi, z) - \varepsilon\right)^+\right) \\ &+ s\left(h(\xi, w) + \tfrac{\partial V}{\partial \xi}(x, z)g(\xi, \theta) + W\theta + Qd\right) \\ &+ a\tfrac{|d|^2 + \left(\left(|\theta| - b - \lambda(\exp(z))\right)^+\right)^2}{1+\kappa(\exp(z))} \\ &+ s\left(\tilde{g}(\xi, w, \theta)\bar{k}(\xi, w, z) - \tfrac{\partial k}{\partial \xi}(\xi, z)\left(f(\xi) + g(\xi, \theta)w\right)\right) \end{aligned} \tag{6.52}$$

Using (6.23), (6.26), (6.52) and the fact that $|v^+ - b^+| \leq |v - b|$ for all $v, b \in \mathbb{R}$, we get

$$\begin{aligned}
\Lambda &\leq -cV(\xi,z) + |s| \left|\tfrac{\partial k}{\partial z}(\xi,z)\right| \Gamma \exp(-z) \left(\bar{V}(\xi,w,z) - \varepsilon\right)^+ - M(\xi,w,z)s^2 \\
&+ \left|\tfrac{\partial V}{\partial z}(\xi,z)\right| \Gamma \exp(-z) \left|\bar{V}(\xi,w,z) - V(\xi,z)\right| + a\tfrac{|d|^2 + \left((|\theta| - b - \lambda(\exp(z)))^+\right)^2}{1+\kappa(\exp(z))} \\
&+ |s| \left(|h(\xi,w)| + \left|\tfrac{\partial V}{\partial \xi}(\xi,z)\right| |g(\xi,\theta)| + |W|\,|\theta| + \left|\tfrac{\partial k}{\partial \xi}(\xi,z)\right| |f(\xi) + g(\xi,\theta)w| + |Q|\,|d|\right)
\end{aligned} \tag{6.53}$$

Using the inequality

$$|s|\,|Q|\,|d| \leq \frac{a\,|d|^2}{1+\kappa(\exp(z))} + \frac{1+\kappa(\exp(z))}{4a} |Q|^2 s^2 \tag{6.54}$$

definition (6.25), the triangle inequality $|f(\xi) + g(\xi,\theta)w| \leq |f(\xi)| + |g(\xi,\theta)|\,|w|$ and inequality (6.24) we get from (6.53)

$$\begin{aligned}
\Lambda &\leq -cV(\xi,z) + |s| \left|\tfrac{\partial k}{\partial z}(\xi,z)\right| \Gamma \exp(-z) \left(\bar{V}(\xi,w,z) - \varepsilon\right)^+ + \tfrac{1}{2}\left|\tfrac{\partial V}{\partial z}(\xi,z)\right| \Gamma \exp(-z) s^2 \\
&+ |s| \left(|h(\xi,w)| + \left|\tfrac{\partial k}{\partial \xi}(\xi,z)\right| |f(\xi)| + \mu(\xi)\left|\tfrac{\partial V}{\partial \xi}(\xi,z)\right|\right) \\
&+ a\tfrac{2|d|^2 + \left((|\theta| - b - \lambda(\exp(z)))^+\right)^2}{1+\kappa(\exp(z))} + |s|\,|\theta| \left(\left|\tfrac{\partial V}{\partial \xi}(\xi,z)\right| \mu(\xi) + \left|\tfrac{\partial k}{\partial \xi}(\xi,z)\right| \mu(\xi)\,|w| + |W|\right) \\
&+ |s|\,\mu(\xi) \left|\tfrac{\partial k}{\partial \xi}(\xi,z)\right| |w| - \left(M(\xi,w,z) - \tfrac{1+\kappa(\exp(z))}{4a} |Q|^2\right) s^2
\end{aligned} \tag{6.55}$$

The triangle inequalities $|W| \leq \left|\frac{\partial k}{\partial \xi}(\xi,z)\Phi(\xi)\right| + |\varphi(\xi,w)|$ and $|w| \leq |s| + |k(\xi,z)|$ (recall that $s = w - k(\xi,z)$), inequalities (6.44), (6.55), and the fact that $\left(\bar{V}(\xi,w,z) - \varepsilon\right)^+ \leq \bar{V}(\xi,w,z)$ give

$$\begin{aligned}
\Lambda &\leq -cV(\xi,z) + |s| \left|\tfrac{\partial k}{\partial z}(\xi,z)\right| \Gamma \exp(-z)\bar{V}(\xi,w,z) + s^2 |\theta| \left(\left|\tfrac{\partial k}{\partial \xi}(\xi,z)\right| \mu(\xi) + \rho(\xi,w)\right) \\
&+ a\tfrac{2|d|^2 + \left((|\theta| - b - \lambda(\exp(z)))^+\right)^2}{1+\kappa(\exp(z))} - \left(M(\xi,w,z) - \tfrac{1+\kappa(\exp(z))}{4a} |Q|^2\right) s^2 \\
&+ \left(\left|\tfrac{\partial k}{\partial \xi}(\xi,z)\right| \mu(\xi) + \rho(\xi,w) + \tfrac{1}{2}\left|\tfrac{\partial V}{\partial z}(\xi,z)\right| \Gamma \exp(-z)\right) s^2 \\
&+ |s|\,|\xi| \left(\left|\tfrac{\partial k}{\partial \xi}(\xi,z)\right| r(\xi) + \rho(\xi,w) + \left(\left|\tfrac{\partial k}{\partial \xi}(\xi,z)\right| \mu(\xi) + \rho(\xi,w) + \mu(\xi)\right) R(\xi,z)\right) \\
&+ |s|\,|\xi|\,|\theta| \left(\rho(\xi,w) + R(\xi,z)\left(1 + \mu(\xi) + \left|\tfrac{\partial k}{\partial \xi}(\xi,z)\right| \mu(\xi) + \rho(\xi,w)\right)\right)
\end{aligned} \tag{6.56}$$

Using the inequalities

$$2\left|\frac{\partial k}{\partial \xi}(\xi,z)\right| \leq 1 + \left|\frac{\partial k}{\partial \xi}(\xi,z)\right|^2 \tag{6.57}$$

$$2\left|\frac{\partial V}{\partial z}(\xi,z)\right| \leq 1 + \left(\frac{\partial V}{\partial z}(\xi,z)\right)^2 \tag{6.58}$$

$$2\left|\frac{\partial k}{\partial z}(\xi,z)\right| \leq 1 + \left(\frac{\partial k}{\partial z}(\xi,z)\right)^2 \tag{6.59}$$

and definition (6.25), we get from (6.56) and (6.47):

$$\begin{aligned}
\Lambda &\leq -cV(\xi,z) + \tfrac{\Gamma\exp(-z)}{2} |s| \left(1 + \left(\tfrac{\partial k}{\partial z}(\xi,z)\right)^2\right) V(\xi,z) \\
&+ \tfrac{\Gamma\exp(-z)}{4} |s|^3 \left(1 + \left(\tfrac{\partial k}{\partial z}(\xi,z)\right)^2\right) + a\tfrac{2|d|^2 + \left((|\theta| - b - \lambda(\exp(z)))^+\right)^2}{1+\kappa(\exp(z))} \\
&+ \left(\tfrac{\mu(\xi)}{2}\left(1 + \left|\tfrac{\partial k}{\partial \xi}(\xi,z)\right|^2\right) + \rho(\xi,w) + \tfrac{\Gamma\exp(-z)}{4}\left(1 + \left(\tfrac{\partial V}{\partial z}(\xi,z)\right)^2\right)\right) s^2 \\
&+ P(\xi,w,z)s^2 |\theta| + |s|\,|\xi|\,P(\xi,w,z)(1 + |\theta|) - \left(M(\xi,w,z) - \tfrac{1+\kappa(\exp(z))}{4a} |Q|^2\right) s^2
\end{aligned} \tag{6.60}$$

The inequalities

$$|s| \leq 1/2 + s^2/2 \tag{6.61}$$

$$|s|\left(1+\left(\frac{\partial k}{\partial z}(\xi,z)\right)^2\right)\Gamma\exp(-z) \leq \frac{c}{2}+\frac{\Gamma^2\exp(-2z)}{2c}s^2\left(1+\left(\frac{\partial k}{\partial z}(\xi,z)\right)^2\right)^2 \tag{6.62}$$

in conjunction with (6.60) give

$$\begin{aligned}
&\Lambda \leq -\tfrac{3c}{4}V(\xi,z)+\tfrac{\Gamma^2\exp(-2z)}{4c}s^2\left(1+\left(\tfrac{\partial k}{\partial z}(\xi,z)\right)^2\right)^2V(\xi,z)+a\tfrac{2|d|^2+\left((|\theta|-b-\lambda(\exp(z)))^+\right)^2}{1+\kappa(\exp(z))}\\
&+\tfrac{\Gamma\exp(-z)}{8}\left(1+s^2\right)\left(1+\left(\tfrac{\partial k}{\partial z}(\xi,z)\right)^2\right)s^2+P(\xi,w,z)\left(b+\lambda\left(\exp(z)\right)\right)s^2\\
&+\left(\tfrac{\mu(\xi)}{2}\left(1+\left|\tfrac{\partial k}{\partial \xi}(\xi,z)\right|^2\right)+\rho(\xi,w)+\tfrac{\Gamma\exp(-z)}{4}\left(1+\left(\tfrac{\partial V}{\partial z}(\xi,z)\right)^2\right)\right)s^2\\
&+|s|\,|\xi|\,P(\xi,w,z)\left(|\theta|-b-\lambda\left(\exp(z)\right)\right)+P(\xi,w,z)s^2\left(|\theta|-b-\lambda\left(\exp(z)\right)\right)\\
&+|s|\,|\xi|\left(1+b+\lambda\left(\exp(z)\right)\right)P(\xi,w,z)-\left(M(\xi,w,z)-\tfrac{1+\kappa(\exp(z))}{4a}|Q|^2\right)s^2
\end{aligned} \tag{6.63}$$

Using the inequalities

$$|\theta|-b-\lambda\left(\exp(z)\right) \leq \left(|\theta|-b-\lambda\left(\exp(z)\right)\right)^+ \tag{6.64}$$

$$\begin{aligned}
&|s|\,|\xi|\,P(\xi,w,z)\left(b+1+\lambda\left(\exp(z)\right)\right)\\
&\leq \tfrac{c}{4\sigma(\xi,z)}|\xi|^2+\tfrac{\sigma(\xi,z)}{c}P^2(\xi,w,z)\left(b+1+\lambda\left(\exp(z)\right)\right)^2s^2
\end{aligned} \tag{6.65}$$

we obtain from (6.63) and (6.46), (6.21)

$$\begin{aligned}
&\Lambda \leq -\tfrac{c}{2}V(\xi,z)-\tfrac{c}{4}s^2+a\tfrac{2|d|^2+\left((|\theta|-b-\lambda(\exp(z)))^+\right)^2}{1+\kappa(\exp(z))}\\
&+P(\xi,w,z)s^2\left(|\theta|-b-\lambda\left(\exp(z)\right)\right)^+ +|s|\,|\xi|\,P(\xi,w,z)\left(|\theta|-b-\lambda\left(\exp(z)\right)\right)^+\\
&-\tfrac{1+\kappa(\exp(z))}{2a}P^2(\xi,w,z)\left(s^2+|\xi|^2\right)s^2
\end{aligned} \tag{6.66}$$

Using (6.25), (6.66) and the inequalities

$$\begin{aligned}
&s^2P(\xi,w,z)\left(|\theta|-b-\lambda\left(\exp(z)\right)\right)^+\\
&\leq \tfrac{a}{2}\tfrac{\left((|\theta|-b-\lambda(\exp(z)))^+\right)^2}{1+\kappa(\exp(z))}+\tfrac{1+\kappa(\exp(z))}{2a}P^2(\xi,w,z)s^4
\end{aligned} \tag{6.67}$$

$$\begin{aligned}
&|s|\,|\xi|\,P(\xi,w,z)\left(|\theta|-b-\lambda\left(\exp(z)\right)\right)^+\\
&\leq \tfrac{a}{2}\tfrac{\left((|\theta|-b-\lambda(\exp(z)))^+\right)^2}{1+\kappa(\exp(z))}+\tfrac{1+\kappa(\exp(z))}{2a}|\xi|^2P^2(\xi,w,z)s^2
\end{aligned} \tag{6.68}$$

we are in a position to conclude that (6.28) holds.

Finally, we proceed to show the validity of (6.27) with $\bar{\sigma}(x,y,z)$ defined by (6.45). Using the triangle inequality $|w| \leq |w-k(\xi,z)|+|k(\xi,z)|$ and the fact that $\left(|w-k(\xi,z)|+|k(\varsigma,z)|\right)^2 \leq 2\,|w-k(\xi,z)|^2+2\,|k(\xi,z)|^2$ we get

$$|\xi|^2+|w|^2 \leq |\xi|^2+2\,|w-k(\xi,z)|^2+2\,|k(\xi,z)|^2 \tag{6.69}$$

The above inequality in conjunction with definition (6.25) and inequalities (6.44) gives

$$|\xi|^2+|w|^2 \leq \left(1+2R^2(\xi,z)\right)|\xi|^2+4\bar{V}(\xi,w,z) \tag{6.70}$$

Using (6.25) and inequalities (6.21), (6.70) we obtain

$$|\xi|^2+|w|^2 \leq \left(\left(1+2R^2(\xi,z)\right)\sigma(\xi,z)+4\right)\bar{V}(\xi,w,z) \tag{6.71}$$

Combining (6.71) and (6.45) we obtain (6.27). ◁

6.4 ▪ Explicit DADS Backstepping Design for a Strict-Feedback Class

In this section we present an explicit control design algorithm for the simplified but representative class of strict-feedback systems with unknown parameters and disturbance, given by

$$\dot{w}_i = w_{i+1} + \varphi_i'(\bar{w}_i)\theta + \alpha_i'(\bar{w}_i)d, \qquad i = 1, ..., n \tag{6.72}$$

where

$$w_{n+1} = u \tag{6.73}$$

and the smooth functions φ_i satisfy $\varphi_i(0) = 0$. Note that this system is nothing but (6.11) with h_i taken as zero and $g_i = 1$, for the sake of focus on how the method handles *uncertainties*. The more general case, where h_i is not zero and g_i is not unity, is already covered in Theorem 6.2.

While a DADS design for this class can be produced by successive application of Lemma 6.1, such a design is not given explicitly in the proof of Theorem 6.2 but only outlined. In this section, we work out the details. To do so, we don't fully specialize the process in the proof of Theorem 6.2, reliant on the construction in the proof of Lemma 6.1, but make some small deviations, dictated by a desire for having a simple recursive definition of the feedback law and notational compactness.

While our DADS design is focused on ensuring robustness in the presence of disturbance d, we are also interested in ensuring, when possible, a regulation to the origin $w = 0$. To make this possible, it is necessary for the nonlinearities multiplied by the unknown parameter θ to vanish at the origin. It is for that reason that we assume that $\varphi_i(0) = 0$, and this proves important in the design we present next.

In the exposition that follows, we *derive* a recursive algorithm for deadzone-adapted disturbance suppression—a "DADS backstepping" algorithm—for system (6.72). Parts of this algorithm are possible to write concisely up front, while other parts, which deal with the robustification, come out of a somewhat complicated Lyapunov design process. We start with a list of definitions in which the key ingredients of the design recursion are collected:

$$\zeta_i = w_i - k_{i-1}(\bar{w}_{i-1}, z) \tag{6.74}$$

$$V_i(\bar{w}_i, z) = \frac{1}{2}\sum_{j=1}^{i} \zeta_i^2 \tag{6.75}$$

$$\pi_i'(\bar{w}_i, z) = \varphi_i' - \sum_{j=1}^{i-1} \frac{\partial k_{i-1}}{\partial w_j}\varphi_j' =: p_i'(\bar{w}_i, z)\bar{\zeta}_i \tag{6.76}$$

$$\rho_i'(\bar{w}_i, z) = \alpha_i' - \sum_{j=1}^{i-1} \frac{\partial k_{i-1}}{\partial w_j}\alpha_j' \tag{6.77}$$

$$\sigma_i(\bar{w}_i, z) = \frac{\partial k_{i-1}}{\partial z}\Gamma \exp(-z)\left(V_i - \frac{\varepsilon^2}{2}\right)^+ \tag{6.78}$$

$$k_i(\bar{w}_i, z) = q_i(\bar{w}_i, z) - c_j\zeta_i - \zeta_{i-1} + \sum_{j=1}^{i-1} \frac{\partial k_{i-1}}{\partial w_j} w_{j+1} \tag{6.79}$$

where p_i is a matrix of dimension $\dim(\theta) \times i$, with each element a smooth function of $(\bar{w}_i, z)$, and $\bar{\zeta}_i := (\zeta_1, ..., \zeta_i)$. The definition (6.74) represents the backstepping transformation, (6.75) is the ith control Lyapunov function, and (6.79) represents the ith stabilizing function, within which the function q_i is yet to be designed.

The actual control law is the last of the stabilizing functions, namely,

$$u = k_n(w, z) \tag{6.80}$$

As we shall see, due to the aforementioned assumption that $\varphi_i(0) = 0$, the functions q_i shall have the property $q_i(0, z) \equiv 0$ and, by a careful examination of the above expressions, one can see that this results in the stabilizing functions having also the property of $k_i(0, z) \equiv 0$, the backstepping transformations having also the property that $(\zeta_1, ..., \zeta_i) = 0$ whenever $\bar{w}_i = 0$, regardless of the value of z, and, consequently, that $V_i(0, z) \equiv 0$ and $\pi_i(0, z) \equiv 0$. It is from the latter property that the factoring $\pi_i' = p_i'\zeta_i$ at the right end of (6.76) follows. We shall exploit this latter property in the design.

The last of the Lyapunov functions, V_i for $i = n$, namely, V_n, will be used in the update law for the gain,

$$\dot{z} = \Gamma \exp(-z) \left(V_n - \frac{\varepsilon^2}{2} \right)^+ \tag{6.81}$$

The key challenge in the design of the DADS backstepping controller is, as we shall see, that this gain-updating law depends, through V_n, on the entire state $(w_1, ..., w_n)$, whereas the stabilizing functions $k_i(\bar{w}_i, z)$, which depend on z and therefore give rise to the V_n-dependent $\dot{z}$ in the recursion, may depend on the components of w only up to w_i, through the term $\sigma_i(\bar{w}_i, z)$ in k_i. This challenge is always encountered in adaptive forms of backstepping and examples of how to address it are shown in [89, 95]. The original solution is in the 1992 paper [85]. Without this challenge, namely, if θ were absent from (6.72) and only d present, the design (robust but not adaptive) would complete in several lines—at most a page. With adaptation present, the design is always complex and we shall see how, in order for the backstepping transformation ζ_i to maintain a triangular dependence on $(w_1, ..., w_i)$, along with the dependence on z, an intricate approach to avoiding the cancellation of the effect of $\dot{z} = \Gamma \exp(-z) \left(V_n - \frac{\varepsilon^2}{2} \right)^+$, which involves not only a dependence on $(w_1, ..., w_i)$ but also on $(w_{i+1}, ..., w_n)$, is conducted, letting, *roughly speaking*, the dependence of $\dot{z}$ on w_{i+1} be dealt with by k_{i+1}, the dependence on w_{n-1} by k_{n-1}, and the portion of $\dot{z}$ uncompensated until the step $n - 1$ absorbed into the final control, $u = k_n$.

Before we start our Lyapunov design, using the definitions of $\zeta_i, \pi_i, \rho_i, k_i$ above, taking the time derivative of ζ_i and performing several substitutions, we get the model in the transformed backstepping coordinates,

$$\dot{\zeta}_i = -c_i\zeta_i - \zeta_{i-1} + \zeta_{i+1} + q_i + \pi_i'\theta + \rho_i'd - \frac{\partial k_{i-1}}{\partial z}\dot{z} \tag{6.82}$$

where $\zeta_0 = \zeta_{n+1} := 0$.

For $i = 1, ..., n-1$, taking the time derivative of $V_i = \frac{1}{2}\sum_{j=1}^{i} \zeta_i^2$,

$$\dot{V}_i = -\sum_{j=1}^{i} c_j\zeta_j^2 + \zeta_i\zeta_{i+1} + \sum_{j=1}^{i} \zeta_j \left(q_j + \pi_j'\theta + \rho_j'd - \frac{\partial k_{j-1}}{\partial z}\dot{z} \right) \tag{6.83}$$

where $k_0 := 0$. Recall, we are designing the stabilizing functions $q_j(\bar{w}_j, z)$, $j = 1, ..., i$ in (6.83). Noting from (6.78) and (6.81) that

$$\begin{aligned} -\sum_{j=1}^{i} \zeta_j \frac{\partial k_{j-1}}{\partial z}\dot{z} &= -\sum_{j=1}^{i} \zeta_j\sigma_j \\ &\quad + \sum_{j=2}^{i} \zeta_j \frac{\partial k_{j-1}}{\partial z} \Gamma \exp(-z) \left[\left(V_j - \frac{\varepsilon^2}{2} \right)^+ - \left(V_n - \frac{\varepsilon^2}{2} \right)^+ \right] \end{aligned} \tag{6.84}$$

we get

$$\dot{V}_i = -\sum_{j=1}^{i} c_j\zeta_j^2 + \zeta_i\zeta_{i+1} + \sum_{j=1}^{i}\zeta_j\left(q_j + \pi_j'\theta + \rho_j'd - \sigma_j\right)$$
$$+\Gamma\exp(-z)\sum_{j=2}^{i}\zeta_j\frac{\partial k_{j-1}}{\partial z}\left[\left(V_j - \frac{\varepsilon^2}{2}\right)^+ - \left(V_n - \frac{\varepsilon^2}{2}\right)^+\right] \tag{6.85}$$

First, we concentrate on the terms $-\zeta_j\sigma_j$, where σ_j is defied in (6.78). This term can be canceled with q_j but this would make q_j nonsmooth since the $(\cdot)^+$ operation within σ_j is nonsmooth. Hence, instead of cancellation, we have to employ domination. First we note that

$$\left(V_j - \frac{\varepsilon^2}{2}\right)^+ \leq V_j = \frac{1}{2}|\bar{\zeta}_j|^2 \tag{6.86}$$

It then follows that

$$-\zeta_j\sigma_j \leq \frac{\Gamma}{2}\exp(-z)\left|\frac{\partial k_{j-1}}{\partial z}\right||\zeta_j|\sum_{l=1}^{i}\zeta_l^2$$
$$\leq \sum_{l=1}^{j} g_l\zeta_l^2 + \zeta_j^2\frac{\Gamma^2}{4}\exp(-2z)\left(\frac{\partial k_{j-1}}{\partial z}\right)^2\sum_{l=1}^{i}\frac{1}{4g_l}\zeta_l^2 \tag{6.87}$$

where $g_j > 0$. Focusing on the double sum

$$\sum_{j=1}^{i}\sum_{l=1}^{j} g_l\zeta_l^2 = \sum_{j=1}^{i}(i-j+1)g_j\zeta_j^2 \leq \sum_{j=1}^{i}(n-j+1)g_j\zeta_j^2 \tag{6.88}$$

we turn to summing (6.87) over j and obtain, for the resulting double sum,

$$-\sum_{j=1}^{i}\zeta_j\sigma_j \leq \sum_{j=1}^{i}(n-j+1)g_j\zeta_j^2$$
$$+\sum_{j=1}^{i}\zeta_j^2\frac{\Gamma^2}{4}\exp(-2z)\left(\frac{\partial k_{j-1}}{\partial z}\right)^2\sum_{l=1}^{j}\frac{1}{4g_l}\zeta_l^2 \tag{6.89}$$

Let us now denote

$$R_j(\bar{w}_j, z) = (n-j+1)g_j + \frac{\Gamma^2}{4}\exp(-2z)\left(\frac{\partial k_{j-1}}{\partial z}\right)^2\sum_{l=1}^{j}\frac{1}{4g_l}\zeta_l^2 \tag{6.90}$$

Inequality (6.89) becomes

$$-\sum_{j=1}^{i}\zeta_j\sigma_j \leq \sum_{j=1}^{i}R_j(\bar{w}_j, z)\zeta_j^2 \tag{6.91}$$

This then turns (6.85) into

$$\dot{V}_i \leq -\sum_{j=1}^{i} c_j\zeta_j^2 + \zeta_i\zeta_{i+1} + \sum_{j=1}^{i}\zeta_j\left(q_j + \pi_j'\theta + \rho_j'd + R_j(\bar{w}_j, z)\zeta_j\right)$$
$$+\Gamma\exp(-z)\sum_{j=2}^{i}\zeta_j\frac{\partial k_{j-1}}{\partial z}\left[\left(V_j - \frac{\varepsilon^2}{2}\right)^+ - \left(V_n - \frac{\varepsilon^2}{2}\right)^+\right] \tag{6.92}$$

Next, by adding and subtracting $(V_{j+1}-\varepsilon)^+, \dots, (V_{n-1}-\varepsilon)^+$ in the last bracket, we get

$$\begin{aligned}\dot{V}_i \leq &-\sum_{j=1}^{i} c_j\zeta_j^2 + \zeta_i\zeta_{i+1} + \sum_{j=1}^{i} \zeta_j\left(q_j + \pi_j'\theta + \rho_j'd + R_j(\bar{w}_j, z)\zeta_j\right) \\ &+\Gamma\exp(-z)\sum_{j=2}^{i}\zeta_j\frac{\partial k_{j-1}}{\partial z}\left[\left(V_i - \frac{\varepsilon^2}{2}\right)^+ - \left(V_n - \frac{\varepsilon^2}{2}\right)^+\right] \\ &+\Gamma\exp(-z)\left[\left(V_i - \frac{\varepsilon^2}{2}\right)^+ - \left(V_n - \frac{\varepsilon^2}{2}\right)^+\right]\sum_{j=2}^{i}\zeta_j\frac{\partial k_{j-1}}{\partial z} \end{aligned} \tag{6.93}$$

where we have split apart the sums depending on $(V_{j+1}-\varepsilon^2/2)^+, \dots, (V_i-\varepsilon^2/2)^+$ from the sums depending on $(V_i-\varepsilon^2/2)^+, \dots, (V_n-\varepsilon^2/2)^+$. We focus on the inner sum in the second row of (6.93) and recall the inequality

$$\left|\left(V_l - \frac{\varepsilon^2}{2}\right)^+ - \left(V_{l+1} - \frac{\varepsilon^2}{2}\right)^+\right| \leq |V_l - V_{l+1}| = \frac{1}{2}\zeta_{l+1}^2 \tag{6.94}$$

We thus obtain

$$\begin{aligned}\dot{V}_i \leq &-\sum_{j=1}^{i} c_j\zeta_j^2 + \zeta_i\zeta_{i+1} + \sum_{j=1}^{i} \zeta_j\left(q_j(\bar{w}_j, z) + \pi_j'\theta + \rho_j'd + R_j(\bar{w}_j, z)\zeta_j\right) \\ &+\frac{\Gamma}{2}\exp(-z)\sum_{j=2}^{i}|\zeta_j|\left|\frac{\partial k_{j-1}}{\partial z}\right|\sum_{l=j+1}^{i}\zeta_l^2 \\ &+\Gamma\exp(-z)\left[\left(V_i - \frac{\varepsilon^2}{2}\right)^+ - \left(V_n - \frac{\varepsilon^2}{2}\right)^+\right]\sum_{j=2}^{i}\zeta_j\frac{\partial k_{j-1}}{\partial z} \end{aligned} \tag{6.95}$$

Since changing the order of nested summations results in

$$\sum_{j=2}^{i}|\zeta_j|\left|\frac{\partial k_{j-1}}{\partial z}\right|\sum_{l=j+1}^{i}\zeta_l^2 = \sum_{j=3}^{i}\zeta_j^2\sum_{l=2}^{j-1}|\zeta_l|\left|\frac{\partial k_{l-1}}{\partial z}\right| \tag{6.96}$$

we get

$$\begin{aligned}\dot{V}_i \leq &-\sum_{j=1}^{i} c_j\zeta_j^2 + \zeta_i\zeta_{i+1} + \sum_{j=1}^{i} \zeta_j\left(q_j(\bar{w}_j, z) + \pi_j'\theta + \rho_j'd + R_j(\bar{w}_j, z)\zeta_j\right) \\ &+\frac{\Gamma}{2}\exp(-z)\sum_{j=3}^{i}\zeta_j^2\sum_{l=2}^{j-1}|\zeta_l|\left|\frac{\partial k_{l-1}}{\partial z}\right| \\ &+\Gamma\exp(-z)\left[\left(V_i - \frac{\varepsilon^2}{2}\right)^+ - \left(V_n - \frac{\varepsilon^2}{2}\right)^+\right]\sum_{j=2}^{i}\zeta_j\frac{\partial k_{j-1}}{\partial z} \\ = &-\sum_{j=1}^{i} c_j\zeta_j^2 + \zeta_i\zeta_{i+1} \\ &+\sum_{j=1}^{i}\zeta_j\left[q_j(\bar{w}_j, z) + \pi_j'\theta + \rho_j'd + R_j(\bar{w}_j, z)\zeta_j + \frac{\Gamma}{2}\exp(-z)\zeta_j\sum_{l=2}^{j-1}|\zeta_l|\left|\frac{\partial k_{l-1}}{\partial z}\right|\right] \\ &+\Gamma\exp(-z)\left[\left(V_i - \frac{\varepsilon^2}{2}\right)^+ - \left(V_n - \frac{\varepsilon^2}{2}\right)^+\right]\sum_{j=2}^{i}\zeta_j\frac{\partial k_{j-1}}{\partial z} \end{aligned} \tag{6.97}$$

where we use the convention that $k_{-1} = k_0 = 0$. The terms $\frac{\Gamma}{2}\exp(-z)|\zeta_l|\left|\frac{\partial k_{l-1}}{\partial z}\right|$ can be smoothly bounded in a number of ways, including

$$|\zeta_l|\left|\frac{\partial k_{l-1}}{\partial z}\right| \leq \gamma_l + \frac{\Gamma^2}{4}\exp(-2z)\frac{1}{4\gamma_l}\zeta_l^2\left(\frac{\partial k_{l-1}}{\partial z}\right)^2 =: s_l(\bar{w}_l, z) \tag{6.98}$$

with $\gamma_i > 0$. Substituting (6.98) into (6.97), we get

$$\begin{aligned}\dot{V}_i \leq & -\sum_{j=1}^{i} c_j\zeta_j^2 + \zeta_i\zeta_{i+1} \\ & +\sum_{j=1}^{i}\zeta_j\left[q_j(\bar{w}_j, z) + \underbrace{\pi_j'\theta + \rho_j'd}+R_j(\bar{w}_j, z)\zeta_j + \zeta_j\sum_{l=2}^{j-1}s_l(\bar{w}_l, z)\right] \\ & +\Gamma\exp(-z)\left[\left(V_i - \frac{\varepsilon^2}{2}\right)^+ - \left(V_n - \frac{\varepsilon^2}{2}\right)^+\right]\sum_{j=2}^{i}\zeta_j\frac{\partial k_{j-1}}{\partial z}\end{aligned} \tag{6.99}$$

What is next to be bounded are the two uncertain underbraced terms in (6.99). First, we note, with Young's inequality, that

$$\zeta_j\rho_j'd \leq \beta_j\left(1+\exp(z)\right)\zeta_j^2|\rho_j|^2 + \frac{1}{4\beta_j\left(1+\exp(z)\right)}|d|^2 \tag{6.100}$$

where $\beta_j > 0$. Second, we get

$$\begin{aligned}\zeta_j\pi_j'\theta \leq & \frac{\left((|\theta|-1-\exp(z))^+\right)^2}{4\beta_j\left(1+\exp(z)\right)} + \beta_j\left(1+\exp(z)\right)\zeta_j^2|\pi_j|^2 \\ & +\left(1+\exp(z)\right)|\zeta_j|\,|\pi_j|\end{aligned} \tag{6.101}$$

Recalling from (6.76) that $\pi_j = \bar{\zeta}_j' p_j$, Young's inequality yields

$$\left(1+\exp(z)\right)|\zeta_j|\,|\pi_j| \leq b_j|\bar{\zeta}_j|^2 + \frac{\left(1+\exp(z)\right)^2}{4b_j}|p_j|^2\zeta_j^2 \tag{6.102}$$

where $b_j > 0$ and $|p_j|$ denotes the Frobenius norm of the matrix p_j of dimension $\dim(\theta) \times j$. We take the Frobenius norm, rather than the matrix 2-norm, because, with the Frobenius norm, $|p_j|^2$ is a smooth function of its arguments, whereas with the 2-norm $|p_j|^2$ is only continuous. The last two inequalities yield

$$\begin{aligned}\zeta_j\pi_j'\theta \leq & \frac{\left((|\theta|-1-\exp(z))^+\right)^2}{4\beta_j\left(1+\exp(z)\right)} + \beta_j\left(1+\exp(z)\right)\zeta_j^2|\pi_j|^2 \\ & +b_j|\bar{\zeta}_j|^2 + \frac{\left(1+\exp(z)\right)^2}{4b_j}|p_j|^2\zeta_j^2\end{aligned} \tag{6.103}$$

namely,

$$\begin{aligned}\zeta_j\pi_j'\theta \leq & \, b_j|\bar{\zeta}_j|^2 + \left(1+\exp(z)\right)\left[\beta_j|\pi_j|^2 + \frac{1+\exp(z)}{4b_j}|p_j|^2\right]\zeta_j^2 \\ & +\frac{\left((|\theta|-1-\exp(z))^+\right)^2}{4\beta_j\left(1+\exp(z)\right)}\end{aligned} \tag{6.104}$$

To summarize, by adding (6.100) and (6.104), we get

$$\begin{aligned}\zeta_j\rho_j'd+\zeta_j\pi_j'\theta &\leq \beta_j\left(1+\exp(z)\right)\zeta_j^2|\rho_j|^2+\frac{1}{4\beta_j\left(1+\exp(z)\right)}|d|^2\\ &\quad+b_j|\bar{\zeta}_j|^2+(1+\exp(z))\left[\beta_j|\pi_j|^2+\frac{1+\exp(z)}{4b_j}|p_j|^2\right]\zeta_j^2\\ &\quad+\frac{\left((|\theta|-1-\exp(z))^+\right)^2}{4\beta_j\left(1+\exp(z)\right)}\\ &=b_j|\bar{\zeta}_j|^2+(1+\exp(z))\left[\beta_j|\rho_j|^2+\beta_j|\pi_j|^2+\frac{1+\exp(z)}{4b_j}|p_j|^2\right]\zeta_j^2\\ &\quad+\frac{\left((|\theta|-1-\exp(z))^+\right)^2+|d|^2}{4\beta_j\left(1+\exp(z)\right)}\end{aligned}\tag{6.105}$$

Substituting the bound (6.105) for the underbraced term in (6.99), we get

$$\begin{aligned}\dot{V}_i &\leq -\sum_{j=1}^{i}c_j\zeta_j^2+\zeta_i\zeta_{i+1}+\sum_{j=1}^{i}b_j|\bar{\zeta}_j|^2\\ &\quad+\sum_{j=1}^{i}\zeta_j\left[q_j(\bar{w}_j,z)+R_j(\bar{w}_j,z)\zeta_j\right.\\ &\quad\left.+(1+\exp(z))\left(\beta_j|\rho_j|^2+\beta_j|\pi_j|^2+\frac{1+\exp(z)}{4b_j}|p_j|^2\right)\zeta_j+\zeta_j\sum_{l=2}^{j-1}s_l(\bar{w}_l,z)\right]\\ &\quad+\frac{\left((|\theta|-1-\exp(z))^+\right)^2+|d|^2}{4\beta_0\left(1+\exp(z)\right)}\\ &\quad+\Gamma\exp(-z)\left[\left(V_i-\frac{\varepsilon^2}{2}\right)^+-\left(V_n-\frac{\varepsilon^2}{2}\right)^+\right]\sum_{j=2}^{i}\zeta_j\frac{\partial k_{j-1}}{\partial z}\end{aligned}\tag{6.106}$$

where

$$\beta_0=\left(\sum_{j=1}^{i}\frac{1}{\beta_j}\right)^{-1}\tag{6.107}$$

Focusing on the term $\sum_{j=1}^{i}b_j|\bar{\zeta}_j|^2$, we get

$$\sum_{j=1}^{i}b_j|\bar{\zeta}_j|^2=\sum_{j=1}^{i}b_{ji}\zeta_j^2\leq\sum_{j=1}^{i}b_{jn}\zeta_j^2\tag{6.108}$$

where

$$b_{ji}=\sum_{l=j}^{i}b_l\leq\sum_{l=j}^{n}b_l=b_{jn}\tag{6.109}$$

Substituting (6.108) into (6.106), we arrive at

$$
\begin{aligned}
\dot V_i \le & -\sum_{j=1}^{i} c_j\zeta_j^2 + \zeta_i\zeta_{i+1} + \sum_{j=1}^{i}\zeta_j\Bigg\{ q_j(\bar w_j, z) + R_j(\bar w_j, z)\zeta_j \\
& + \Bigg[b_{jn} + (1+\exp(z))\left(\beta_j|\rho_j|^2 + \beta_j|\pi_j|^2 + \frac{1+\exp(z)}{4b_j}|p_j|^2\right) \\
& + \frac{\Gamma}{2}\exp(-z)\sum_{l=2}^{j-1} s_l(\bar w_l, z)\Bigg]\zeta_j\Bigg\} \\
& + \frac{\left((|\theta| - 1 - \exp(z))^+\right)^2 + |d|^2}{4\beta_0\left(1+\exp(z)\right)} \\
& + \Gamma\exp(-z)\left[\left(V_i - \frac{\varepsilon^2}{2}\right)^+ - \left(V_n - \frac{\varepsilon^2}{2}\right)^+\right]\sum_{j=2}^{i}\zeta_j\frac{\partial k_{j-1}}{\partial z}
\end{aligned}
\tag{6.110}
$$

Let us now denote

$$
\begin{aligned}
S_j(\bar w_j, z) = & \; b_{jn} + (1+\exp(z))\left(\beta_j|\rho_j|^2 + \beta_j|\pi_j|^2 + \frac{1+\exp(z)}{4b_j}|p_j|^2\right) \\
& + \sum_{l=2}^{j-1}\left(\gamma_l + \frac{\Gamma^2}{4}\exp(-2z)\frac{1}{4\gamma_l}\zeta_l^2\left(\frac{\partial k_{l-1}}{\partial z}\right)^2\right)
\end{aligned}
\tag{6.111}
$$

This results in

$$
\begin{aligned}
\dot V_i \le & -\sum_{j=1}^{i} c_j\zeta_j^2 + \zeta_i\zeta_{i+1} + \sum_{j=1}^{i}\zeta_j\left\{ q_j(\bar w_j, z) + R_j(\bar w_j, z)\zeta_j + S_j(\bar w_i, z)\zeta_j\right\} \\
& + \frac{\left((|\theta| - 1 - \exp(z))^+\right)^2 + |d|^2}{4\beta_0\left(1+\exp(z)\right)} \\
& + \Gamma\exp(-z)\left[\left(V_i - \frac{\varepsilon^2}{2}\right)^+ - \left(V_n - \frac{\varepsilon^2}{2}\right)^+\right]\sum_{j=2}^{i}\zeta_j\frac{\partial k_{j-1}}{\partial z}
\end{aligned}
\tag{6.112}
$$

Let us now add R_j from (6.90) and S_j from (6.111),

$$
\begin{aligned}
R_j(\bar w_j, z) + S_j(\bar w_j, z) = & \;(n-j+1)g_j + \frac{\Gamma^2}{4}\exp(-2z)\left(\frac{\partial k_{j-1}}{\partial z}\right)^2\sum_{l=1}^{j}\frac{1}{4g_l}\zeta_l^2 \\
& + \sum_{l=j}^{n} b_l + (1+\exp(z))\left(\beta_j|\rho_j|^2 + \beta_j|\pi_j|^2 + \frac{1+\exp(z)}{4b_j}|p_j|^2\right) \\
& + \sum_{l=2}^{j-1}\gamma_l + \frac{\Gamma^2}{4}\exp(-2z)\sum_{l=2}^{j-1}\frac{1}{4\gamma_l}\zeta_l^2\left(\frac{\partial k_{l-1}}{\partial z}\right)^2
\end{aligned}
\tag{6.113}
$$

regroup the terms, and relabel the sum as T_j,

$$\begin{aligned}
T_j(\bar{w}_j, z) &= R_j(\bar{w}_j, z) + S_j(\bar{w}_j, z) \\
&= \sum_{l=2}^{j-1} \gamma_l + \sum_{l=j}^{n} b_l + (n - j + 1) g_j \\
&\quad + (1 + \exp(z)) \left(\beta_j |\rho_j|^2 + \beta_j |\pi_j|^2 + \frac{1 + \exp(z)}{4b_j} |p_j|^2 \right) \\
&\quad + \frac{\Gamma^2}{4} \exp(-2z) \left[\left(\frac{\partial k_{j-1}}{\partial z} \right)^2 \sum_{l=1}^{j} \frac{1}{4g_l} \zeta_l^2 + \sum_{l=2}^{j-1} \frac{1}{4\gamma_l} \zeta_l^2 \left(\frac{\partial k_{l-1}}{\partial z} \right)^2 \right]
\end{aligned} \tag{6.114}$$

For the convenience of interpretation of the roles of different terms in T_j, we label its three components as

$$T_j(\bar{w}_j, z) = \eta_j + \kappa_j(\bar{w}_j, z) + \mu_j(\bar{w}_j, z) \tag{6.115}$$

$$\eta_j = \sum_{l=2}^{j-1} \gamma_l + \sum_{l=j}^{n} b_l + (n - j + 1) g_j \tag{6.116}$$

$$\kappa_j(\bar{w}_j, z) = (1 + \exp(z)) \left(\beta_j |\rho_j|^2 + \beta_j |\pi_j|^2 + \frac{1 + \exp(z)}{4b_j} |p_j|^2 \right) \tag{6.117}$$

$$\mu_j(\bar{w}_j, z) = \frac{\Gamma^2}{4} \exp(-2z) \left[\left(\frac{\partial k_{j-1}}{\partial z} \right)^2 \sum_{l=1}^{j} \frac{1}{4g_l} \zeta_l^2 + \sum_{l=2}^{j-1} \frac{1}{4\gamma_l} \zeta_l^2 \left(\frac{\partial k_{l-1}}{\partial z} \right)^2 \right] \tag{6.118}$$

Returning to (6.112), we now simply pick

$$q_j(\bar{w}_j, z) = -T_j(\bar{w}_j, z)\zeta_j, \quad j = 1, ..., n \tag{6.119}$$

and obtain

$$\begin{aligned}
\dot{V}_i &\le -\sum_{j=1}^{i} c_j \zeta_j^2 + \zeta_i \zeta_{i+1} + \frac{\left((|\theta| - 1 - \exp(z))^+ \right)^2 + |d|^2}{4\beta_0 (1 + \exp(z))} \\
&\quad + \Gamma \exp(-z) \left[\left(V_i - \frac{\varepsilon^2}{2} \right)^+ - \left(V_n - \frac{\varepsilon^2}{2} \right)^+ \right] \sum_{j=2}^{i} \zeta_j \frac{\partial k_{j-1}}{\partial z}
\end{aligned} \tag{6.120}$$

The inequality (6.120) was derived for $i = 1, ..., n - 1$ but it is also valid for $i = n$, except that $\zeta_{n+1} = 0$ because of (6.73). Hence, setting $i = n$, and noting specifically that this renders the second line in (6.120) zero, we get

$$\dot{V}_n \le -\sum_{j=1}^{n} c_j \zeta_j^2 + \frac{\left((|\theta| - 1 - \exp(z))^+ \right)^2 + |d|^2}{4\beta_0 (1 + \exp(z))} \tag{6.121}$$

Introducing

$$c_0 = \min_{i=1,...,n} c_i \tag{6.122}$$

we get

$$\dot{V}_n \le -2c_0 V_n + \frac{\left((|\theta| - 1 - \exp(z))^+\right)^2 + |d|^2}{4\beta_0 (1 + \exp(z))} . \tag{6.123}$$

With (6.121), using Theorem 2.8 we get the following result, under an adaptive controller summarized in Table 6.1.

Recursive design definitions: $i = 1, \ldots, n$

$$\zeta_i = w_i - k_{i-1}(\bar{w}_{i-1}, z) \tag{6.124}$$

$$V_i(\bar{w}_i, z) = \frac{1}{2}\sum_{j=1}^{i} \zeta_i^2 \tag{6.125}$$

$$\pi_i'(\bar{w}_i, z) = \varphi_i' - \sum_{j=1}^{i-1} \frac{\partial k_{i-1}}{\partial w_j}\varphi_j' =: p_i'(\bar{w}_i, z)\bar{\zeta}_i \tag{6.126}$$

$$\rho_i'(\bar{w}_i, z) = \alpha_i' - \sum_{j=1}^{i-1} \frac{\partial k_{i-1}}{\partial w_j}\alpha_j' \tag{6.127}$$

$$\begin{aligned} k_i(\bar{w}_i, z) = &- \left[c_i + \eta_i + \kappa_i(\bar{w}_i, z) + \mu_i(\bar{w}_i, z)\right]\zeta_i \\ &-\zeta_{i-1} + \sum_{j=1}^{i-1} \frac{\partial k_{i-1}}{\partial w_j} w_{j+1} \end{aligned} \tag{6.128}$$

$$\eta_i = \sum_{j=2}^{i-1} \gamma_j + \sum_{j=i}^{n} b_j + (n - i + 1) g_i \tag{6.129}$$

$$\kappa_i(\bar{w}_i, z) = (1 + \exp(z))\left(\beta_i|\rho_i|^2 + \beta_i|\pi_i|^2 + \frac{1 + \exp(z)}{4b_i}|p_i|^2\right) \tag{6.130}$$

$$\mu_i(\bar{w}_i, z) = \frac{\Gamma^2}{16}\exp(-2z)\left[\left(\frac{\partial k_{i-1}}{\partial z}\right)^2 \sum_{j=1}^{i} \frac{1}{g_j}\zeta_j^2 + \sum_{j=2}^{i-1} \frac{1}{\gamma_j}\left(\frac{\partial k_{j-1}}{\partial z}\right)^2 \zeta_j^2\right] \tag{6.131}$$

Controller:

$$u = k_n(w, z) \tag{6.132}$$

Update law:

$$\dot{z} = \Gamma \exp(-z)\left(V_n - \frac{\varepsilon^2}{2}\right)^+ \tag{6.133}$$

Table 6.1. *Explicit DADS backstepping design for strict-feedback system* (6.72), (6.73), *with properties stated in Theorem 6.3.*

Theorem 6.3 (Explicit DADS backstepping control): *Consider the system* (6.72), (6.73) *in feedback with the control law defined by* (6.132), (6.128), (6.124), (6.126), (6.127), (6.129), (6.130), (6.131), *along with the gain update law defined by* (6.133), (6.125). *For all positive constants* $\varepsilon, \Gamma, c_i, b_{ji}, \beta_i, g_i, \gamma_i$, *all the properties in Theorem 6.2 hold, with* $b = 1$, $c = 2c_0$,

$a = \frac{1}{4\beta_0}$, *and* $\lambda = \kappa$ *identity functions. In particular, for all initial conditions* (w_0, z_0), *all values of the unknown parameter vector* θ, *and all essentially bounded disturbance signals* d, *the unique solution of the closed-loop system exists and, moreover,* $V_n(t)$ *satisfies a p-IOS bound, the system's output state* $w_1(t)$ *satisfies*

$$\limsup_{t\to+\infty} |w_1(t)| \leq \varepsilon \tag{6.134}$$

and there exists a continuous function B *such that the gain* $z(t)$ *satisfies a p-UBIBS bound*

$$z_0 \leq z(t) \leq \lim_{s\to+\infty} z(s) \leq B\left(w_0, z_0, \|d\|_\infty, \|\theta\|\right) \tag{6.135}$$

for all $t \geq 0$.

Before we move on, let us examine the gains (6.129), (6.130), (6.131), which appear in the stabilizing function k_i in (6.128),

- The constant gain η_i is positive and can be chosen arbitrarily low, through choices of low values for γ_i, b_i, g_i. Additionally, $c_i >$ can be chosen arbitrarily low, which means that the overall linear gain $c_i + \eta_i > 0$ is arbitrarily low.
- The nonlinear damping $\kappa_i(\bar{w}, z)$ has a role of robustification against the disturbance d and arises in both the case when the unknown parameter θ is matched (such as in the scalar example in Chapter 4) and when it is unmatched. While dependent on the adapted gain z, $\kappa_i(\bar{w}, z)$ is independent of the update rate for z (it is not dependent on the adaptation gain Γ).
- The nonlinear damping $\mu_i(\bar{w}, z)$ arises only when θ is *unmatched* and its role is to dominate the adverse effect of fast adaptation, namely, the adaptation rate $\dot{z}$ acting as a disturbance, as evident from the last term in (6.82).

6.5 ▪ Illustrative Examples

Our first example does not contain uncertain constant parameters θ. The example deals with a simple linear controllable system and shows that even for this case the DADS control scheme can offer remarkable advantages.

Example 6.1 (A disturbed chain of two integrators—revisited): Our first example deals with the linear time-invariant system

$$\begin{aligned} \dot{w}_1 &= w_2 + d \\ \dot{w}_2 &= u \\ w &= (w_1, w_2)' \in \mathbb{R}^2, u \in \mathbb{R}, d \in \mathbb{R} \end{aligned} \tag{6.136}$$

System (6.136) has already been studied in the previous chapter (Example 5.3) where it was shown that the zero p-OAG property $\limsup_{t\to+\infty}(|w(t)|) \leq \alpha$ with a constant $\alpha > 0$ cannot be achieved for the entire state $w = (w_1, w_2)$ of the system (6.136) (the constant parameter θ is irrelevant for this example); no matter what is the applied controller.

Of course, it is possible to achieve the ISS property for system (6.136). Any stabilizing linear feedback guarantees the ISS property for the closed-loop system. However, it is not possible to assign the gain function for system (6.136); no matter what is the applied linear controller $u = -k_1 w_1 - k_2 w_2$ with $k_1, k_2 > 0$. To see this, notice that the application of a constant disturbance $d \in \mathbb{R}$ for the closed-loop system (6.136) with any stabilizing linear controller gives $\lim_{t\to+\infty}(w_2(t)) = -d$, $\lim_{t\to+\infty}(w_1(t)) = \frac{k_2}{k_1}d$ and, consequently, we get $\limsup_{t\to+\infty}(|w(t)|) \geq \|d\|_\infty \sqrt{1+\frac{k_2^2}{k_1^2}}$ for constant disturbances. Therefore, it is not possible to assign the gain function (or the asymptotic gain function) by applying a linear stabilizing feedback.

For this example, the DADS control scheme can provide an alternative *nonlinear dynamical feedback controller* that can guarantee the zero p-OAG property $\limsup_{t\to+\infty}(|w_1(t)|) \leq \alpha$, for the output $y = w_1$, with a constant $\alpha > 0$. More specifically, for every $a, c, \Gamma, \varepsilon > 0$ the controller

$$\begin{aligned}
u &= -\left(1+\tfrac{\Gamma}{2a}\left(\tfrac{1}{2}w_1^2+\tfrac{1}{2}(w_2+\rho w_1)^2-\varepsilon\right)^+\right)w_1-\rho w_2-\left(c+(\rho-c)\rho^2\right)(w_2+\rho w_1)\\
\rho &= c+\tfrac{1+\exp(z)}{2a}\\
\dot{z} &= \Gamma\exp(-z)\left(\tfrac{1}{2}w_1^2+\tfrac{1}{2}(w_2+\rho w_1)^2-\varepsilon\right)^+
\end{aligned} \tag{6.137}$$

guarantees that there exists a function $B \in C^0\left(\mathbb{R}^2 \times \mathbb{R} \times \mathbb{R}_+\right)$ such that for every $(w_0, z_0) \in \mathbb{R}^2 \times \mathbb{R}$, $d \in L^\infty(\mathbb{R}_+;\mathbb{R})$ the unique solution of (6.136), (6.137) with $(w(0), z(0)) = (w_0, z_0)$ exists and satisfies the following estimates for all $t \geq 0$:

$$\begin{aligned}
&\limsup_{t\to+\infty}\left(w_1^2(t)+\left(w_2(t)+\left(c+\tfrac{1+\exp(z(t))}{2a}\right)w_1(t)\right)^2\right)\\
&\leq \min\left(2\varepsilon,\ \chi\left(\limsup_{t\to+\infty}(|d(t)|),\exp\left(\lim_{t\to+\infty}(z(t))\right)\right)\right)
\end{aligned} \tag{6.138}$$

$$\begin{aligned}
&w_1^2(t)+\left(w_2(t)+\left(c+\tfrac{1+\exp(z(t))}{2a}\right)w_1(t)\right)^2\\
&\leq \exp(-2ct)\left(w_1^2(0)+\left(w_2(0)+\left(c+\tfrac{1+\exp(z(0))}{2a}\right)w_1(0)\right)^2\right)+\chi\left(\|d\|_\infty,\exp(z_0)\right)
\end{aligned} \tag{6.139}$$

$$|w(t)| \leq B\left(w_0, z_0, \|d\|_\infty\right) \tag{6.140}$$

$$z_0 \leq z(t) \leq \lim_{s\to+\infty}(z(s)) \leq B\left(w_0, z_0, \|d\|_\infty\right) \tag{6.141}$$

where $\chi(s_1, s_2) = \frac{a s_1^2}{c(1+s_2)}$ for all $s_1, s_2 \geq 0$. Furthermore, if $\lim_{t\to+\infty}(d(t)) = 0$ then

$$\lim_{t\to+\infty}(w(t)) = 0. \tag{6.142}$$

The DADS controller (6.137) is derived by applying the backstepping methodology that was used in Lemma 6.1. We start with the Lyapunov function $V_1(w_1) = \frac{1}{2}w_1^2$ for which the feedback law $w_2 = -\left(c+\frac{1+\exp(z)}{2a}\right)w_1$ satisfies the inequality $\dot{V}_1 \leq -2cV_1(w_1) + \frac{ad^2}{2(1+\exp(z))}$. Then we continue by adding one integrator and repeating all steps in the proof of Lemma 6.1 (with no parameter θ): the feedback law (6.137) guarantees the differential inequality $\dot{V} \leq -2cV(w,z) + \frac{ad^2}{1+\exp(z)}$ for the function $V(w,z) = \frac{1}{2}w_1^2 + \frac{1}{2}\left(w_2 + cw_1 + \frac{1+\exp(z)}{2a}w_1\right)^2$. The stability estimates (6.138), (6.139), (6.140), (6.141) are given by Theorem 2.8.

We next use controller (6.137) with

$$c = a = 1, \Gamma = 12, \varepsilon = 0.02 \tag{6.143}$$

and we make comparisons with the linear controller

$$u = -11w_1 - 7w_2 \tag{6.144}$$

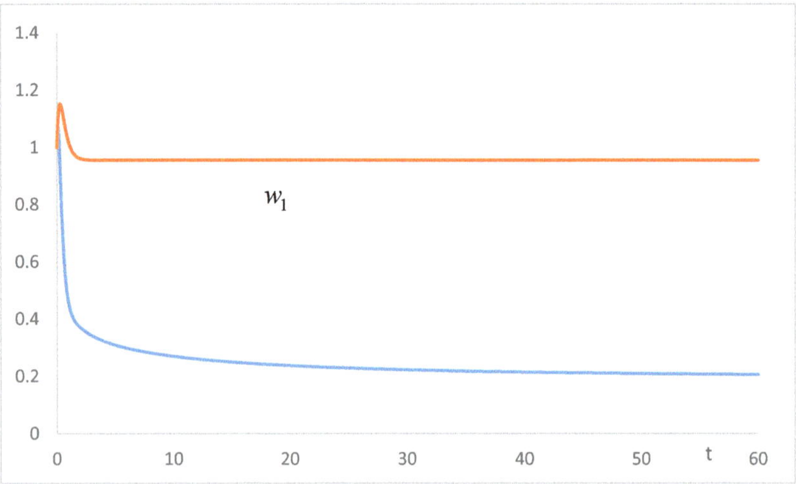

Figure 6.1. *The evolution of the state w_1 for the solutions of the closed-loop systems* (6.136), (6.137) *(blue line—DADS) and* (6.136), (6.144) *(red line—linear feedback) with $d \equiv 1.5$.*

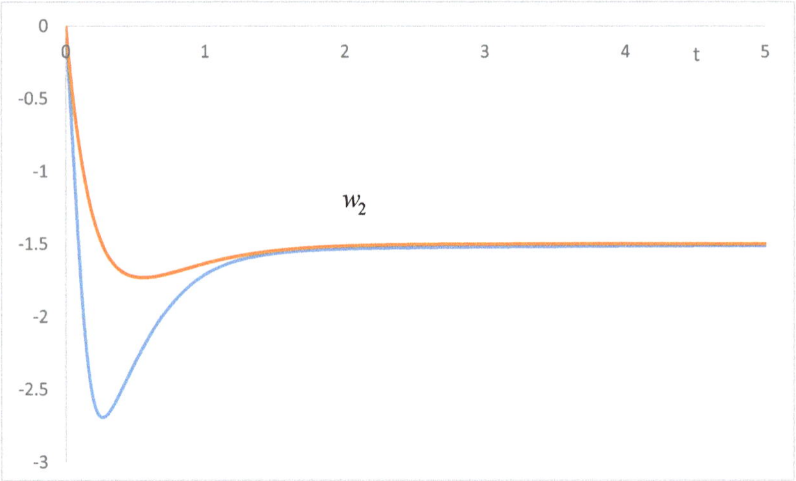

Figure 6.2. *The evolution of the state w_2 for the solutions of the closed-loop systems* (6.136), (6.137) *(blue line-DADS) and* (6.136), (6.144) *(red line-linear feedback) with $d \equiv 1.5$.*

which guarantees the inequality $\dot{U} \leq -2U(w) + \frac{1}{2}d^2$ for the function

$$U(w) = \frac{1}{2}w_1^2 + \frac{1}{2}(w_2 + 2w_1)^2 . \tag{6.145}$$

Figure 6.1 and Figure 6.2 show the evolution of the states w_1, w_2 for the two closed-loop systems with the constant disturbance $d \equiv 1.5$ and initial condition

$$w_1(0) = 1, w_2(0) = 0, z(0) = -\ln(5) \tag{6.146}$$

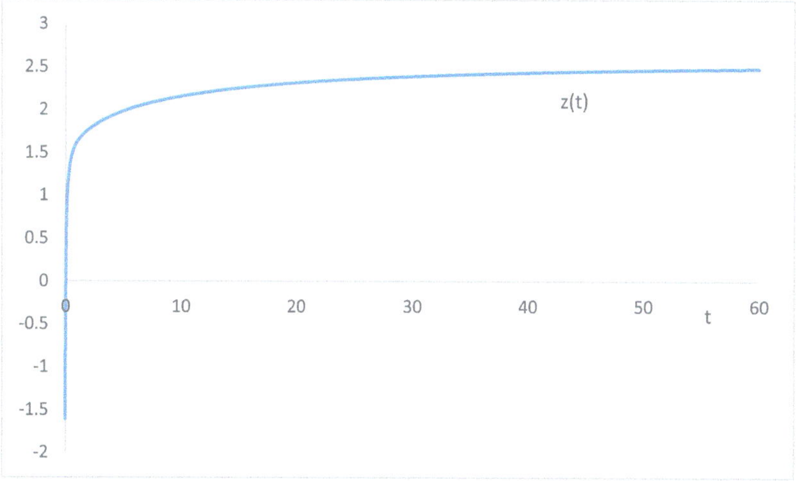

Figure 6.3. *The evolution of $z(t)$ for the solution of the closed-loop system* (6.136), (6.137) *(DADS controller) with $d \equiv 1.5$.*

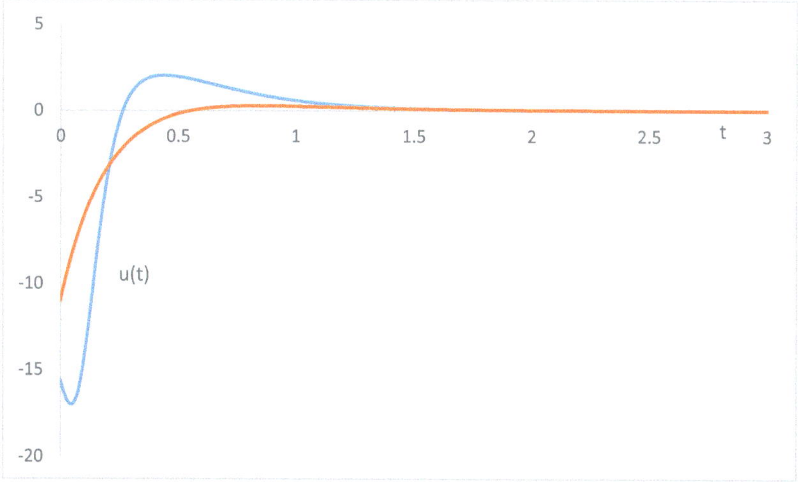

Figure 6.4. *The evolution of the control input u for the solutions of the closed-loop systems* (6.136), (6.137) *(blue line—DADS) and* (6.136), (6.144) *(red line—linear feedback) with $d \equiv 1.5$.*

It is clear that the DADS controller (6.137) guarantees the inequality $\limsup_{t \to +\infty} (|w_1(t)|) \leq 0.2$ although the convergence to the set $|w_1| \leq 0.2$ is slow. Similarly, the state $z(t)$ converges slowly to a value close to 2.5 (see Figure 6.3). On the other hand, the convergence of w_2 to the value -1.5 is fast. The controller (6.144) cannot guarantee the inequality $\limsup_{t \to +\infty} (|w_1(t)|) \leq 0.2$ and the state $w_1(t)$ converges rapidly to the value $21/22 = 0.9545$. Figure 6.4 shows that the DADS controller (6.137) does not require a larger control effort than the linear controller (6.144).

Therefore, the DADS control scheme may be used as an alternative to linear feedback control even for time-invariant, controllable linear systems. ◁

Our second example shows an application of the results of the present chapter to the stabilization of a jet engine.

Example 6.2 (Jet engine): The following model was proposed on pages 67–72 in [87] for the no-stall part of a jet engine compression system:

$$\begin{aligned}&\dot{\xi} = -\tfrac{3}{2}\xi^2 - \tfrac{1}{2}\xi^3 + w\\&\dot{w} = u\\&(\xi, w)' \in \mathbb{R}^2, u \in \mathbb{R}\end{aligned} \tag{6.147}$$

Since model (6.147) is obtained after many approximations (including a three-state Galerkin approximation of a nonlinear PDE model and use of empirical compressor and throttle characteristics equations), it is reasonable to assume that the coefficients $\frac{3}{2}$ and $\frac{1}{2}$ appearing in the first equation of (6.147) are not constant but unknown and possibly time-varying coefficients. Moreover, it is also possible that an additive uncertainty is present in the second equation of (6.147). Therefore, it is reasonable to study the stabilization problem for the following uncertain jet engine model:

$$\begin{aligned}&\dot{\xi} = \theta_1\xi^2 + \theta_2\xi^3 + w\\&\dot{w} = u + d\\&(\xi, w)' \in \mathbb{R}^2, \theta = (\theta_1, \theta_2)' \in \mathbb{R}^2, u \in \mathbb{R}, d \in \mathbb{R}\end{aligned} \tag{6.148}$$

Model (6.148) is a system of the form (6.35) with

$$\begin{aligned}&f(\xi) = 0, g(\xi) = 1, G(\xi) = 0, \Phi(\xi) = \left(\xi^2, \xi^3\right)\\&h(\xi, w) = 0, \varphi'(\xi, w) = (0, 0), \alpha'(\xi, w) = 1\end{aligned} \tag{6.149}$$

Let $c, a > 0$ be given positive constants. Using the inequalities

$$\begin{aligned}&\theta_1\xi^3 + \theta_2\xi^4 \leq |\theta_1|\,\xi^2\,|\xi| + |\theta_2|\,\xi^4\\&\leq |\theta|\,\xi^2\,|\xi| + |\theta|\,\xi^4\\&\leq (|\theta| - 1 - \exp(z))^+\,\xi^2\,|\xi| + (|\theta| - 1 - \exp(z))^+\,\xi^4\\&+ (1 + \exp(z))\,\xi^2\,|\xi| + (1 + \exp(z))\,\xi^4\\&\leq (|\theta| - 1 - \exp(z))^+\,\xi^2\,|\xi| + (|\theta| - 1 - \exp(z))^+\,\xi^4\\&+ (1 + \exp(z))\,\xi^2\left(1 + \xi^2\right) + (1 + \exp(z))\,\xi^4\\&\leq \tfrac{1+\exp(z)}{2a}\xi^6 + a\tfrac{\left((|\theta|-1-\exp(z))^+\right)^2}{2(1+\exp(z))} + \tfrac{1+\exp(z)}{2a}\xi^8 + a\tfrac{\left((|\theta|-1-\exp(z))^+\right)^2}{2(1+\exp(z))}\\&+ (1 + \exp(z))\,\xi^2 + 2\,(1 + \exp(z))\,\xi^4\\&= (1 + \exp(z))\left(\tfrac{1}{2a}\xi^4 + \tfrac{1}{2a}\xi^6 + 1 + 2\xi^2\right)\xi^2 + a\tfrac{\left((|\theta|-1-\exp(z))^+\right)^2}{1+\exp(z)}\end{aligned} \tag{6.150}$$

we conclude that inequality (6.37) holds with

$$\begin{aligned}&V(\xi) = \tfrac{1}{2}\xi^2, \kappa(s) = \lambda(s) = s, b = 1\\&k(\xi, z) = -\tfrac{c}{2}\xi - (1 + \exp(z))\left(\tfrac{1}{2a}\xi^4 + \tfrac{1}{2a}\xi^6 + 1 + 2\xi^2\right)\xi\end{aligned} \tag{6.151}$$

Since (6.151) implies that $\frac{\partial k}{\partial \xi}(\xi, z) = -\frac{c}{2} - (1 + \exp(z))\left(\frac{5}{2a}\xi^4 + \frac{7}{2a}\xi^6 + 1 + 6\xi^2\right)$, we get

$$\begin{aligned}&\left|\tfrac{\partial k}{\partial \xi}(\xi, z)\left(\xi^2, \xi^3\right)\right| = \left(\tfrac{c}{2} + (1 + \exp(z))\left(\tfrac{5}{2a}\xi^4 + \tfrac{7}{2a}\xi^6 + 1 + 6\xi^2\right)\right)\xi^2\sqrt{1 + \xi^2}\\&\leq \left(\tfrac{c}{2} + (1 + \exp(z))\left(\tfrac{5}{2a}\xi^4 + \tfrac{7}{2a}\xi^6 + 1 + 6\xi^2\right)\right)|\xi|\,\sqrt{2 + 2\xi^2}\sqrt{V(\xi)}\end{aligned} \tag{6.152}$$

and we conclude that inequality (6.36) holds with

$$\begin{aligned}&r_1(\xi, w, z) = \left(\tfrac{c}{2} + (1 + \exp(z))\left(\tfrac{5}{2a}\xi^4 + \tfrac{7}{2a}\xi^6 + 1 + 6\xi^2\right)\right)|\xi|\,\sqrt{2 + 2\xi^2}\\&r_2(\xi, w, z) = 0\end{aligned} \tag{6.153}$$

It is straightforward to apply Corollary 6.1 to system (6.148) and obtain a robust adaptive feedback controller for the jet engine compression system. $\triangleleft$

Our last example shows that the DADS control scheme can be applied to systems which are not in the form (6.1), (6.2) or in the form (6.11).

Example 6.3 (A linear stabilizable uncertain system—revisited): Consider the system

$$\begin{aligned}&\dot{\xi}_1 = -\xi_1 + d, \quad \dot{\xi}_2 = \theta\xi_1 + \xi_3, \quad \dot{\xi}_3 = u \\ &\xi = (\xi_1, \xi_2, \xi_3)' \in \mathbb{R}^3, u \in \mathbb{R}, \theta \in \mathbb{R}, d \in \mathbb{R}\end{aligned} \tag{6.154}$$

System (6.154) has already been studied in the previous chapter (Example 5.4) where it was shown that the p-OAG property $\limsup\limits_{t\to+\infty}(|\xi(t)|) \leq \alpha(\|d\|_\infty)$ from the input d to the entire plant state $\xi = (\xi_1, \xi_2, \xi_3)$ with a non-decreasing function $\alpha : \mathbb{R}_+ \to (0, +\infty)$ independent of the constant parameter θ cannot be achieved for system (6.154); no matter what is the applied controller. Moreover, system (6.154) is not in the form (6.1), (6.2) or in the form (6.11). However, we notice that the transformation

$$\begin{aligned}w_1 &= \xi_2 \\ w_2 &= \xi_3 \\ w_3 &= \xi_1 \\ \tilde{d} &= \theta\xi_1\end{aligned} \tag{6.155}$$

recasts system (6.154) as the cascade connection of the following subsystems:

$$\begin{aligned}&\dot{w}_1 = w_2 + \tilde{d} \\ &\dot{w}_2 = u \\ &w = (w_1, w_2)' \in \mathbb{R}^2, u \in \mathbb{R}, \tilde{d} \in \mathbb{R}\end{aligned} \tag{6.156}$$

$$\begin{aligned}&\dot{w}_3 = -w_3 + d \\ &w_3 \in \mathbb{R}, d \in \mathbb{R}\end{aligned} \tag{6.157}$$

Exploiting the explicit formula for the solution of (6.157) $w_3(t) = \exp(-(t-t_0))\,w_3(t_0) + \int_{t_0}^{t} \exp(-(t-s))d(s)ds$, we obtain (using (6.66)) the following estimates for all $d \in L^\infty(\mathbb{R}_+; \mathbb{R})$ and for every solution of (6.154):

$$|\xi_1(t)| \leq \exp(-t)\,|\xi_1(0)| + \|d\|_\infty \tag{6.158}$$

$$\limsup_{t\to+\infty}(|\xi_1(t)|) \leq \limsup_{t\to+\infty}(|d(t)|) \tag{6.159}$$

On the other hand, subsystem (6.157) is identical to system (6.136) that was studied in Example 6.1. Therefore, using (6.158) and (6.159), we conclude that for every $a, c, \Gamma, \varepsilon > 0$ the nonlinear adaptive controller

$$\begin{aligned}&u = -\left(1 + \tfrac{\Gamma}{2a}\left(\tfrac{1}{2}\xi_2^2 + \tfrac{1}{2}(\xi_3 + \rho\xi_2)^2 - \varepsilon\right)^+\right)\xi_2 - \rho\xi_3 - \left(c + (\rho - c)\rho^2\right)(\xi_3 + \rho\xi_2) \\ &\rho = c + \tfrac{1+\exp(z)}{2a} \\ &\dot{z} = \Gamma\exp(-z)\left(\tfrac{1}{2}\xi_2^2 + \tfrac{1}{2}(\xi_3 + \rho\xi_2)^2 - \varepsilon\right)^+\end{aligned} \tag{6.160}$$

guarantees that there exists a function $B \in C^0\left(\mathbb{R}^4 \times \mathbb{R}_+ \times \mathbb{R}_+\right)$ such that for every $(\xi_0, z_0) \in \mathbb{R}^3 \times \mathbb{R}$, $d \in L^\infty(\mathbb{R}_+;\mathbb{R})$, $\theta \in \mathbb{R}$ the unique solution of (6.154), (6.160) with $(\xi(0), z(0)) = (\xi_0, z_0)$ exists and satisfies the following estimates for all $t \geq 0$:

$$\begin{aligned}&\limsup_{t\to+\infty}\left(\xi_2^2(t) + \left(\xi_3(t) + \left(c + \tfrac{1+\exp(z(t))}{2a}\right)\xi_2(t)\right)^2\right)\\&\leq \min\left(2\varepsilon\,,\, \chi\left(|\theta|\limsup_{t\to+\infty}(|d(t)|)\,,\exp\left(\lim_{t\to+\infty}(z(t))\right)\right)\right)\end{aligned} \tag{6.161}$$

$$\begin{aligned}&\xi_2^2(t) + \left(\xi_3(t) + \left(c + \tfrac{1+\exp(z(t))}{2a}\right)\xi_2(t)\right)^2\\&\leq \exp(-2ct)\left(\xi_2^2(0) + \left(\xi_3(0) + \left(c + \tfrac{1+\exp(z(0))}{2a}\right)\xi_2(0)\right)^2\right)\\&+\chi\left(|\theta|\,|\xi_1(0)| + \|d\|_\infty\,, \exp(z_0)\right)\end{aligned} \tag{6.162}$$

$$|\xi(t)| \leq B\left(\xi_0, z_0, \|d\|_\infty, |\theta|\right) \tag{6.163}$$

$$z_0 \leq z(t) \leq \lim_{s\to+\infty}(z(s)) \leq B\left(\xi_0, z_0, \|d\|_\infty, |\theta|\right) \tag{6.164}$$

where $\chi(s_1, s_2) = \frac{as_1^2}{c(1+s_2)}$ for all $s_1, s_2 \geq 0$. Furthermore, if $\lim_{t\to+\infty}(d(t)) = 0$ then

$$\lim_{t\to+\infty}(\xi(t)) = 0 \tag{6.165}$$

Therefore, we can understand why the p-OAG property $\limsup_{t\to+\infty}(|\xi(t)|) \leq \alpha(\|d\|_\infty)$ with a non-decreasing function $\alpha : \mathbb{R}_+ \to (0, +\infty)$ independent of the constant parameter θ cannot be achieved for the entire plant state $\xi = (\xi_1, \xi_2, \xi_3)$ of the system (6.154) but only for the portion $y = (\xi_1, \xi_2)$ of the state. The problem is the state component ξ_3. The DADS controller (6.160) guarantees the estimate $\limsup_{t\to+\infty}(|\xi_2(t)|) \leq \sqrt{2\varepsilon}$ (recall (6.72)) and estimate $\limsup_{t\to+\infty}(|\xi_1(t)|) \leq \|d\|_\infty$ (recall (6.69)) and cannot provide such estimates for the state component ξ_3. This is expected: equation $\dot{\xi}_2 = \theta\xi_1 + \xi_3$ (recall (6.65)) shows that in order to "push and keep" the state component ξ_2 in the region $|\xi_2| \leq \sqrt{2\varepsilon}$, the state component ξ_3 has to cancel the effect of the "disturbance" $\theta\xi_1$, which depends heavily on the constant parameter θ. ◁

6.6 ▪ Proofs

In this section we provide the proofs of the main results of the chapter.

Proof of Theorem 6.1: By virtue of Theorem 2.8, it suffices to show that for every tuple of constants $b, \Gamma, \varepsilon, a, c > 0$ and for every $\kappa, \lambda \in K_\infty \cap C^\infty(\mathbb{R}_+)$ there exist $k, V \in C^\infty\left(\mathbb{R}^n \times \mathbb{R}^{m+1}\right)$ with $V(0,0,z) = k(0,0,z) = 0$ for $z \in \mathbb{R}$, $V(\xi, w, z) > 0$ for $(x, y) \neq 0$, and a constant $M > 0$ such that

(Ia) for each $\tilde{M} > 0$ the set $\left\{(\xi, w, z) \in \mathbb{R}^n \times \mathbb{R}^m \times \left[-\tilde{M}, \tilde{M}\right] \;:\; V(\xi, w, z) \leq \tilde{M}\right\}$ is bounded, and

(IIa) the following inequalities hold:

$$|\xi|^2 + w_1^2 \leq MV(\xi, w, z), \ \textit{for all} \ (\xi, w, z) \in \mathbb{R}^n \times \mathbb{R}^m \times \mathbb{R} \tag{6.166}$$

$$\begin{aligned}
&\sum_{i=1}^{n-1} \tfrac{\partial V}{\partial \xi_i}(\xi, w, z)\xi_{i+1} + \sum_{j=1}^{m-1} \tfrac{\partial V}{\partial w_j}(\xi, w, z)h_j(\xi, w_1, ..., w_j) \\
&+\sum_{j=1}^{m-1} \tfrac{\partial V}{\partial w_j}(\xi, w, z)\left(g_j(\xi, w_1, ..., w_j, \theta)w_{j+1} + \varphi_j'(\xi, w_1, ..., w_j)\theta + \alpha_j'(\xi, w_1, ..., w_j)d\right) \\
&+\tfrac{\partial V}{\partial w_m}(\xi, w, z)\left(h_m(\xi, w) + g_m(\xi, w, \theta)k(\xi, w, z) + \varphi_m'(\xi, w)\theta + \alpha_m'(\xi, w)d\right) \\
&+\tfrac{\partial V}{\partial \xi_n}(\xi, w, z)w_1 + \tfrac{\partial V}{\partial z}(\xi, w, z)\Gamma\exp(-z)\left(V(\xi, w, z) - \tfrac{\varepsilon^2}{2M}\right)^+ \\
&\leq -cV(\xi, w, z) + \tfrac{a}{M}\tfrac{|d|^2 + \left((|\theta| - b - \lambda(\exp(z)))^+\right)^2}{1+\kappa(\exp(z))} \\
&\textit{for all} \ (\xi, w, z) \in \mathbb{R}^n \times \mathbb{R}^m \times \mathbb{R}, \ d \in \mathbb{R}^l, \ \theta \in \Theta
\end{aligned} \tag{6.167}$$

Let arbitrary constants $b, \Gamma, \varepsilon, a, c > 0$ and arbitrary functions $\kappa, \lambda \in K_\infty \cap C^\infty(\mathbb{R}_+)$ be given. The functions $k, V \in C^\infty(\mathbb{R}^n \times \mathbb{R}^m \times \mathbb{R})$ are constructed inductively by using Lemma 6.1.

Define the matrices

$$A = \{a_{i,j} : i, j = 1, ..., n\}, \bar{b} = \begin{bmatrix} 0 & \cdots & 0 & 1 \end{bmatrix}' \tag{6.168}$$

where

$$a_{i,j} = 1 \text{ for } j = i+1, \ i = 1, ..., n-1 \text{ and } a_{i,j} = 0 \text{ if otherwise} \tag{6.169}$$

Notice that equations (6.1) and definitions (6.168), (6.169) and the fact that $x_{n+1} = y_1$ imply the equation

$$\dot{\xi} = A\xi + \bar{b}w_1 \tag{6.170}$$

Since the pair of matrices $(A, \bar{b})$ is a controllable pair there exists a vector $\omega \in \mathbb{R}^n$ and a positive definite, symmetric matrix $P \in \mathbb{R}^{n \times n}$ such that

$$P(A + \bar{b}\omega') + (A + \bar{b}\omega')'P \leq -2^m cP \tag{6.171}$$

Next, we notice that the functions

$$V_1(\xi, w_1, z) := \xi' P\xi + \frac{1}{2}(w_1 - \omega'\xi)^2, k_1(\xi, w_1, z) := -\frac{G(\xi, w_1, z)}{\eta_1(\xi, w_1)}(w_1 - \omega'\xi) \tag{6.172}$$

with

$$\begin{aligned}
G(\xi, w_1, z) &:= |\omega'\bar{b}| + 2^{m-2}c + r(\xi, w_1)(1 + b + \lambda(\exp(z))) \\
&+\tfrac{M}{2^{m+1}c}\left(K + r(\xi, w_1)(1 + |\omega|)(1 + b + \lambda(\exp(z)))\right)^2 \\
&+\tfrac{M(1+\kappa(\exp(z)))}{2^{3-m}a}\left(|\alpha_1(\xi, w_1)|^2 + r^2(\xi, w_1)\left(|\xi|^2 + w_1^2\right)\right)
\end{aligned} \tag{6.173}$$

where $K = |2\bar{b}'P - \omega' A - \omega'\bar{b}\omega'|$, η_1 is the function involved in (6.3), and $r : \mathbb{R}^n \times \mathbb{R} \to (0, +\infty)$ is a smooth function that satisfies for all $(\xi, w_1) \in \mathbb{R}^n \times \mathbb{R}$

$$|h_1(\xi, w_1)| + |\varphi_1(\xi, w_1)| \leq r(\xi, w_1)(|\xi| + |w_1|) \tag{6.174}$$

satisfy the following inequality for all $(x, y_1, z) \in \mathbb{R}^n \times \mathbb{R} \times \mathbb{R}$, $d \in \mathbb{R}^l$, $\theta \in \Theta$:

$$\begin{aligned}
&\sum_{i=1}^{n-1} \tfrac{\partial V_1}{\partial \xi_i}(\xi, w_1, z)\xi_{i+1} + \tfrac{\partial V_1}{\partial \xi_n}(\xi, w_1, z)y_1 + \tfrac{\partial V_1}{\partial z}(\xi, w_1, z)\Gamma\exp(-z)\left(V(\xi, w_1, z) - \tfrac{\varepsilon^2}{2M}\right)^+ \\
&+\tfrac{\partial V_1}{\partial w_1}(\xi, w_1, z)\left(h_1(\xi, w_1) + g_1(\xi, w_1, \theta)k_1(\xi, w_1, z) + \varphi_1'(\xi, w_1)\theta + \alpha_1'(\xi, w_1)d\right) \\
&\leq -2^{m-1}cV_1(x, y_1, z) + \tfrac{2^{1-m}a}{M}\tfrac{|d|^2 + \left((|\theta| - b - \lambda(\exp(z)))^+\right)^2}{1+\kappa(\exp(z))}
\end{aligned} \tag{6.175}$$

where $M > 0$ is any constant for which the following inequality holds for all $(\xi, w_1) \in \mathbb{R}^n \times \mathbb{R}$:

$$M\left(\xi' P\xi + \frac{1}{2}(w_1 - \omega'\xi)^2\right) \geq |\xi|^2 + w_1^2 \tag{6.176}$$

The existence of a constant $M > 0$ for which (6.176) holds follows from (6.172) and the fact that $\xi' P\xi + \frac{1}{2}(w_1 - \omega'\xi)^2$ is a positive definite quadratic form on $\mathbb{R}^{n+1}$.

Inequality (6.175) is a direct consequence of definitions (6.172), (6.173), inequalities (6.3), (6.171), (6.174), and (6.176) (which implies the inequality $M\xi' P\xi \geq |\xi|^2$ for all $\xi \in \mathbb{R}^n$) as well as the inequalities

$$|\xi| + |w_1| \leq (1 + |\omega|)|\xi| + |w_1 - \omega'\xi| \tag{6.177}$$

$$\begin{aligned} &|w_1 - \omega'\xi|\,|\alpha_1(\xi, w_1)|\,|d| \\ &\leq \tfrac{2^{1-m}a|d|^2}{M(1+\kappa(\exp(z)))} + \tfrac{M(1+\kappa(\exp(z)))}{2^{3-m}a}|\alpha_1(\xi, w_1)|^2 (w_1 - \omega'\xi)^2 \end{aligned} \tag{6.178}$$

$$\begin{aligned} &|w_1 - \omega'\xi|\, r(\xi, w_1)\,(|\xi| + |w_1|)\,(|\theta| - b - \lambda(\exp(z)))^+ \\ &\leq 2^{1-m}a\tfrac{\left((|\theta|-b-\lambda(\exp(z)))^+\right)^2}{M(1+\kappa(\exp(z)))} + \tfrac{M(1+\kappa(\exp(z)))}{2^{3-m}a} r^2(\xi, w_1)\,(|\xi| + |w_1|)^2 (w_1 - \omega'\xi)^2 \end{aligned} \tag{6.179}$$

$$(|\xi| + |w_1|)^2 \leq 2|\xi|^2 + 2w_1^2 \tag{6.180}$$

$$|\theta| - b - \lambda(\exp(z)) \leq (|\theta| - b - \lambda(\exp(z)))^+ \tag{6.181}$$

$$\begin{aligned} &|w_1 - \omega'\xi|\,|\xi|\,(K + r(\xi, w_1)(1 + |\omega|)(1 + b + \lambda(\exp(z)))) \\ &\leq 2^{m-1}c\tfrac{|\xi|^2}{M} + \tfrac{M}{2^{m+1}c}(K + r(\xi, w_1)(1 + |\omega|)(1 + b + \lambda(\exp(z))))^2 (w_1 - \omega'\xi)^2 \end{aligned} \tag{6.182}$$

which give inequality (6.175).

Applying Lemma 6.1 exactly $m-1$ times, we construct functions $k_j \in C^\infty\left(\mathbb{R}^n \times \mathbb{R}^j \times \mathbb{R}\right)$ for $j = 2, ..., m$ with $k_j(0, 0, z) = 0$ for $z \in \mathbb{R}$, such that (6.167) holds for

$$V(\xi, w, z) = x' Px + \frac{1}{2}(w_1 - \omega'\xi)^2 + \frac{1}{2}\sum_{j=2}^{m}(w_j - k_{j-1}(\xi, w_1, ..., w_{j-1}, z))^2 \tag{6.183}$$

Notice that (6.176), (6.183) guarantee the validity of (6.166). ◁

Finally, we provide the proof of Theorem 6.2.

Proof of Theorem 6.2: By virtue of Theorem 2.8, it suffices to show that for every tuple of constants $b, \Gamma, \varepsilon, a, c > 0$ and for every $\kappa, \lambda \in K_\infty \cap C^\infty(\mathbb{R}_+)$ there exist functions $k, V \in C^\infty(\mathbb{R}^n \times \mathbb{R})$ with $V(0, z) = k(0, z) = 0$ for $z \in \mathbb{R}$, $V(w, z) > 0$ for $w \neq 0$, such that

(Ib) for each $\tilde{M} > 0$ the set $\left\{(w, z) \in \mathbb{R}^n \times \left[-\tilde{M}, \tilde{M}\right] : V(w, z) \leq \tilde{M}\right\}$ is bounded, and

(IIb) the following inequalities hold:

$$\frac{1}{2}w_1^2 \leq V(w, z) \text{ for all } (w, z) \in \mathbb{R}^n \times \mathbb{R} \tag{6.184}$$

$$\begin{aligned} &\textstyle\sum_{i=1}^{n-1}\frac{\partial V}{\partial w_i}(w, z)\,(h_i(w_1, ..., w_i) + g_i(w_1, ..., w_i, \theta)x_{i+1} + \varphi_i'(w_1, ..., w_i)\theta + \alpha_i'(w_1, ..., w_i)d) \\ &+\tfrac{\partial V}{\partial w_n}(w, z)\,(h_n(w) + g_n(w, \theta)k(w, z) + \varphi_n'(w)\theta + \alpha_n'(w)d) \\ &+\tfrac{\partial V}{\partial z}(w, z)\Gamma\exp(-z)\left(V(w, z) - \tfrac{\varepsilon^2}{2}\right)^+ \\ &\leq -cV(w, z) + a\tfrac{|d|^2 + \left((|\theta|-b-\lambda(\exp(z)))^+\right)^2}{1+\kappa(\exp(z))} \\ &\text{for all } (w, z) \in \mathbb{R}^n \times \mathbb{R},\ d \in \mathbb{R}^l,\ \theta \in \Theta \end{aligned} \tag{6.185}$$

Let arbitrary constants $b, \Gamma, \varepsilon, a, c > 0$ and arbitrary functions $\kappa, \lambda \in K_\infty \cap C^\infty(\mathbb{R}_+)$ be given. The functions $k, V \in C^\infty(\mathbb{R}^n \times \mathbb{R})$ are constructed inductively by using Lemma 6.1.

Indeed, we notice that the functions

$$V_1(w_1, z) := \frac{1}{2} w_1^2,\ k_1(w_1, z) := -\frac{M(w_1, z)}{\eta_1(w_1)} w_1 \tag{6.186}$$

with

$$\begin{aligned} M(w_1, z) &:= (b + 1 + \lambda(\exp(z)))\, r(w_1) \\ &+ \tfrac{1+\kappa(\exp(z))}{2^{3-n}a} \left(|\alpha_1(w_1)|^2 + r^2(w_1) w_1^2 \right) + 2^{n-2} c \end{aligned} \tag{6.187}$$

where η_1 is the function involved in (6.12) and $r : \mathbb{R} \to (0, +\infty)$ is a smooth function that satisfies

$$|h_1(w_1)| + |\varphi_1(w_1)| \le r(w_1)\, |w_1| \text{ for all } w_1 \in \mathbb{R}, \tag{6.188}$$

satisfy the following inequality for all $(w_1, z) \in \mathbb{R} \times \mathbb{R}$, $d \in \mathbb{R}^l$, $\theta \in \Theta$:

$$\begin{aligned} &\tfrac{\partial V_1}{\partial w_1}(w_1, z) \left(h_1(w_1) + g_1(w_1, \theta) k_1(w_1, z) + \varphi_1'(w_1)\theta + \alpha_1'(w_1) d \right) \\ &+ \tfrac{\partial V_1}{\partial z}(w_1, z) \Gamma \exp(-z) \left(V(w_1, z) - \tfrac{\varepsilon^2}{2} \right)^+ \\ &\le -2^{n-1} c V_1(w_1, z) + 2^{1-n} a \tfrac{|d|^2 + \left((|\theta| - b - \lambda(\exp(z)))^+ \right)^2}{1+\kappa(\exp(z))} \end{aligned} \tag{6.189}$$

Inequality (6.189) is a direct consequence of definitions (6.186), (6.187), inequalities (6.12) and (6.188), as well as the inequalities

$$|w_1|\, |\alpha_1(w_1)|\, |d| \le 2^{1-n} a \frac{|d|^2}{1 + \kappa(\exp(z))} + \frac{1 + \kappa(\exp(z))}{2^{3-n} a} |\alpha_1(w_1)|^2 w_1^2 \tag{6.190}$$

$$|\theta| - b - \lambda(\exp(z)) \le (|\theta| - b - \lambda(\exp(z)))^+ \tag{6.191}$$

$$\begin{aligned} &r(w_1) w_1^2 (|\theta| - b - \lambda(\exp(z)))^+ \\ &\le \tfrac{1+\kappa(\exp(z))}{2^{3-n}a} r^2(w_1) w_1^4 + 2^{1-n} a \tfrac{\left((|\theta| - b - \lambda(\exp(z)))^+ \right)^2}{1+\kappa(\exp(z))} \end{aligned} \tag{6.192}$$

which give inequality (6.189).

Applying Lemma 6.1 exactly $n - 1$ times, we construct functions $k_i \in C^\infty\left(\mathbb{R}^i \times \mathbb{R}\right)$ for $i = 2, ..., n$ with $k_i(0, z) = 0$ for $z \in \mathbb{R}$, such that (6.185) holds for the function

$$V(w, z) = \frac{1}{2} w_1^2 + \frac{1}{2} \sum_{i=2}^{n} (w_i - k_{i-1}(w_1, ..., w_{i-1}, z))^2 \tag{6.193}$$

Notice that equality (6.193) guarantees the validity of (6.184). The proof is complete. ◃

Chapter 7

L^{∞}-Robust Identifier: Batch Least Squares

7.1 ▪ Introduction

The advantage of using direct adaptive controllers is the fact that the identifier can focus on the task of identification only without worrying about the task of regulation. This is an important advantage because the task of identification of unknown constant parameters is not easy. Particularly, in the case where disturbances are present, the task of robust (with respect to the disturbances) identification is a challenge that only rarely has been addressed in the literature. An exception is the paper [31] whose results were extended in [60], [98], [120] and were applied to the aircraft icing problem in [104]. The identifier proposed in [31] is derived by means of game-theoretic arguments (which involve a least-squares cost) and guarantees a prespecified level of robustness when there is sufficient excitation. The identifier proposed in [31] can handle disturbances of class $L^2(\mathbb{R}_+)$.

In this section, we extend the ideas that were presented in [70], [72], [74] and were analyzed in Chapter 3 about the construction of robust identifiers. The proposed robust identifier can handle disturbances of class $L^{\infty}(\mathbb{R}_+)$.

The robust identifier that we construct in this chapter can work in parallel with every direct adaptive controller (not only DADS). It is based on the idea of least-squares and provides characteristics that no other identifier can provide. Moreover, it can be applied to a wide class of nonlinear systems (systems which have right-hand sides that are affine with respect to the unknown parameters and to the disturbances). The proposed identifier uses batches of data obtained in an interval of specific length in order to provide or not a new parameter estimate. The parameter estimate is provided only if the level of excitation of the measured outputs is sufficient to dominate the effect of the disturbance. In this way, the identifier can guarantee a prespecified level of robustness when there is sufficient excitation.

7.2 ▪ A Robust Least-Squares Identifier

We are given the control system

$$\dot{y} = f(y,z) + \Phi(y,z)\theta + g(y,z)d \tag{7.1}$$

where $y \in \mathbb{R}^n$ is the state which is measured, $z \in \mathbb{R}^m$ is an input that is measured, $\theta \in \Theta \subseteq \mathbb{R}^p$ is an unknown constant parameter, and $d \in \mathbb{R}^l$ is an input that is not measured (a disturbance). We assume that the set $\Theta \subseteq \mathbb{R}^p$ where the constant parameter takes values is a known closed convex set. We also assume that the mappings $f : \mathbb{R}^n \times \mathbb{R}^m \to \mathbb{R}^n$, $\Phi : \mathbb{R}^n \times \mathbb{R}^m \to \mathbb{R}^{n\times p}$, $g : \mathbb{R}^n \times \mathbb{R}^m \to \mathbb{R}^{n\times l}$ are locally Lipschitz mappings.

We next assume that for every $d \in L^\infty_{loc}\left(\mathbb{R}_+;\mathbb{R}^l\right)$, $x_0 \in \mathbb{R}^n$ there exists $z \in C^0\left(\mathbb{R}_+;\mathbb{R}^m\right)$ such that the solution $x(t) \in \mathbb{R}^n$ of (7.1) with initial condition $x(0) = x_0$ exists for all $t \geq 0$.

Let $y(t) \in \mathbb{R}^n$ be an arbitrary solution of (7.1) that corresponds to certain inputs $z \in C^0\left(\mathbb{R}_+;\mathbb{R}^m\right)$, $d \in L^\infty\left(\mathbb{R}_+;\mathbb{R}^l\right)$.

Integrating (7.1) we get for all $t, s \geq 0$

$$y(t)-y(s) = \int_s^t f(y(\tau), z(\tau))d\tau+\left(\int_s^t \Phi(y(\tau), z(\tau))d\tau\right)\theta+\int_s^t g(y(\tau), z(\tau))d(\tau)d\tau \quad (7.2)$$

We define for all $t, s \geq 0$

$$\begin{aligned} a(t,s) &:= y(t) - y(s) - \textstyle\int_s^t f(y(\tau), z(\tau))d\tau \\ b(t,s) &:= \textstyle\int_s^t \Phi(y(\tau), z(\tau))d\tau \\ D(t,s) &:= \textstyle\int_s^t g(y(\tau), z(\tau))d(\tau)d\tau \end{aligned} \quad (7.3)$$

Thus, we obtain from (7.2) and definitions (7.3) the following equation for all $t, s \geq 0$:

$$a(t,s) = b(t,s)\theta + D(t,s) \quad (7.4)$$

Let a constant $T > 0$, $r \geq 0$ be given. We define the function

$$h(\vartheta; r) = \int_r^{T+r} \int_r^{T+r} |a(t,s) - b(t,s)\vartheta|^2 \, dsdt \text{ for all } \vartheta \in \mathbb{R}^p \quad (7.5)$$

It is clear that in the disturbance-free case where $d = 0$, the mapping $\mathbb{R}^p \ni \vartheta \to h(\vartheta; r)$ has a global minimum at $\vartheta = \theta$. In the general case, if the mapping $\mathbb{R}^p \ni \vartheta \to h(\vartheta; r)$ has a minimum at $\vartheta^* \in \mathbb{R}^p$ then (by virtue of Fermat's theorem) we obtain

$$A(r) = B(r)\vartheta^* \quad (7.6)$$

where

$$\begin{aligned} A(r) &:= \textstyle\int_r^{T+r}\int_r^{T+r} b'(t,s)a(t,s)dsdt \\ B(r) &:= \textstyle\int_r^{T+r}\int_r^{T+r} b'(t,s)b(t,s)dsdt \end{aligned} \quad (7.7)$$

Definition (7.7) implies that the square matrix $B(r) \in \mathbb{R}^{p\times p}$ is symmetric and positive semidefinite (since $\xi' B(r)\xi = \int_r^{T+r}\int_r^{T+r} |b(t,s)\xi|^2 \, dsdt$ for all $\xi \in \mathbb{R}^p$). Therefore, $\det(B(T)) \geq 0$. Moreover, using (7.3) and (7.7) we get the following characterization:

"$\det(B(r)) = 0$ if and only if
there exists $\xi \in \mathbb{R}^p$, $\xi \neq 0$ with $\Phi(y(t), z(t))\xi = 0$ for all $t \in [r, T+r]$"

To see this, notice that $\det(B(r)) = 0$ implies the existence of $\xi \in \mathbb{R}^p$, $\xi \neq 0$ with $B(r)\xi = 0$. Since $\xi' B(r)\xi = \int_r^{T+r}\int_r^{T+r} |b(t,s)\xi|^2 \, dsdt$ we get (by continuity of b; recall (7.3)) that $b(t,s)\xi = \int_s^t \Phi(y(\tau), z(\tau))\xi d\tau = 0$ for all $t, s \in [r, T+r]$. Fixing $s = r$ and differentiating with respect to t we obtain that $\Phi(y(t), z(t))\xi = 0$ for all $t \in [r, T+r]$. Conversely, if there exists $\xi \in \mathbb{R}^p$, $\xi \neq 0$ with $\Phi(y(t), z(t))\xi = 0$ for $t \in [r, T+r]$ a.e. then we get $b(t,s)\xi = \int_s^t \Phi(y(\tau), z(\tau))\xi d\tau = 0$ for all $t, s \in [r, T+r]$. Since $\xi' B(r)\xi = \int_r^{T+r}\int_r^{T+r} |b(t,s)\xi|^2 \, dsdt$ we conclude that $\xi' B(r)\xi = 0$. Consequently, the positive semidefinite matrix $B(r) \in \mathbb{R}^{p\times p}$ is not positive definite. Therefore, $\det(B(r)) = 0$.

However, it should be noticed that in general, we have no guarantee that $B(r) \in \mathbb{R}^{p\times p}$ is an invertible matrix.

Assuming that $\det(B(r)) > 0$ we obtain from (7.6)

$$\vartheta^* = (B(r))^{-1} A(r) \tag{7.8}$$

Using definitions (7.7) and equation (7.4) we get

$$\vartheta^* = \theta + (B(r))^{-1} \int_r^{T+r} \int_r^{T+r} b'(t,s) D(t,s) ds dt \tag{7.9}$$

In this case, using (7.4), (7.5), (7.7) and (7.9) we obtain the following equation:

$$(\vartheta^* - \theta)' B(r) (\vartheta^* - \theta) = \int_r^{T+r} \int_r^{T+r} |D(t,s)|^2 ds dt - h(\vartheta^*; r) \tag{7.10}$$

Definition (7.3) implies the following estimate for all $t, s \in [r, T+r]$:

$$|D(t,s)| \leq |t-s| \max_{\tau \in [r, T+r]} (|g(y(\tau), z(\tau))|) \|d\|_\infty \tag{7.11}$$

Therefore, we obtain from (7.10) and (7.11) the following estimate that holds if $\det(B(r)) > 0$:

$$\lambda_{\min}(B(r)) \left| (B(r))^{-1} A(r) - \theta \right|^2 \leq \frac{T^4}{6} \left(\max_{\tau \in [r, T+r]} (|g(y(\tau), z(\tau))|) \right)^2 \|d\|_\infty^2 \tag{7.12}$$

We are now ready to give our robust least-squares identifier. Let $P_\Theta : \mathbb{R}^p \to \Theta$ denote the projection mapping on the closed, convex set $\Theta \subseteq \mathbb{R}^p$. For $T, \eta, \sigma > 0$ arbitrary constants, we consider next the following identifier with (arbitrary) initial condition $\hat{\theta}(0) = \hat{\theta}_0 \in \Theta$:

$$\psi_0 = \sigma T^2 \tag{7.13}$$

$$\psi_{i+1} = \max\left(\psi_i, \frac{\sqrt{6\lambda_{\min}(B(iT))}}{\eta + \max_{\tau \in [iT, (i+1)T]} (|g(y(\tau), z(\tau))|)} \right) \text{ for } i \in \mathbb{N} \tag{7.14}$$

$$\begin{gathered} \hat{\theta}((i+1)T) = P_\Theta\left((B(iT))^{-1} A(iT) \right) \\ \text{for } i \in \mathbb{N} \text{ and } \frac{\sqrt{6\lambda_{\min}(B(iT))}}{\eta + \max_{\tau \in [iT, (i+1)T]} (|g(y(\tau), z(\tau))|)} \geq \psi_i \end{gathered} \tag{7.15}$$

$$\hat{\theta}((i+1)T) = \hat{\theta}(iT) \text{ for } i \in \mathbb{N} \text{ and } \frac{\sqrt{6\lambda_{\min}(B(iT))}}{\eta + \max_{\tau \in [iT, (i+1)T]} (|g(y(\tau), z(\tau))|)} < \psi_i \tag{7.16}$$

$$\hat{\theta}(t) = \hat{\theta}(iT) \text{ for } i \in \mathbb{N} \text{ and } t \in [iT, (i+1)T) \tag{7.17}$$

Equation (7.14) implies that $\psi_{i+1} \geq \psi_i$ for all $i \in \mathbb{N}$. Consequently, we get $\psi_i \geq \psi_0 = \sigma T^2$ for all $i \in \mathbb{N}$. Furthermore, we get (by induction)

$$\psi_{i+1} = \max\left(\sigma T^2, \max_{j=0,\ldots,i}\left(\frac{\sqrt{6\lambda_{\min}\left(B(jT)\right)}}{\eta + \max\limits_{\tau\in[jT,(j+1)T]}\left(|g(y(\tau),z(\tau))|\right)}\right)\right) \tag{7.18}$$

for all $i \in \mathbb{N}$.

The parameter update rule (7.15), (7.16) with (7.14) is driven by the following logic. Let us refer to the matrix $B(iT)$ informally as the "parameter's batch regressor". Let us also refer to $\max\limits_{\tau\in[iT,(i+1)T]}\left(|g(y(\tau),z(\tau))|\right)$ as the "disturbance's batch gain". The logic of (7.16) is to not update $\hat{\theta}$ when the disturbance's batch gain dominates the parameter regressor. And, conversely, the logic of (7.15) is the opposite: update the parameter estimate $\hat{\theta}$ when the parameter's batch regressor dominates the disturbance's batch gain. Finally, the logic of (7.14) is to keep raising the bar for updating $\hat{\theta}$. Overall, updating only when the disturbance does not dominate offers a hope for robustness of parameter estimation.

Using (7.9), (7.12), (7.15) and the fact that $\Theta \subseteq \mathbb{R}^p$ is a closed convex set (which implies that $\left|P_\Theta\left(\hat{\theta}\right) - P_\Theta(\theta)\right| \le \left|\hat{\theta} - \theta\right|$ for all $\hat{\theta}, \theta \in \mathbb{R}^p$) we get the following estimate when $\frac{\sqrt{6\lambda_{\min}(B(iT))}}{\eta + \max\limits_{\tau\in[iT,(i+1)T]}(|g(y(\tau),z(\tau))|)} \ge \psi_i$:

$$\begin{aligned}
&\left|\hat{\theta}((i+1)T) - \theta\right| = \left|P_\Theta\left(\left(B(iT)\right)^{-1}A(iT)\right) - \theta\right| \\
&= \left|P_\Theta\left(\left(B(iT)\right)^{-1}A(iT)\right) - P_\Theta(\theta)\right| \\
&\le \left|\left(B(iT)\right)^{-1}A(iT) - \theta\right| \\
&\le \frac{T^2}{\sqrt{6\lambda_{\min}(B(iT))}} \max_{\tau\in[iT,(i+1)T]}\left(|g(y(\tau),z(\tau))|\right)\|d\|_\infty \\
&\le \frac{T^2}{\psi_i}\|d\|_\infty \le \frac{1}{\sigma}\|d\|_\infty
\end{aligned} \tag{7.19}$$

The above estimate justifies the idea behind the batch identifier (7.14), (7.15), (7.16), (7.17): we use the most recent estimate $P_\Theta\left(\left(B(iT)\right)^{-1}A(iT)\right)$ that minimizes the gain

$$\frac{T^2}{\sqrt{6\lambda_{\min}\left(B(iT)\right)}} \max_{\tau\in[iT,(i+1)T]}\left(|g(y(\tau),z(\tau))|\right) \tag{7.20}$$

of the disturbance d to the estimation error $\left|P_\Theta\left(\left(B(iT)\right)^{-1}A(iT)\right) - \theta\right|$, and it is only when that gain does not exceed $1/\sigma$ that we update $\hat{\theta}$ to $P_\Theta\left(\left(B(iT)\right)^{-1}A(iT)\right)$.

Therefore, we have proved the following result for the hybrid identifier (7.14), (7.15), (7.16), (7.17).

Proposition 7.1: *If there exists $i \in \mathbb{N}$ with $\frac{\sqrt{6\lambda_{\min}(B(iT))}}{\eta + \max\limits_{\tau\in[iT,(i+1)T]}(|g(y(\tau),z(\tau))|)} \ge \psi_i$ then the following estimate holds for all $t \ge (i+1)T$:*

$$\left|\hat{\theta}(t) - \theta\right| \le \frac{1}{\sigma}\|d\|_\infty \tag{7.21}$$

The hybrid identifier (7.14), (7.15), (7.16), (7.17) guarantees robustness with respect to the disturbance d with assignable gain (equal to $1/\sigma$) of the disturbance to the estimation error. Moreover, the identifier (7.14), (7.15), (7.16), (7.17) gives a specific indication, namely

$$\frac{\sqrt{6\lambda_{\min}\left(B(iT)\right)}}{\eta + \max\limits_{\tau\in[iT,(i+1)T]}\left(|g(y(\tau),z(\tau))|\right)} \ge \psi_i \tag{7.22}$$

if and when robust identification has occurred. If there is no such $i \in \mathbb{N}$ then there is no $i \in \mathbb{N}$ with $\frac{\sqrt{6\lambda_{\min}(B(iT))}}{\eta + \max\limits_{\tau \in [iT,(i+1)T]} (|g(y(\tau),z(\tau))|)} \geq \sigma T^2$ and the equation $\hat{\theta}(t) = \hat{\theta}_0$ holds for all $t \geq 0$, i.e., the identifier has stuck to the initial estimate since the effect of the disturbance is too strong.

The bound (7.21) guarantees that the parameter estimate $\hat{\theta}(t)$ is always bounded (even when disturbances are present) and exact estimation in finite time of the unknown parameter $\theta \in \mathbb{R}$ when there exists sufficient excitation of the state and the disturbance is absent. Moreover, the bound (7.21) guarantees an ultimate estimation error which is proportional to the magnitude of the disturbance when there exists sufficient excitation of the state: this feature brings to mind the identifier proposed in [74] but it cannot be provided by other adaptive schemes in the literature. The gain of the disturbance for the estimation error (equal to $1/\sigma$) indicates a *trade-off* between "learning" and "robustness": when the update threshold $\sigma > 0$ is large then the gain from the disturbance to parameter error is small (indicating strong robustness) but this implies that it is less likely the "learning" process to be complete (since it is less likely that there exists $i \in \mathbb{N}$ with $\frac{\sqrt{6\lambda_{\min}(B(iT))}}{\eta + \max\limits_{\tau \in [iT,(i+1)T]} (|g(y(\tau),z(\tau))|)} \geq \sigma T^2$). On the other hand, when $\sigma > 0$ is small then the gain of the disturbance is large (indicating lack of robustness) but this implies that it is more possible the "learning" process to be complete.

The analysis above showed that when there exists $i \in \mathbb{N}$ with $\frac{\sqrt{6\lambda_{\min}(B(iT))}}{\eta + \max\limits_{\tau \in [iT,(i+1)T]} (|g(y(\tau),z(\tau))|)} \geq \psi_i$ then we have $\left|\hat{\theta}((i+1)T) - \theta\right| \leq \frac{T^2}{\psi_i} \|d\|_\infty$. Consequently, when identification occurs then we get $\limsup\limits_{t \to +\infty} \left(\left|\hat{\theta}(t) - \theta\right|\right) \leq \frac{T^2}{\lim\limits_{i \to +\infty}(\psi_i)} \|d\|_\infty$ and the actual asymptotic gain of the disturbance to the estimation error is even less than $1/\sigma$: it is equal to $\frac{T^2}{\lim\limits_{i \to +\infty}(\psi_i)} \leq \frac{1}{\sigma}$. The quantity $\frac{T^2}{\lim\limits_{i \to +\infty}(\psi_i)}$ is known to the user, though not a priori but only in the course of running the identifier. Therefore, the user not only knows if and when identification occurs but also knows what the estimated asymptotic gain of the disturbance to the estimation error is when identification occurs.

The parameter $T > 0$ plays a significant role in the robust identifier (7.14), (7.15), (7.16), (7.17). If we select $T > 0$ to be large, then it is less likely that there exists $i \in \mathbb{N}$ with $\frac{\sqrt{6\lambda_{\min}(B(iT))}}{\eta + \max\limits_{\tau \in [iT,(i+1)T]} (|g(y(\tau),z(\tau))|)} \geq \sigma T^2$. This happens because of the conservatism of estimates (7.11), (7.12): when $T > 0$ is large both inequalities (7.11) and (7.12) become "very conservative" estimates. On the other hand, a very small value of $T > 0$ creates numerical problems.

The parameter $\eta > 0$ should be selected as small as possible in order to reduce conservatism. This parameter is simply introduced in order to guarantee that $\eta + \max\limits_{\tau \in [iT,(i+1)T]} (|g(y(\tau), z(\tau))|)$ is always positive. If the quantity $\max\limits_{\tau \in [iT,(i+1)T]} (|g(y(\tau), z(\tau))|)$ is always positive, such as in the common case of an additive disturbance, then one can even select η to be equal to zero.

The implementation of the identifier (7.14), (7.15), (7.16), (7.17) can be performed by means of the following system of $(1+p)(2n+p)$ ODEs:

$$\begin{aligned}
&\dot{q}(t) = \Phi(y(t), z(t)) \\
&\dot{\zeta}(t) = f(y(t), z(t)) \\
&\dot{\chi}(t) = q(t) \\
&\dot{Q}(t) = (q(t))' \, q(t) \\
&\dot{G}(t) = (q(t))' \, (y(t) - \zeta(t)) \\
&\dot{\nu}(t) = \zeta(t) - y(t) \\
&\text{for } i \in \mathbb{N} \text{ and } t \in [iT, (i+1)T)
\end{aligned} \tag{7.23}$$

with

$$\begin{array}{l} q(iT) = \chi(iT) = 0 \in \mathbb{R}^{n\times p} \\ \zeta(iT) = \nu(iT) = 0 \in \mathbb{R}^{n} \\ Q(iT) - 0 \in \mathbb{R}^{p\times p} \\ G(iT) = 0 \in \mathbb{R}^{p} \\ \text{for } i \in \mathbb{N} \end{array} \tag{7.24}$$

and

$$\begin{array}{l} A(iT) = 2T \lim\limits_{t\to(i+1)T^-} (G(t)) + 2 \lim\limits_{t\to(i+1)T^-} \left((\chi(t))' \nu(t)\right) \\ B(iT) = 2T \lim\limits_{t\to(i+1)T^-} (Q(t)) - 2 \lim\limits_{t\to(i+1)T^-} \left((\chi(t))' \chi(t)\right) \\ \text{for } i \in \mathbb{N} \end{array} \tag{7.25}$$

Formulas (7.23), (7.24) give

$$\begin{array}{l} q(t) = \int_{iT}^{t} \Phi(y(\tau), z(\tau))d\tau \\ \zeta(t) = \int_{iT}^{t} f(y(\tau), z(\tau))d\tau \\ \chi(t) = \int_{iT}^{t} q(s)ds \\ Q(t) = \int_{iT}^{t} (q(s))' q(s)ds \\ G(t) = \int_{iT}^{t} (q(s))' (y(s) - \zeta(s))\, ds \\ \nu(t) = \int_{iT}^{t} (\zeta(s) - y(s))\, ds \\ \text{for } i \in \mathbb{N} \text{ and } t \in [iT, (i+1)T) \end{array} \tag{7.26}$$

and consequently, we can verify the following identities for the quantities a, b defined by (7.3):

$$\begin{array}{l} b(t, s) = q(t) - q(s) \\ a(t, s) = y(t) - y(s) - \zeta(t) + \zeta(s) \end{array} \tag{7.27}$$

for $t, s \in [iT, (i+1)T)$. The above identities in conjunction with (7.7) and (7.26) give us equations (7.25).

In practice less than $(1+p)(2n+p)$ ODEs are needed for the implementation of the identifier (7.14), (7.15), (7.16), (7.17), because one can take into account that the matrix $Q(t) \in \mathbb{R}^{p\times p}$ is symmetric.

The major disadvantage of the identifier (7.14), (7.15), (7.16), (7.17) is that we cannot be sure a priori that for any given $T, \eta, \sigma > 0$ there will be $i \in \mathbb{N}$ with $\frac{\sqrt{6\lambda_{\min}(B(iT))}}{\eta + \max\limits_{\tau\in[iT,(i+1)T]}(|g(y(\tau), z(\tau))|)} \geq \sigma T^2$. Therefore, we cannot be sure that identification will indeed take place. However, no identifier in adaptive control can guarantee that identification will indeed take place *without a persistence of excitation assumption*, which is hardly ever a priori verifiable. Moreover, no identifier in adaptive control *without a persistence of excitation assumption* can inform us *if and when* identification has occurred. Finally, no identifier in adaptive control can guarantee a robustness property like (7.21).

7.3 ▪ A Scalar Example

We study again the scalar example that was studied in Chapter 4 under DADS control, namely, the closed-loop system:

$$\dot{y} = \theta y - \left((1 + \exp(z)) \frac{1 + y^2}{4a} + c + \frac{1}{2c} (1 + \exp(z))^2 \right) y + d \tag{7.28}$$

$$\dot{z} = \Gamma \exp(-z) \left(\frac{1}{2} y^2 - \varepsilon \right)^+ \tag{7.29}$$

where $y, z, d, \theta \in \mathbb{R}$ and $c, a, \Gamma, \varepsilon > 0$ are constants. System (7.28), (7.29) corresponds to the case (7.1) with $n = p = 1$, $\Theta = \mathbb{R}$, $f(y,z) = -\left((1+\exp(z))\frac{1+y^2}{4a} + c + \frac{1}{2c}(1+\exp(z))^2\right)y$, $\Phi(y,z) = y$, and $g(y,z) = 1$. The system was studied for all cases that were shown in the simulation section (Section 4.10) in Chapter 4. We implemented the robust identifier (7.14), (7.15), (7.16), (7.17) with $T = 0.01$ and $\eta = 0$. Moreover, equations (7.26) were implemented in a sampled-data fashion by exploiting the trapezoid rule (see [131]) for an integer $N \geq 1$:

$$
\begin{aligned}
&q(iT+j\delta) \\
&= \tfrac{\delta}{2}\textstyle\sum_{l=0}^{j-1}\left(\Phi\left(y\left(iT+l\delta\right),z\left(iT+l\delta\right)\right)+\Phi\left(y\left(iT+(l+1)\delta\right),z\left(iT+(l+1)\delta\right)\right)\right)\\
&\zeta(iT+j\delta)\\
&= \tfrac{\delta}{2}\textstyle\sum_{l=0}^{j-1}\left(f\left(y\left(iT+l\delta\right),z\left(iT+l\delta\right)\right)+f\left(y\left(iT+(l+1)\delta\right),z\left(iT+(l+1)\delta\right)\right)\right)\\
&\chi(iT+j\delta) = \tfrac{\delta}{2}\textstyle\sum_{l=0}^{j-1}\left(q\left(iT+l\delta\right)+q\left(iT+(l+1)\delta\right)\right)\\
&Q(iT+j\delta) = \tfrac{\delta}{2}\textstyle\sum_{l=0}^{j-1}\left(q'\left(iT+l\delta\right)q\left(iT+l\delta\right)+q'\left(iT+(l+1)\delta\right)q\left(iT+(l+1)\delta\right)\right)\\
&G(iT+j\delta) = \tfrac{\delta}{2}\textstyle\sum_{l=0}^{j-1}q'\left(iT+l\delta\right)\left(y\left(iT+l\delta\right)-\zeta\left(iT+l\delta\right)\right)\\
&\qquad +\tfrac{\delta}{2}\textstyle\sum_{l=0}^{j-1}q'\left(iT+(l+1)\delta\right)\left(y\left(iT+(l+1)\delta\right)-\zeta\left(iT+(l+1)\delta\right)\right)\\
&\nu(iT+j\delta) = \tfrac{\delta}{2}\textstyle\sum_{l=0}^{j-1}\left(\zeta\left(iT+l\delta\right)-y\left(iT+l\delta\right)+\zeta\left(iT+(l+1)\delta\right)-y\left(iT+(l+1)\delta\right)\right)\\
&\text{for } i\in\mathbb{N} \text{ and } j=1,\dots,N-1
\end{aligned}
\tag{7.30}
$$

$$
\begin{aligned}
&\lim_{t\to(i+1)T^-}(q(t))\\
&= \tfrac{\delta}{2}\textstyle\sum_{l=0}^{N-1}\left(\Phi\left(y\left(iT+l\delta\right),z\left(iT+l\delta\right)\right)+\Phi\left(y\left(iT+(l+1)\delta\right),z\left(iT+(l+1)\delta\right)\right)\right)\\
&\lim_{t\to(i+1)T^-}(\zeta(t))\\
&= \tfrac{\delta}{2}\textstyle\sum_{l=0}^{N-1}\left(f\left(y\left(iT+l\delta\right),z\left(iT+l\delta\right)\right)+f\left(y\left(iT+(l+1)\delta\right),z\left(iT+(l+1)\delta\right)\right)\right)\\
&\lim_{t\to(i+1)T^-}(\chi(t)) = \tfrac{\delta}{2}\textstyle\sum_{l=0}^{N-1}\left(q\left(iT+l\delta\right)+q\left(iT+(l+1)\delta\right)\right)\\
&\lim_{t\to(i+1)T^-}(Q(t))\\
&= \tfrac{\delta}{2}\textstyle\sum_{l=0}^{N-1}\left(q'\left(iT+l\delta\right)q\left(iT+l\delta\right)+q'\left(iT+(l+1)\delta\right)q\left(iT+(l+1)\delta\right)\right)\\
&\lim_{t\to(i+1)T^-}(G(t)) = \tfrac{\delta}{2}\textstyle\sum_{l=0}^{N-1}q'\left(iT+l\delta\right)\left(y\left(iT+l\delta\right)-\zeta\left(iT+l\delta\right)\right)\\
&\qquad +\tfrac{\delta}{2}\textstyle\sum_{l=0}^{N-1}q'\left(iT+(l+1)\delta\right)\left(y\left(iT+(l+1)\delta\right)-\zeta\left(iT+(l+1)\delta\right)\right)\\
&\lim_{t\to(i+1)T^-}(\nu(t))\\
&= \tfrac{\delta}{2}\textstyle\sum_{l=0}^{N-1}\left(\zeta\left(iT+l\delta\right)-y\left(iT+l\delta\right)+\zeta\left(iT+(l+1)\delta\right)-y\left(iT+(l+1)\delta\right)\right)\\
&\text{for } i\in\mathbb{N}
\end{aligned}
\tag{7.31}
$$

where $\delta = T/N$. We used $N = 10$ in our simulations.

It should be noted that formulas (7.30), (7.31) are not exact but numerical approximations. Therefore, it is expected that formulas (7.30), (7.31) will introduce an additional component in the estimation error: the component of the discretization error.

The initial condition for the parameter estimate was $\hat{\theta}_0 = 0$. When we used a large threshold $\sigma = 1$, for the robust identifier (7.14), (7.15), (7.16), (7.17), then no identification occurred. On the other hand, when we used $\sigma = 0.1$ for the robust identifier (7.14), (7.15), (7.16), (7.17), then identification occurred. The results are shown in Figure 7.1, Figure 7.2, Figure 7.3, and Figure 7.4. It is clearly seen that in all cases identification occurred in the interval $t \in [0, 0.2]$. In the disturbance-free case the resulting small estimation error is due to the discretization, i.e., due to the use of formulas (7.30), (7.31). The estimated asymptotic gain $\frac{T^2}{\lim\limits_{i\to+\infty}(\psi_i)}$ in all cases was less than $2 < 1/\sigma = 10$, while the actual estimation error was less than 27%.

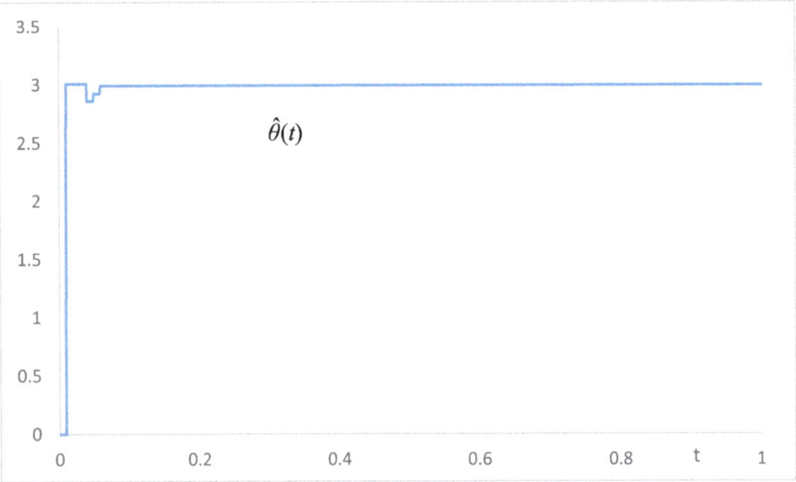

Figure 7.1. *The evolution of $\hat{\theta}$ for system* (7.28), (7.29) *with* $d \equiv 0$, $\theta = 3$, $c = a = 1, \Gamma = 100, \varepsilon = 0.00125$ *and* $y(0) = 0.5$, $z(0) = -\ln(10)$. *Limit value for $\hat{\theta}(t)$: 2.9926. Estimated asymptotic gain: 1.9576.*

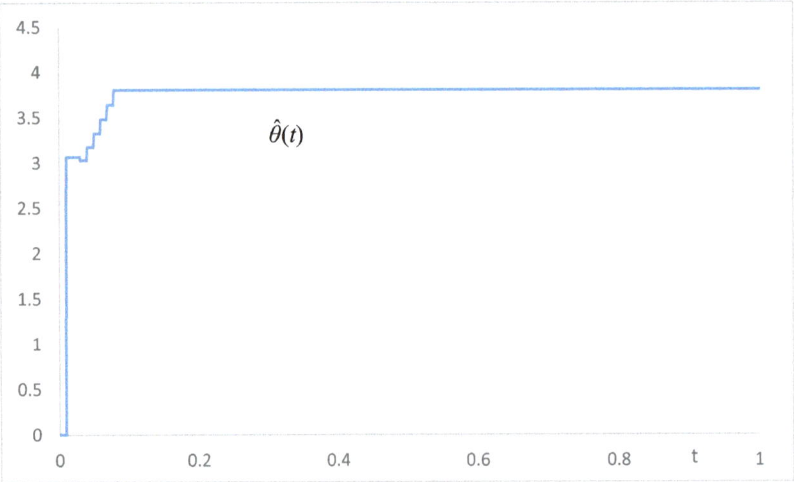

Figure 7.2. *The evolution of $\hat{\theta}$ for system* (7.28), (7.29) *with* $d = 2\sin(\pi t)$, $\theta = 3$, $c = a = 1, \Gamma = 100, \varepsilon = 0.00125$ *and* $y(0) = 0.5$, $z(0) = -\ln(10)$. *Limit value for $\hat{\theta}(t)$: 3.8044. Estimated asymptotic gain: 1.9143.*

7.4 ▪ Notes and comments

1. The results of the chapter can give the impression that the identifier can be directly used in a certainty equivalence fashion (as in [70], [72]) in order to design a robust adaptive controller. Unfortunately, this is not the case. There are two obstacles for achieving robustness by using the proposed identifier in a certainty equivalence way. The first obstacle is that our least-squares identifier does not always achieve identification. As the scalar example showed, there are cases in which no identification occurs (due to lack of excitation). The second obstacle is the fact that our identifier does not guarantee a certain level of parameter

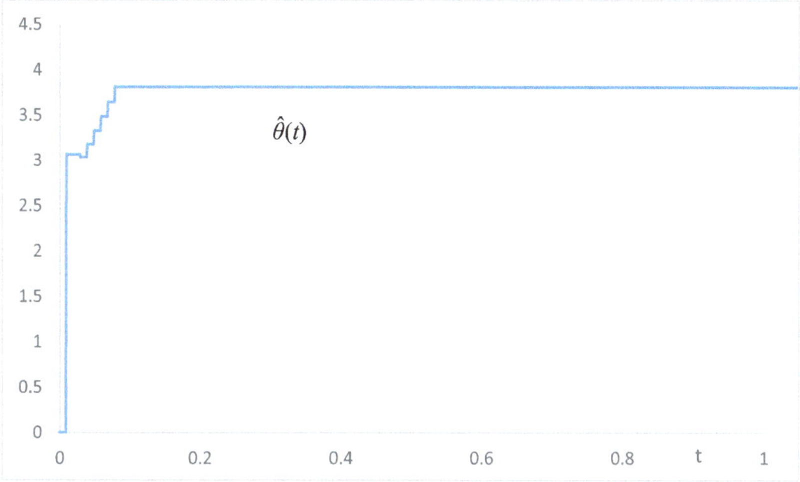

Figure 7.3. *The evolution of* $\hat{\theta}$ *for system* (7.28), (7.29) *with* $d = 2\sin(\pi t)$, $\theta = 3$, $c = a = 1, \Gamma = 100, \varepsilon = 0.00005$ *and* $y(0) = 0.5$, $z(0) = -\ln(10)$. *Limit value for* $\hat{\theta}(t)$*: 3.8096. Estimated asymptotic gain: 1.9155.*

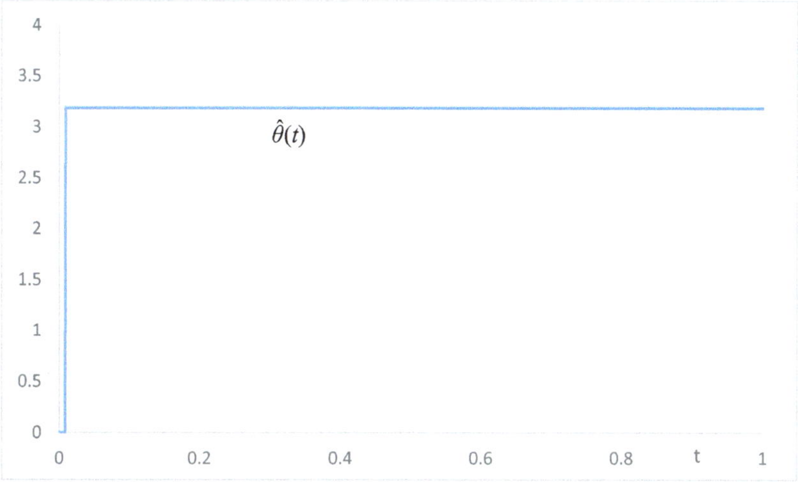

Figure 7.4. *The evolution of* $\hat{\theta}$ *for system* (7.28), (7.29) *with* $d = 2\sin(\pi t)$, $\theta = 3$, $c = a = 1, \Gamma = 1000, \varepsilon = 0.00005$ *and* $y(0) = 0.5$, $z(0) = -\ln(10)$. *Limit value for* $\hat{\theta}(t)$*: 3.1844. Estimated asymptotic gain: 1.9749.*

estimation error; it guarantees a certain level of the gain of the disturbance to the parameter estimation error. Inequality (7.21) shows that the parameter estimation error is bounded by $\sigma^{-1}\|d\|_\infty$ (when identification occurs). While the threshold $\sigma > 0$ is assignable, the norm of the disturbance is not known and may be (arbitrarily) large.

2. It is clear that the proposed identifier has similar features with the identifier proposed in [70], [72]. However, there are important differences between the proposed identifier and the results in [70], [72]: in [70], [72] the identifier works in an event-triggered fashion (with regulation-based event triggering) while here the identifier works in a sampled-data

fashion where the size of the sampling period (T) is a design parameter that can be selected optimally.

3. The implementation of the identifier (7.14), (7.15), (7.16), (7.17) by means of (7.24), (7.25), and (7.30) gives an important advantage: the values of the state $y(t)$ and input $z(t)$ need not be measured continuously; instead we can have sampled measurements only. Of course, as we pointed out above, this kind of implementation introduces an additional source of error (discretization error) whose level can be controlled by reducing appropriately the discretization time step $\delta = T/N$ (sampling period) at least in the case that the state, the input, the disturbance, and the unknown parameter take values in a compact set.

4. When the number of unknown parameters p and the number of states n are large numbers or when the set $\Theta \subseteq \mathbb{R}^p$ is very complicated, then the computations involved in (7.14), (7.15), (7.16), (7.17), (7.24), (7.25) and (7.30), (7.31) (and more specifically the computation of the vector $P_\Theta\left((B(iT))^{-1} A(iT)\right)$ and the computation of an accurate lower bound for the quantity $\frac{\sqrt{6\lambda_{\min}(B(iT))}}{\eta + \max\limits_{\tau\in[iT,(i+1)T]}(|g(y(\tau),z(\tau))|)}$) can be very demanding. Therefore, another possibility arises: the use of machine learning tools for the computation of $\hat{\theta}((i+1)T)$ and ψ_{i+1}. This possibility is discussed in Chapter 9.

5. The idea of using batches of data and not all the available data is a key point that enables us to allow disturbances in $L^\infty(\mathbb{R}_+)$. Indeed, the L^2 norm of the disturbance arises naturally in the equations obtained by a least-squares analysis (recall (7.10) and (7.3)). However, due to the fact that there is a finite sampling period we can obtain (7.12), which involves the sup norm of the disturbance. The authors of [31] pioneered the study of robust identification for nonlinear ODEs with disturbances. The identifier in [31] uses all available data up to time t, and employs an advanced form of a forgetting factor for robustification and, consequently, can handle disturbances in $L^2(\mathbb{R}_+)$ but not in $L^\infty(\mathbb{R}_+)$.

Chapter 8

The Aircraft Wing Rock Instability

8.1 ▪ Introduction

The aircraft wing rock phenomenon is an oscillation of the roll angle in an aircraft and is essentially a self-sustained oscillatory motion around the longitudinal aircraft axis. Usually, it occurs in high-performance aircraft when flying at a high angle of attack, at subsonic speeds. The wing rock is a substantially nonlinear and undesired phenomenon, which may become dangerous when the amplitude of the oscillation is large. Engineers have worked on the suppression of the wing rock phenomenon without degrading the maneuverability of the aircraft.

A variety of mathematical models and controllers have been proposed for the description/suppression of wing rock (see [4], [5], [19], [22], [32], [37], [38], [39], [49], [63], [87], [99], [128], [163]). The models are usually empirical, containing a number of parameters that depend on many aircraft characteristics. Here we use the mathematical model that is based on wind tunnel tests at NASA Langley Research Center and was also used in [87]. The model is given by the equations

$$\begin{aligned}
&\dot{x}_1 = x_2 \\
&\dot{x}_2 = \theta_1 x_1 + \theta_2 x_2 + \theta_3 |x_1| x_2 + \theta_4 |x_2| x_2 + (\beta + \theta_5) x_3 + d_1 \\
&\dot{x}_3 = -\mu x_3 + \mu u + d_2 \\
&x = (x_1, x_2, x_3)' \in \mathbb{R}^3, u \in \mathbb{R}, d = (d_1, d_2)' \in \mathbb{R}^2, \theta = (\theta_1, \theta_2, \theta_3, \theta_4, \theta_5)' \in \Theta
\end{aligned} \tag{8.1}$$

where $x \in \mathbb{R}^3$ is the state, $d \in \mathbb{R}^2$ is the disturbance, $u \in \mathbb{R}$ is the control input, $\theta = (\theta_1, \theta_2, \theta_3, \theta_4, \theta_5)'$ is the vector of unknown constant parameters that takes values in the closed convex (but unbounded) set $\Theta = \mathbb{R}^4 \times \mathbb{R}_+$ (i.e., θ_5 must be non-negative), and $\beta, \mu > 0$ are known constants. In this model, x_1 is the roll angle, x_2 is the roll rate, and x_3 is the aileron deflection angle. The aileron dynamics is first-order. The disturbances d_1, d_2 are used for the description of unknown forces acting on the wings and the aileron, respectively. The control objective is the suppression of the wing rock, which mathematically is equivalent to the stabilization of the output

$$y = (x_1, x_2)' \in \mathbb{R}^2 \tag{8.2}$$

for the control system (8.1).

All simulations shown in this chapter were performed by Dr. Dionysios Theodosis. The authors would like to thank Dr. Dionysios Theodosis for his help.

8.2 ▪ Construction of the Adaptive Controller

Since system (8.1) does not have smooth functions in its right-hand sides, it is not covered by Theorem 6.1. However, as remarked in Chapter 6, the backstepping Lemma 6.1 can also be proved for locally Lipschitz functions in the right-hand sides. Therefore, in order to design a DADS controller for system (8.1) we apply backstepping by using the same steps as in the proof of Lemma 6.1 in Chapter 6. We next present the steps of the design in detail for pedagogical purposes. All relations that follow are valid for all $x \in \mathbb{R}^3$, $d \in \mathbb{R}^2$, $z \in \mathbb{R}$, $\theta \in \Theta$ but we omit referring to this fact for simplicity.

1st Step: We start with the scalar subsystem

$$\dot{x}_1 = x_2 \tag{8.3}$$

and the function

$$V_1(x,z) = \frac{1}{2}x_1^2 \tag{8.4}$$

The feedback law

$$k_1(x,z) = -cx_1 \tag{8.5}$$

where $c > 0$ is a constant, gives the equation

$$\dot{V}_1(x,z,d) = -cx_1^2 + x_1(x_2 - k_1(x,z)) \tag{8.6}$$

The first step is complete.

2nd Step: We consider the planar subsystem

$$\begin{aligned} &\dot{x}_1 = x_2 \\ &\dot{x}_2 = \theta_1 x_1 + \theta_2 x_2 + \theta_3 |x_1| x_2 + \theta_4 |x_2| x_2 + (\beta + \theta_5) x_3 + d_1 \end{aligned} \tag{8.7}$$

and the function

$$V_2(x,z) = V_1(x,z) + \frac{1}{2}(x_2 - k_1(x,z))^2 = \frac{1}{2}x_1^2 + \frac{1}{2}(x_2 + cx_1)^2 \tag{8.8}$$

We use the feedback law

$$k_2(x,z) = -\beta^{-1}M(x,z)(x_2 + cx_1) \tag{8.9}$$

where $M(x,z)$ is a positive smooth function to be selected.

Using (8.6), (8.7), (8.8), and (8.9) we get

$$\begin{aligned} &\dot{V}_2(x,z,d) = -cx_1^2 + (x_1 + cx_2)(x_2 + cx_1) \\ &+ (x_2 + cx_1)(\theta_1 x_1 + \theta_2 x_2 + \theta_3 |x_1| x_2 + \theta_4 |x_2| x_2) \\ &+ (x_2 + cx_1) d_1 - (\beta + \theta_5)\beta^{-1}M(x,z)(x_2 + cx_1)^2 \\ &+ (x_2 + cx_1)(\beta + \theta_5)(x_3 - k_2(x,z)) \end{aligned} \tag{8.10}$$

The inequality

$$(x_2 + cx_1) d_1 \le \frac{a}{1 + \exp(z)} d_1^2 + \frac{1 + \exp(z)}{4a}(x_2 + cx_1)^2 \tag{8.11}$$

where $a > 0$ is a constant and the fact that

$$\beta + \theta_5 \ge \beta > 0, M(x,z) > 0, |\theta_i| \le |\theta|, i = 1, ..., 4 \tag{8.12}$$

in conjunction with (8.10) provide the following estimate:

$$\begin{aligned}&\dot{V}_2(x,z,d) \le -cx_1^2 - \left(c^2-1\right)x_1\left(x_2+cx_1\right)\\&+|\theta|\left|x_2+cx_1\right|\left(|x_1|+|x_2|+|x_1||x_2|+|x_2|^2\right)\\&+\tfrac{a}{1+\exp(z)}d_1^2+\left(c+\tfrac{1+\exp(z)}{4a}-M(x,z)\right)\left(x_2+cx_1\right)^2\\&+\left(x_2+cx_1\right)\left(\beta+\theta_5\right)\left(x_3-k_2(x,z)\right)\end{aligned} \tag{8.13}$$

The inequalities

$$-\left(c^2-1\right)x_1\left(x_2+cx_1\right) \le \frac{c}{6}x_1^2+\frac{3}{2c}\left(c^2-1\right)^2\left(x_2+cx_1\right)^2 \tag{8.14}$$

$$|\theta| \le 1+\exp(z)+\left(|\theta|-1-\exp(z)\right)^+ \tag{8.15}$$

combined with (8.13) allow us to derive the following inequality:

$$\begin{aligned}&\dot{V}_2(x,z,d) \le -\tfrac{5c}{6}x_1^2+\left(1+\exp(z)\right)\left|x_2+cx_1\right|\left(|x_1|+|x_2|+|x_1||x_2|+|x_2|^2\right)\\&+\left(|\theta|-1-\exp(z)\right)^+\left|x_2+cx_1\right|\left(|x_1|+|x_2|+|x_1||x_2|+|x_2|^2\right)\\&+\tfrac{a}{1+\exp(z)}d_1^2+\left(c+\tfrac{3}{2c}\left(c^2-1\right)^2+\tfrac{1+\exp(z)}{4a}-M(x,z)\right)\left(x_2+cx_1\right)^2\\&+\left(x_2+cx_1\right)\left(\beta+\theta_5\right)\left(x_3-k_2(x,z)\right)\end{aligned} \tag{8.16}$$

Using the inequalities

$$\begin{aligned}&\left(|\theta|-1-\exp(z)\right)^+\left|x_2+cx_1\right||x_1|\\&\le \tfrac{a}{4(1+\exp(z))}\left(\left(|\theta|-1-\exp(z)\right)^+\right)^2+\tfrac{1+\exp(z)}{a}x_1^2\left(x_2+cx_1\right)^2\\&\left(|\theta|-1-\exp(z)\right)^+\left|x_2+cx_1\right||x_2|\\&\le \tfrac{a}{4(1+\exp(z))}\left(\left(|\theta|-1-\exp(z)\right)^+\right)^2+\tfrac{1+\exp(z)}{a}x_2^2\left(x_2+cx_1\right)^2\\&\left(|\theta|-1-\exp(z)\right)^+\left|x_2+cx_1\right||x_1||x_2|\\&\le \tfrac{a}{4(1+\exp(z))}\left(\left(|\theta|-1-\exp(z)\right)^+\right)^2+\tfrac{1+\exp(z)}{a}x_1^2x_2^2\left(x_2+cx_1\right)^2\\&\left(|\theta|-1-\exp(z)\right)^+\left|x_2+cx_1\right||x_2|^2\\&\le \tfrac{a}{4(1+\exp(z))}\left(\left(|\theta|-1-\exp(z)\right)^+\right)^2+\tfrac{1+\exp(z)}{a}x_2^4\left(x_2+cx_1\right)^2\end{aligned} \tag{8.17}$$

as well as the fact that (triangle inequality)

$$|x_2| \le \left|x_2+cx_1\right|+c|x_1| \tag{8.18}$$

we get from (8.16)

$$\begin{aligned}&\dot{V}_2(x,z,d) \le -\tfrac{5c}{6}x_1^2+a\tfrac{d_1^2+\left(\left(|\theta|-1-\exp(z)\right)^+\right)^2}{1+\exp(z)}\\&+\tfrac{1+\exp(z)}{a}\left(x_1^2+x_2^2\right)\left(1+x_2^2\right)\left(x_2+cx_1\right)^2\\&+\left(1+\exp(z)\right)\left(1+|x_2|\right)\left(x_2+cx_1\right)^2\\&+\left(1+\exp(z)\right)\left|x_2+cx_1\right|(1+c)\left(1+|x_2|\right)|x_1|\\&+\left(c+\tfrac{3}{2c}\left(c^2-1\right)^2+\tfrac{1+\exp(z)}{4a}-M(x,z)\right)\left(x_2+cx_1\right)^2\\&+\left(x_2+cx_1\right)\left(\beta+\theta_5\right)\left(x_3-k_2(x,z)\right)\end{aligned} \tag{8.19}$$

Using the inequalities

$$1+|x_2| \leq 2\left(1+x_2^2\right) \tag{8.20}$$

$$\begin{aligned}&(1+\exp(z))\,|x_2+cx_1|\,(1+c)\,(1+|x_2|)\,|x_1|\\&\leq \tfrac{c}{6}x_1^2+\tfrac{3(1+c)^2}{2c}\,(1+\exp(z))^2\,(1+|x_2|)^2\,(x_2+cx_1)^2\end{aligned} \tag{8.21}$$

we obtain from (8.19)

$$\begin{aligned}&\dot V_2(x,z,d) \leq -\tfrac{2c}{3}x_1^2 - c\,(x_2+cx_1)^2 + a\tfrac{d_1^2+\left((|\theta|-1-\exp(z))^+\right)^2}{1+\exp(z)}\\&+\tfrac{1+\exp(z)}{a}\left(x_1^2+x_2^2+2a\right)\left(1+x_2^2\right)(x_2+cx_1)^2\\&+\tfrac{6(1+c)^2}{c}\,(1+\exp(z))^2\left(1+x_2^2\right)^2(x_2+cx_1)^2\\&+\left(2c+\tfrac{3}{2c}\left(c^2-1\right)^2+\tfrac{1+\exp(z)}{4a}-M(x,z)\right)(x_2+cx_1)^2\\&+(x_2+cx_1)\,(\beta+\theta_5)\,(x_3-k_2(x,z))\end{aligned} \tag{8.22}$$

By selecting

$$\begin{aligned}&M(x,z)=2c+\tfrac{3}{2c}\left(c^2-1\right)^2+\tfrac{1+\exp(z)}{4a}\\&+\tfrac{1+\exp(z)}{a}\left(x_1^2+x_2^2+2a\right)\left(1+x_2^2\right)\\&+\tfrac{6(1+c)^2}{c}\,(1+\exp(z))^2\left(1+x_2^2\right)^2\end{aligned} \tag{8.23}$$

we obtain from (8.22)

$$\begin{aligned}&\dot V_2(x,z,d) \leq -\tfrac{2c}{3}x_1^2 - c\,(x_2+cx_1)^2 + a\tfrac{d_1^2+\left((|\theta|-1-\exp(z))^+\right)^2}{1+\exp(z)}\\&\quad+(x_2+cx_1)\,(\beta+\theta_5)\,(x_3-k_2(x,z))\end{aligned} \tag{8.24}$$

3rd Step: This step is the final step where we consider the whole system (8.1). Since it is the final step, we do not have to care for the regularity properties of the feedback law; a locally Lipschitz feedback law will suffice. Therefore, in this step we do not proceed with domination of all terms (as in the previous step) but (for simplicity) we use cancelation of all terms that can be canceled. We use the function

$$\begin{aligned}V(x,z)&=V_2(x,z)+\tfrac{1}{2}\left(x_3-k_2(x,z)\right)^2\\&=\tfrac{1}{2}x_1^2+\tfrac{1}{2}(x_2+cx_1)^2+\tfrac{1}{2}\left(x_3-k_2(x,z)\right)^2\end{aligned} \tag{8.25}$$

Exploiting (8.1), (8.24) and (8.25) as well as the update law

$$\dot z=\Gamma\exp(-z)\,(V(x,z)-\varepsilon)^+ \tag{8.26}$$

where $\Gamma,\varepsilon>0$ are constants, we obtain

$$\begin{aligned}&\dot V(x,z,d) \leq -\tfrac{2c}{3}x_1^2 - c\,(x_2+cx_1)^2\\&+a\tfrac{d_1^2+\left((|\theta|-1-\exp(z))^+\right)^2}{1+\exp(z)} + (x_3-k_2(x,z))\,(\beta\,(x_2+cx_1)-\mu x_3+\mu u)\\&-(x_3-k_2(x,z))\left(\tfrac{\partial k_2}{\partial x_1}(x,z)x_2+\tfrac{\partial k_2}{\partial x_2}(x,z)\beta x_3+\tfrac{\partial k_2}{\partial z}(x,z)\Gamma\exp(-z)\,(V(x,z)-\varepsilon)^+\right)\\&-(x_3-k_2(x,z))\tfrac{\partial k_2}{\partial x_2}(x,z)\,(\theta_1x_1+\theta_2x_2+\theta_3|x_1|\,x_2+\theta_4|x_2|\,x_2+\theta_5x_3)\\&+(x_3-k_2(x,z))\,(x_2+cx_1)\,\theta_5-(x_3-k_2(x,z))\tfrac{\partial k_2}{\partial x_2}(x,z)d_1+(x_3-k_2(x,z))\,d_2\end{aligned} \tag{8.27}$$

In what follows we use the notation

$$X=x_3-k_2(x,z) \tag{8.28}$$

We use the feedback law

$$\begin{aligned}u &= \mu^{-1}\tfrac{\partial k_2}{\partial x_1}(x,z)x_2 + \mu^{-1}\beta\tfrac{\partial k_2}{\partial x_2}(x,z)x_3 + x_3 - \mu^{-1}P(x,z)X \\ &+\mu^{-1}\Gamma\tfrac{\partial k_2}{\partial z}(x,z)\exp(-z)\left(V(x,z)-\varepsilon\right)^+ - \mu^{-1}\beta\left(x_2+cx_1\right)\end{aligned} \tag{8.29}$$

where $P(x,z)$ is a positive smooth function to be selected. Inserting (8.29) in (8.27), we get

$$\begin{aligned}&\dot V(x,z,d) \le -\tfrac{2c}{3}x_1^2 - c\left(x_2+cx_1\right)^2 \\ &+a\tfrac{d_1^2+\left(\left(|\theta|-1-\exp(z)\right)^+\right)^2}{1+\exp(z)} - P(x,z)X^2 \\ &+|\theta|\,|X|\,|x_2+cx_1| \\ &+|\theta|\,|X|\left|\tfrac{\partial k_2}{\partial x_2}(x,z)\right|\left(|x_1|+|x_2|+|x_1|\,|x_2|+|x_2|^2+|x_3|\right) \\ &-X\tfrac{\partial k_2}{\partial x_2}(x,z)d_1 + Xd_2\end{aligned} \tag{8.30}$$

Using the inequalities

$$-X\frac{\partial k_2}{\partial x_2}(x,z)d_1 \le a\frac{d_1^2}{1+\exp(z)} + \frac{1+\exp(z)}{4a}\left(\frac{\partial k_2}{\partial x_2}(x,z)\right)^2 X^2 \tag{8.31}$$

$$Xd_2 \le a\frac{d_2^2}{1+\exp(z)} + \frac{1+\exp(z)}{4a}X^2 \tag{8.32}$$

$$|\theta| \le 1+\exp(z)+\left(|\theta|-1-\exp(z)\right)^+ \tag{8.33}$$

in conjunction with (8.30) we get

$$\begin{aligned}&\dot V(x,z,d) \le -\tfrac{2c}{3}x_1^2 - c\left(x_2+cx_1\right)^2 + a\tfrac{2d_1^2+d_2^2+\left(\left(|\theta|-1-\exp(z)\right)^+\right)^2}{1+\exp(z)} \\ &+\left(|\theta|-1-\exp(z)\right)^+|X|\,|x_2+cx_1| + \left(1+\exp(z)\right)|X|\,|x_2+cx_1| \\ &+\left(|\theta|-1-\exp(z)\right)^+|X|\left|\tfrac{\partial k_2}{\partial x_2}(x,z)\right|\left(|x_1|+|x_2|+|x_1|\,|x_2|+|x_2|^2+|x_3|\right) \\ &+\left(1+\exp(z)\right)|X|\left|\tfrac{\partial k_2}{\partial x_2}(x,z)\right|\left(|x_1|+|x_2|+|x_1|\,|x_2|+|x_2|^2+|x_3|\right) \\ &+\left(\tfrac{1+\exp(z)}{4a}\left(\left(\tfrac{\partial k_2}{\partial x_2}(x,z)\right)^2+1\right)-P(x,z)\right)X^2\end{aligned} \tag{8.34}$$

The inequalities

$$\begin{aligned}&\left(|\theta|-1-\exp(z)\right)^+|X|\,|x_2+cx_1| \\ &\le \tfrac{a\left(\left(|\theta|-1-\exp(z)\right)^+\right)^2}{2(1+\exp(z))} + \tfrac{1+\exp(z)}{2a}\left(x_2+cx_1\right)^2X^2\end{aligned} \tag{8.35}$$

$$\left(1+\exp(z)\right)|X|\,|x_2+cx_1| \le \frac{c}{6}\left(x_2+cx_1\right)^2 + \frac{3}{2c}\left(1+\exp(z)\right)^2X^2 \tag{8.36}$$

$$\begin{aligned}&\left(|\theta|-1-\exp(z)\right)^+|X|\left|\tfrac{\partial k_2}{\partial x_2}(x,z)\right|\left(|x_1|+|x_2|+|x_1|\,|x_2|+|x_2|^2+|x_3|\right) \\ &\le \tfrac{a\left(\left(|\theta|-1-\exp(z)\right)^+\right)^2}{2(1+\exp(z))} \\ &+\tfrac{1+\exp(z)}{2a}\left(\tfrac{\partial k_2}{\partial x_2}(x,z)\right)^2\left(|x_1|+|x_2|+|x_1|\,|x_2|+|x_2|^2+|x_3|\right)^2X^2\end{aligned} \tag{8.37}$$

allow us to derive the following estimate from (8.34):

$$\begin{aligned}&\dot V(x,z,d) \le -\tfrac{2c}{3}x_1^2 - \tfrac{5c}{6}\left(x_2+cx_1\right)^2 + a\tfrac{2d_1^2+d_2^2+2\left(\left(|\theta|-1-\exp(z)\right)^+\right)^2}{1+\exp(z)} \\ &+\tfrac{1+\exp(z)}{2a}\left(\left(x_2+cx_1\right)^2+\left(\tfrac{\partial k_2}{\partial x_2}(x,z)\right)^2\left(|x_1|+|x_2|+|x_1|\,|x_2|+|x_2|^2+|x_3|\right)^2\right)X^2 \\ &+\left(1+\exp(z)\right)|X|\left|\tfrac{\partial k_2}{\partial x_2}(x,z)\right|\left(|x_1|+|x_2|+|x_1|\,|x_2|+|x_2|^2+|x_3|\right) \\ &+\left(\tfrac{1+\exp(z)}{4a}\left(\left(\tfrac{\partial k_2}{\partial x_2}(x,z)\right)^2+1\right)+\tfrac{3(1+\exp(z))^2}{2c}-P(x,z)\right)X^2\end{aligned} \tag{8.38}$$

The triangle inequality in conjunction with (8.9) gives

$$|x_2| \leq |x_2 + cx_1| + c\,|x_1| \tag{8.39}$$

$$|x_3| \leq |X| + \beta^{-1}M(x,z)\,|x_2 + cx_1| \tag{8.40}$$

Using the above inequalities and (8.38) we obtain

$$\begin{aligned}
\dot{V}(x,z,d) &\leq -\tfrac{2c}{3}x_1^2 - \tfrac{5c}{6}(x_2+cx_1)^2 - cX^2 + a\tfrac{2d_1^2+d_2^2+2\left((|\theta|-1-\exp(z))^+\right)^2}{1+\exp(z)} \\
&+\tfrac{1+\exp(z)}{2a}\left((x_2+cx_1)^2 + \left(\tfrac{\partial k_2}{\partial x_2}(x,z)\right)^2\left(|x_1|+|x_2|+|x_1|\,|x_2|+|x_2|^2+|x_3|\right)^2\right)X^2 \\
&+(1+\exp(z))\,|X|\left|\tfrac{\partial k_2}{\partial x_2}(x,z)\right|(1+c)\,(1+|x_2|)\,|x_1| \\
&+(1+\exp(z))\,|X|\left|\tfrac{\partial k_2}{\partial x_2}(x,z)\right|\left(1+|x_2|+\beta^{-1}M(x,z)\right)|x_2+cx_1| \\
&+\left(\tfrac{1+\exp(z)}{4a}\left(\left(\tfrac{\partial k_2}{\partial x_2}(x,z)\right)^2 + 4a\left|\tfrac{\partial k_2}{\partial x_2}(x,z)\right|+1\right) + \tfrac{3(1+\exp(z))^2}{2c} + c - P(x,z)\right)X^2
\end{aligned} \tag{8.41}$$

Finally, we exploit the inequalities

$$\begin{aligned}
&(1+\exp(z))\,|X|\left|\tfrac{\partial k_2}{\partial x_2}(x,z)\right|(1+c)\,(1+|x_2|)\,|x_1| \\
&\leq \tfrac{c}{6}x_1^2 + \tfrac{3(1+c)^2}{2c}(1+\exp(z))^2\left(\tfrac{\partial k_2}{\partial x_2}(x,z)\right)^2(1+|x_2|)^2X^2
\end{aligned} \tag{8.42}$$

$$\begin{aligned}
&(1+\exp(z))\,|X|\left|\tfrac{\partial k_2}{\partial x_2}(x,z)\right|\left(1+|x_2|+\beta^{-1}M(x,z)\right)|x_2+cx_1| \\
&\leq \tfrac{c}{6}(x_2+cx_1)^2 + \tfrac{3}{2c}(1+\exp(z))^2\left(\tfrac{\partial k_2}{\partial x_2}(x,z)\right)^2\left(1+|x_2|+\beta^{-1}M(x,z)\right)^2X^2
\end{aligned} \tag{8.43}$$

in order to get from (8.41) the estimate

$$\begin{aligned}
\dot{V}(x,z,d) &\leq -\tfrac{c}{2}x_1^2 - \tfrac{2c}{3}(x_2+cx_1)^2 - cX^2 + a\tfrac{2d_1^2+d_2^2+2\left((|\theta|-1-\exp(z))^+\right)^2}{1+\exp(z)} \\
&+\tfrac{1+\exp(z)}{2a}\left((x_2+cx_1)^2 + \left(\tfrac{\partial k_2}{\partial x_2}(x,z)\right)^2\left(|x_1|+|x_2|+|x_1|\,|x_2|+|x_2|^2+|x_3|\right)^2\right)X^2 \\
&+\tfrac{3(1+\exp(z))^2}{2c}\left(\tfrac{\partial k_2}{\partial x_2}(x,z)\right)^2\left((1+c)^2(1+|x_2|)^2 + \left(1+|x_2|+\beta^{-1}M(x,z)\right)^2\right)X^2 \\
&+\left(\tfrac{1+\exp(z)}{4a}\left(\left(\tfrac{\partial k_2}{\partial x_2}(x,z)\right)^2 + 4a\left|\tfrac{\partial k_2}{\partial x_2}(x,z)\right|+1\right) + \tfrac{3(1+\exp(z))^2}{2c} + c - P(x,z)\right)X^2
\end{aligned} \tag{8.44}$$

By selecting

$$\begin{aligned}
P(x,z) &= \tfrac{1+\exp(z)}{4a}\left(\left(\tfrac{\partial k_2}{\partial x_2}(x,z)\right)^2 + 4a\left|\tfrac{\partial k_2}{\partial x_2}(x,z)\right|+1\right) + \tfrac{3(1+\exp(z))^2}{2c} + c \\
&+\tfrac{3(1+\exp(z))^2}{2c}\left(\tfrac{\partial k_2}{\partial x_2}(x,z)\right)^2\left((1+c)^2(1+|x_2|)^2 + \left(1+|x_2|+\beta^{-1}M(x,z)\right)^2\right) \\
&+\tfrac{1+\exp(z)}{2a}\left((x_2+cx_1)^2 + \left(\tfrac{\partial k_2}{\partial x_2}(x,z)\right)^2\left(|x_1|+|x_2|+|x_1|\,|x_2|+|x_2|^2+|x_3|\right)^2\right)
\end{aligned} \tag{8.45}$$

we obtain from (8.44) the following estimate:

$$\dot{V}(x,z,d) \leq -\frac{c}{2}\left(x_1^2 + (x_2+cx_1)^2 + X^2\right) + 2a\frac{|d|^2 + \left((|\theta|-1-\exp(z))^+\right)^2}{1+\exp(z)} \tag{8.46}$$

The above estimate combined with definition (8.25) gives

$$\dot{V}(x,z,d) \leq -cV(x,z) + 2a\frac{|d|^2 + \left((|\theta| - 1 - \exp(z))^+\right)^2}{1+\exp(z)} \tag{8.47}$$

Thus, the feedback design process is complete.

For the reader's convenience, Table 8.1 provides all the formulas that constitute the DADS controller for (8.1).

Table 8.1. *The DADS controller for* (8.1).

Constants	$c, a, \Gamma, \varepsilon > 0$
Update law	$\dot{z} = \Gamma\exp(-z)\left(V(x,z) - \varepsilon\right)^+$
Control	$u = \mu^{-1}\frac{\partial k_2}{\partial x_1}(x,z)x_2 + \mu^{-1}\beta\frac{\partial k_2}{\partial x_2}(x,z)x_3 - \mu^{-1}P(x,z)\left(x_3 - k_2(x,z)\right)$ $+x_3 + \mu^{-1}\Gamma\frac{\partial k_2}{\partial z}(x,z)\exp(-z)\left(V(x,z)-\varepsilon\right)^+ - \mu^{-1}\beta\left(x_2 + cx_1\right)$
Auxiliary functions	$P(x,z) := \frac{\rho}{4a}\left(\left(\frac{\partial k_2}{\partial x_2}(x,z)\right)^2 + 4a\left\lvert\frac{\partial k_2}{\partial x_2}(x,z)\right\rvert + 1\right) + \frac{3\rho^2}{2c} + c$ $+\frac{3\rho^2}{2c}\left(\frac{\partial k_2}{\partial x_2}(x,z)\right)^2\left((1+c)^2(1+\lvert x_2\rvert)^2 + \left(1+\lvert x_2\rvert + \beta^{-1}M(x,z)\right)^2\right)$ $+\frac{\rho}{2a}\left(\frac{\partial k_2}{\partial x_2}(x,z)\right)^2\left(\lvert x_1\rvert + \lvert x_2\rvert + \lvert x_1\rvert\lvert x_2\rvert + \lvert x_2\rvert^2 + \lvert x_3\rvert\right)^2$ $+\frac{\rho}{2a}\left(x_2 + cx_1\right)^2$ $k_2(x,z) := -\beta^{-1}M(x,z)\left(x_2 + cx_1\right)$ $M(x,z) := \frac{\rho}{a}\left(x_1^2 + x_2^2 + 2a\right)\left(1 + x_2^2\right) + \frac{\rho}{4a} + 2c$ $+\frac{3}{2c}\left(c^2-1\right)^2 + \frac{6(1+c)^2}{c}\rho^2\left(1+x_2^2\right)^2$ $V(x,z) := \frac{1}{2}x_1^2 + \frac{1}{2}\left(x_2 + cx_1\right)^2 + \frac{1}{2}\left(x_3 - k_2(x,z)\right)^2$ $\rho := 1 + \exp(z)$

It should be noticed that the function V defined by (8.25) satisfies the estimate:

$$x_1^2 + x_2^2 \leq 2\left(c^2+2\right)V(x,z) \tag{8.48}$$

Applying Theorem 2.8 with $b = 1$, $\kappa(s) = \lambda(s) = s$ and using (8.48) we can guarantee that there exists a function $B \in C^0\left(\mathbb{R}^3 \times \mathbb{R} \times \mathbb{R}_+ \times \mathbb{R}_+\right)$ such that for every $(x_0, z_0) \in \mathbb{R}^3 \times \mathbb{R}$, $d \in L^\infty\left(\mathbb{R}_+;\mathbb{R}^2\right)$, $\theta \in L^\infty\left(\mathbb{R}_+;\Theta\right)$ the unique solution of (8.1), (8.26), (8.29) with $(z(0), z(0)) = (x_0, z_0)$ exists and satisfies the following estimates for all $t \geq 0$:

$$\limsup_{t\to+\infty}\left(\lvert x_1(t)\rvert^2 + \lvert x_2(t)\rvert^2\right) \leq 2\left(c^2+2\right)\min\left(\varepsilon, 2a\frac{l_d^2 + \left((l_\theta - 1 - \exp(l_z))^+\right)^2}{c\left(1+\exp(l_z)\right)}\right) \tag{8.49}$$

$$\limsup_{t\to+\infty}\left(V(x(t),z(t))\right)\leq \min\left(\varepsilon, 2a\frac{l_d^2+\left(\left(l_\theta-1-\exp(l_z)\right)^+\right)^2}{c\left(1+\exp(l_z)\right)}\right) \tag{8.50}$$

$$V\left(x(t),z(t)\right)\leq \exp\left(-ct\right)V\left(x_0,z_0\right)+2a\frac{\|d\|_\infty^2+\left(\left(\|\theta\|_\infty-1-\exp(z_0)\right)^+\right)^2}{c\left(1+\exp(z_0)\right)} \tag{8.51}$$

$$|x(t)|\leq B\left(x_0,z_0,\|d\|_\infty,\|\theta\|_\infty\right) \tag{8.52}$$

$$z_0\leq z(t)\leq l_z\leq B\left(x_0,z_0,\|d\|_\infty,\|\theta\|_\infty\right) \tag{8.53}$$

where $l_d = \limsup_{t\to+\infty}(|d(t)|)$, $l_\theta = \limsup_{t\to+\infty}(|\theta(t)|)$, $l_z = \lim_{t\to+\infty}(z(t))$. Moreover, if $l_\theta \leq 1+\exp(l_z)$ and $l_d = 0$ then $\lim_{t\to+\infty}(x(t)) = 0$.

Let us now state the stability properties of the closed-loop system (8.1), (8.26), (8.29) with output $y = (x_1, x_2)$ when $\theta \in \mathbb{R}^5$ is constant:

- estimates (8.52), (8.52) guarantee the p-UBIBS property,
- estimate (8.49) guarantees the zero p-OAG property with asymptotic residual constant equal to $\sqrt{2\left(c^2+2\right)\varepsilon}$,
- estimates (8.48), (8.51) guarantee the p-IOS property with residual constant

$$\alpha = \left(|\theta|-1\right)^+\sqrt{4a\left(c^2+2\right)/c} \tag{8.54}$$

  and gain function

$$\gamma(s) = s\sqrt{4a\left(c^2+2\right)/c} \tag{8.55}$$

When $\theta \in \mathbb{R}^5$ is constant and in the disturbance-free case (i.e., when $d \equiv 0$), the stability properties of the closed-loop system (8.1), (8.26), (8.29) with output $y = (x_1, x_2)$ can also be stated:

- estimates (8.52), (8.53) guarantee the Lagrange stability property,
- the set $\left\{(0,z)' \in \mathbb{R}^3, z \in \mathbb{R}\right\}$ is invariant,
- estimate (8.51) guarantees the p-UGAOS property and residual constant

$$\alpha = \left(|\theta|-1\right)^+\sqrt{4a\left(c^2+2\right)/c} \tag{8.56}$$

- estimate (8.49) shows that system (8.1), (8.26), (8.29) with output $y = (x_1, x_2)$ is p-GOA with asymptotic residual constant equal to $\tilde{\alpha} = \sqrt{2\left(c^2+2\right)}\min\left(\sqrt{\varepsilon}, \left(|\theta|-1\right)^+\sqrt{2a/c}\right) \leq \alpha$,
- at least one of the following holds: (i) $\lim_{t\to+\infty}(x(t)) = 0$; (ii) $|\theta| > 1$ and $z(t) \leq \lim_{s\to+\infty}(z(s)) < \ln\left(|\theta|-1\right)$, for all $t \geq 0$.

8.3 ▪ The Identifier

For system (8.1) we can apply the robust identifier that was described in the previous chapter. It suffices to use the differential equation for x_2 which is the only differential equation that involves the unknown parameters. We first define the projection operator on the set $\Theta = \mathbb{R}^4 \times \mathbb{R}_+$ by means of the formula

$$P_\Theta\left((\theta_1, ..., \theta_4, \theta_5)'\right) = \left(\theta_1, ..., \theta_4, (\theta_5)^+\right)' \tag{8.57}$$

The identifier is given for arbitrary constants $T, \sigma > 0$ by the following equations for arbitrary initial condition $\hat{\theta}(0) = \hat{\theta}_0 \in \Theta$:

$$\psi_0 = \sigma T^2 \tag{8.58}$$

$$\psi_{i+1} = \max\left(\psi_i, \sqrt{6\lambda_{\min}\left(B(iT)\right)}\right) \text{ for } i \in \mathbb{N} \tag{8.59}$$

$$\hat{\theta}((i+1)T) = P_\Theta\left(\left(B\left(iT\right)\right)^{-1} A\left(iT\right)\right) \text{ for } i \in \mathbb{N} \text{ and } \sqrt{6\lambda_{\min}\left(B(iT)\right)} \geq \psi_i \tag{8.60}$$

$$\hat{\theta}((i+1)T) = \hat{\theta}(iT) \text{ for } i \in \mathbb{N} \text{ and } \sqrt{6\lambda_{\min}\left(B(iT)\right)} < \psi_i \tag{8.61}$$

$$\hat{\theta}(t) = \hat{\theta}(iT) \text{ for } i \in \mathbb{N} \text{ and } t \in [iT, (i+1)T) \tag{8.62}$$

$$\begin{aligned}
&\dot{q}(t) = \left[\begin{array}{ccccc} x_1(t) & x_2(t) & |x_1(t)|\, x_2(t) & |x_2(t)|\, x_2(t) & x_3(t) \end{array}\right] \in \mathbb{R}^{1\times 5}\\
&\dot{\zeta}(t) = \beta x_3(t) \in \mathbb{R}\\
&\dot{\chi}(t) = q(t) \in \mathbb{R}^{1\times 5}\\
&\dot{Q}(t) = (q(t))'\, q(t) \in \mathbb{R}^{5\times 5}\\
&\dot{G}(t) = (q(t))'\, (x_2(t) - \zeta(t)) \in \mathbb{R}^{5\times 1}\\
&\dot{\nu}(t) = \zeta(t) - x_2(t) \in \mathbb{R}\\
&\text{for } i \in \mathbb{N} \text{ and } t \in [iT, (i+1)T)
\end{aligned} \tag{8.63}$$

$$\begin{aligned}
&q(iT) = \chi(iT) = 0 \in \mathbb{R}^{1\times 5}\\
&\zeta(iT) = \nu(iT) = 0 \in \mathbb{R}\\
&Q(iT) = 0 \in \mathbb{R}^{5\times 5}\\
&G(iT) = 0 \in \mathbb{R}^5\\
&\text{for } i \in \mathbb{N}
\end{aligned} \tag{8.64}$$

$$\begin{aligned}
&A\left(iT\right) = 2T \lim_{t\to(i+1)T^-}\left(G(t)\right) + 2\left(\lim_{t\to(i+1)T^-}\left(\chi(t)\right)\right)' \lim_{t\to(i+1)T^-}\left(\nu(t)\right)\\
&B\left(iT\right) = 2T \lim_{t\to(i+1)T^-}\left(Q(t)\right) - 2\left(\lim_{t\to(i+1)T^-}\left(\chi(t)\right)\right)' \lim_{t\to(i+1)T^-}\left(\chi(t)\right)\\
&\text{for } i \in \mathbb{N}
\end{aligned} \tag{8.65}$$

Equation (8.63) gives 42 ODEs but some of them are redundant since we can exploit the fact that the matrix $Q(t) \in \mathbb{R}^{5\times 5}$ is symmetric and, consequently, we need only 15 of the 25 ODEs $\dot{Q}(t) = (q(t))'\, q(t) \in \mathbb{R}^{5\times 5}$. Therefore, the identifier (8.58)–(8.65) actually involves 32 ODEs.

The major computational difficulty that is met in equations (8.58)–(8.65) is the computation of $\lambda_{\min}\left(B(iT)\right)$, i.e., the smallest eigenvalue of the symmetric, positive semidefinite matrix $B(iT) \in \mathbb{R}^{5\times 5}$. Its computation can only be performed numerically.

Another issue that should be noted for the identifier (8.58)–(8.65) is the fact that only the wing disturbance d_1 appears in the differential equation for x_2 that was used for the derivation of the identifier. Therefore, the identifier (8.58)–(8.65) is not affected by the aileron disturbance

d_2, and the parameter estimate

$$\left|\hat{\theta}(t) - \theta\right| \leq \frac{1}{\sigma} \|d_1\|_\infty \tag{8.66}$$

ultimately holds if there exists sufficient excitation (i.e., if $\sqrt{6\lambda_{\min}(B(iT))} \geq \sigma T^2$ for some $i \in \mathbb{N}$).

8.4 ▪ Simulations

The following parameter values were used for the simulations (the same as in [87]):

$$\begin{aligned}
\mu &= 15 \\
\beta &= 0.5 \\
\theta_1 &= -26.67 \\
\theta_2 &= 0.76485 \\
\theta_3 &= -2.9225 \\
\theta_4 &= 0 \\
\theta_5 &= 1
\end{aligned} \tag{8.67}$$

The open-loop system (8.1) (i.e., system (8.1) with $u \equiv 0$, $d \equiv 0$) with the above parameter values admits periodic solutions that represent the wing rock oscillation. The wing rock phenomenon is visible in Figure 8.1, Figure 8.2, Figure 8.3, and Figure 8.4, where it is shown that the solution is attracted by a limit cycle.

When the identifier (8.59), (8.60), (8.61), (8.62) is applied to the open-loop system (8.1) then identification is achieved with $\sigma = 10^{-6}$. Larger values for σ do not allow identification. This small value of σ shows that there is a robustness problem for the identification in the open-loop system (8.1). The problem is the fact that the state component x_3 very rapidly to zero; in fact, we get from (8.1) that $x_3(t) = \exp(-\mu t)\, x_3(0)$ for $t \geq 0$. Consequently, it is difficult to obtain a robust estimation of θ_5 (which is multiplied by x_3 in the differential equation for x_2). However, even for such a small value of σ the identifier manages to make exact identification in finite-time for the open-loop system (8.1). This is shown in Figure 8.5 and Figure 8.6.

We have applied the DADS controller (8.26), (8.29) with the following parameters:

$$\begin{aligned}
\Gamma &= 100 \\
a &= c = 1 \\
\varepsilon &= 0.01
\end{aligned} \tag{8.68}$$

The DADS controller achieves convergence of the output $y = (x_1, x_2)' \in \mathbb{R}^2$ to zero even in the presence of persistent disturbances. This is shown in Figure 8.7 and Figure 8.8 for the disturbance-free case and in Figure 8.9 and Figure 8.10 for the case where persistent disturbances are present.

The success of the DADS controller has an unfortunate result. Due to the fast convergence of the state components x_1, x_2 for the closed-loop system (8.1) with (8.26), (8.29), we do not have enough excitation to achieve identification. The identifier (8.59), (8.60), (8.61), (8.62) fails to produce parameter estimates even for very small values of σ. Larger initial values for x_1, x_2 can provide sufficient excitation for an initial transient period but this is not meaningful from a modeling point of view (recall that x_1 is the roll angle, x_2 is the roll rate, and x_3 is the aileron deflection angle).

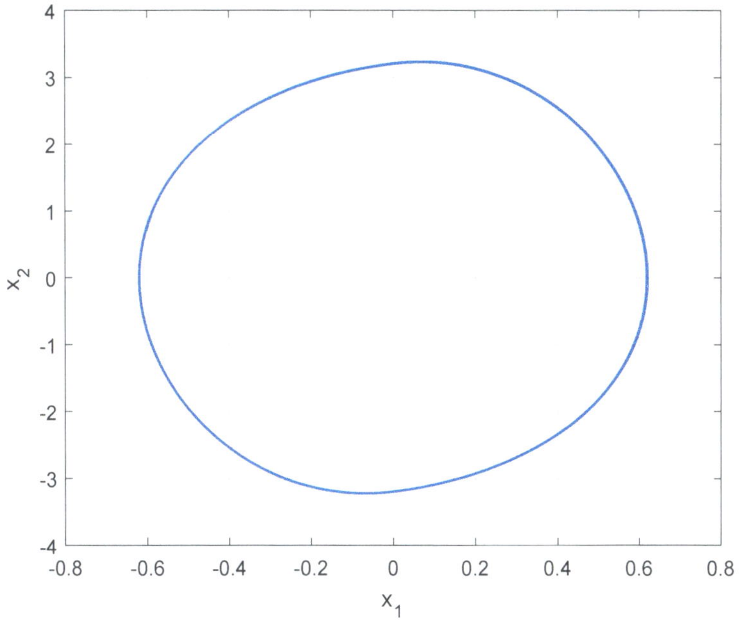

Figure 8.1. *The solution of the open-loop system* (8.1) *with* $d \equiv 0$*, initial condition* $x_1(0) = -0.6$*,* $x_2(0) = 0.7$*,* $x_3(0) = 0.8$ *in the* x_1*-*x_2 *plane.*

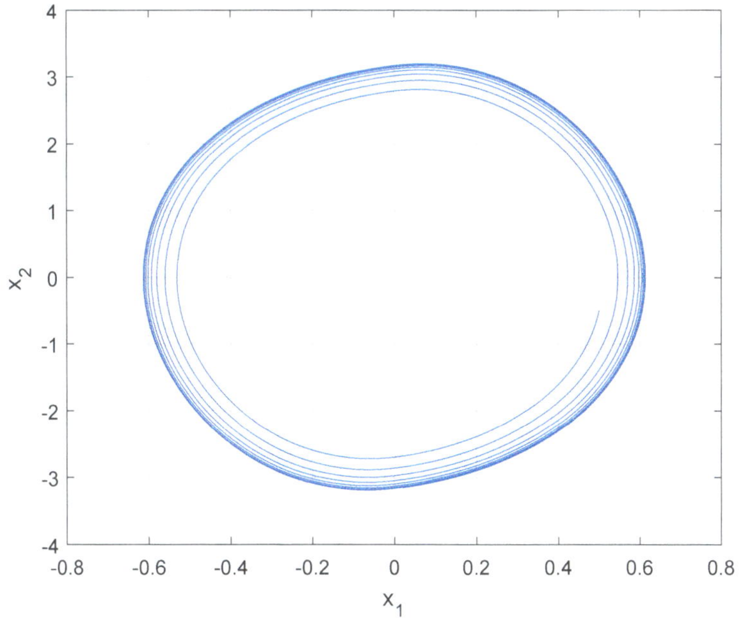

Figure 8.2. *The solution of the open-loop system* (8.1) *with* $d \equiv 0$*, initial condition* $x_1(0) = 0.5$*,* $x_2(0) = -0.5$*,* $x_3(0) = -0.3$ *in the* x_1*-*x_2 *plane.*

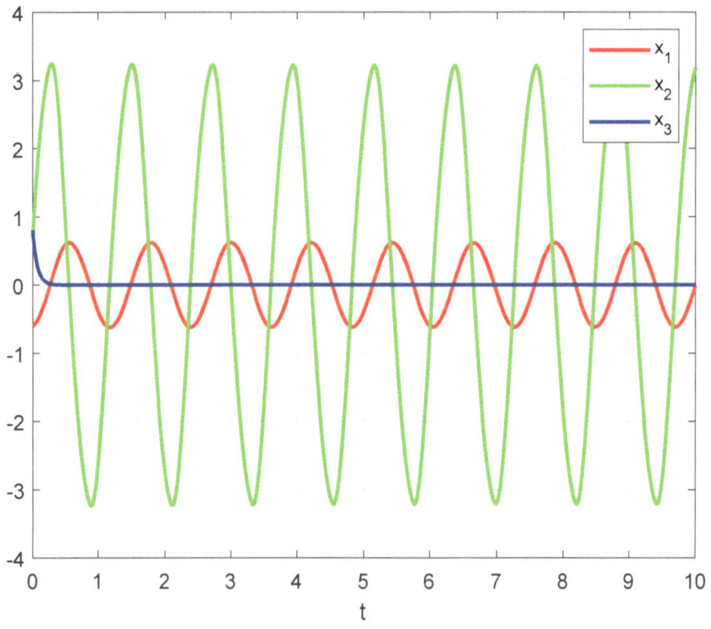

Figure 8.3. *The solution of the open-loop system* (8.1) *with* $d \equiv 0$, *initial condition* $x_1(0) = -0.6$, $x_2(0) = 0.7$, $x_3(0) = 0.8$.

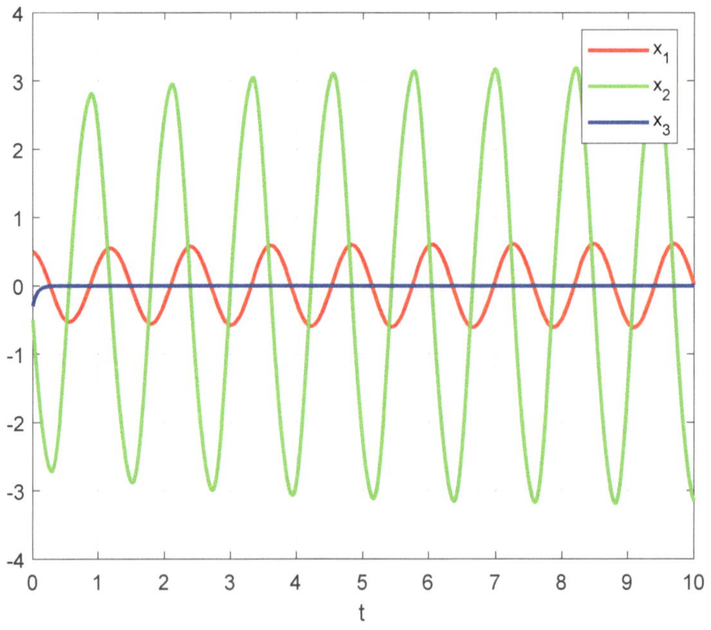

Figure 8.4. *The solution of the open-loop system* (8.1) *with* $d \equiv 0$, *initial condition* $x_1(0) = 0.5$, $x_2(0) = -0.5$, $x_3(0) = -0.3$.

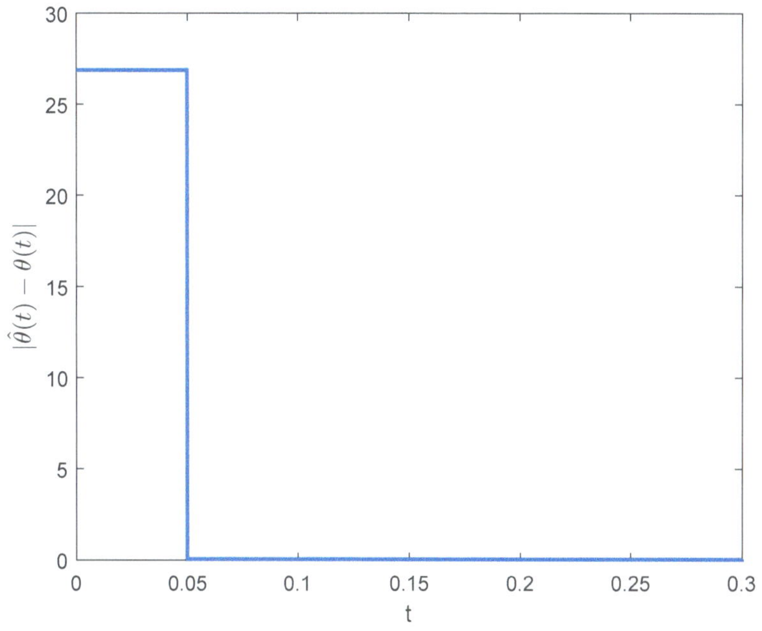

Figure 8.5. *The behavior of the identifier* (8.59), (8.60), (8.61), (8.62) *with* $T = 0.05$, $\sigma = 10^{-6}$ *for the open-loop system* (8.1) *with* $d \equiv 0$, *initial condition* $x_1(0) = -0.6$, $x_2(0) = 0.7$, $x_3(0) = 0.8$, $\hat{\theta}(0) = 0 \in \mathbb{R}^5$. *Estimated asymptotic gain:* $\frac{T^2}{\lim\limits_{i\to+\infty}(\psi_i)} = 8.33 \cdot 10^5$.

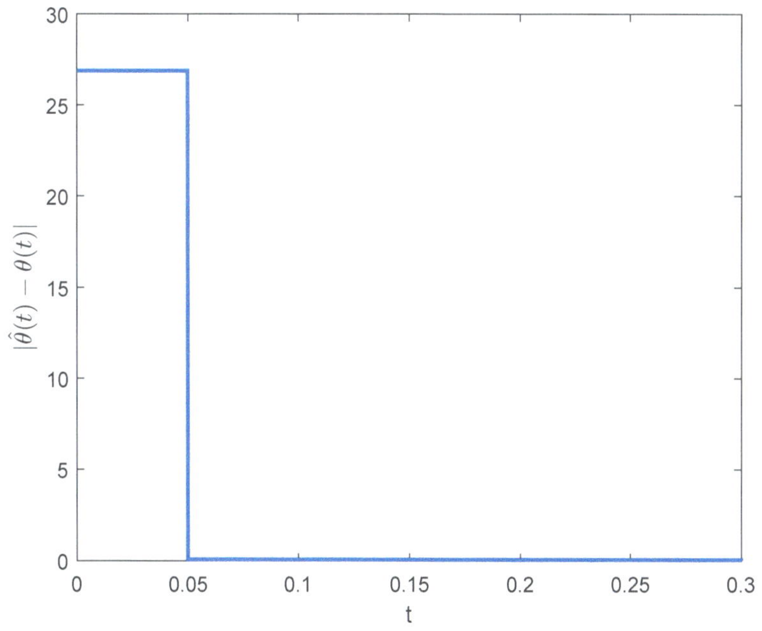

Figure 8.6. *The behavior of the identifier* (8.59), (8.60), (8.61), (8.62) *with* $T = 0.05$, $\sigma = 10^{-6}$ *for the open-loop system* (8.1) *with* $d \equiv 0$, *initial condition* $x_1(0) = 0.5$, $x_2(0) = -0.5$, $x_3(0) = -0.3$, $\hat{\theta}(0) = 0 \in \mathbb{R}^5$. *Estimated asymptotic gain:* $\frac{T^2}{\lim\limits_{i\to+\infty}(\psi_i)} = 8.3 \cdot 10^5$.

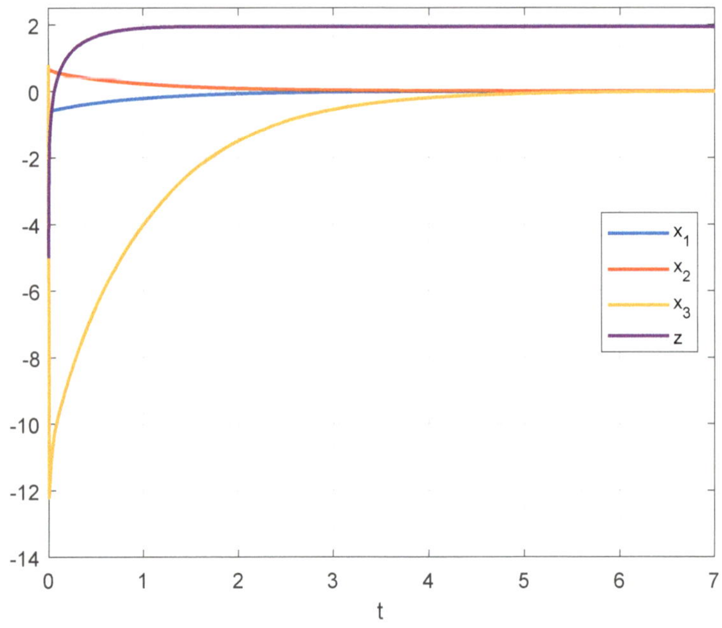

Figure 8.7. *The solution of the closed-loop system* (8.1) *with DADS controller* (8.26), (8.29), $d \equiv 0$, *initial condition* $x_1(0) = -0.6$, $x_2(0) = 0.7$, $x_3(0) = 0.8$, $z(0) = -5$.

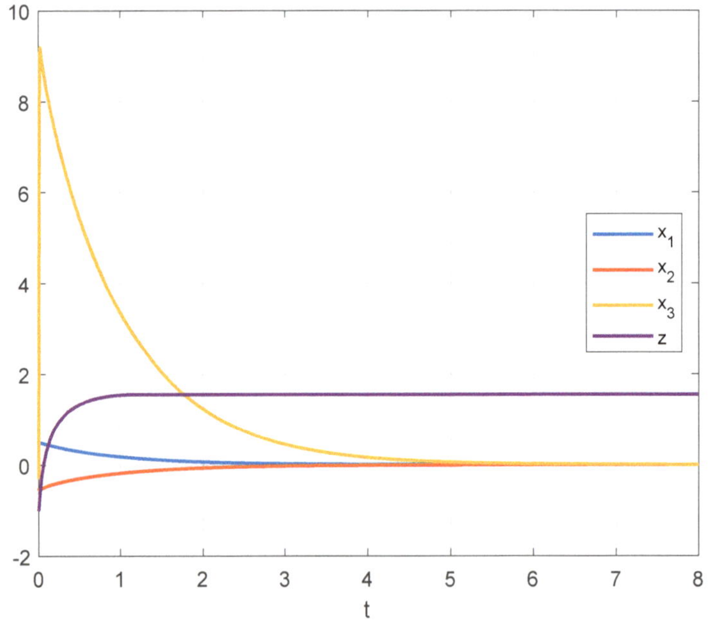

Figure 8.8. *The solution of the closed-loop system* (8.1) *with DADS controller* (8.26), (8.29), $d \equiv 0$, *initial condition* $x_1(0) = 0.5$, $x_2(0) = -0.5$, $x_3(0) = -0.3$, $z(0) = -1$.

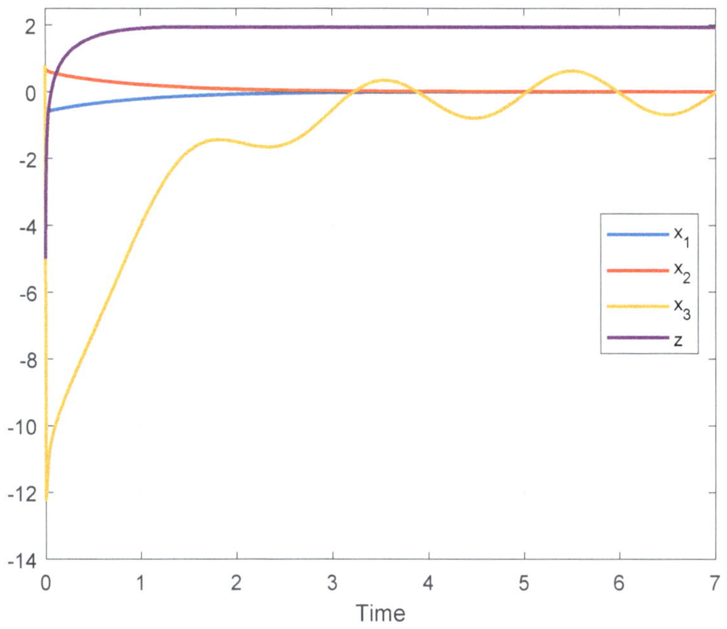

Figure 8.9. *The solution of the closed-loop system* (8.1) *with DADS controller* (8.26), (8.29), $d_1(t) = \sin(\pi t)$, $d_2(t) = 2\sin(\pi t + 0.5)$, *initial condition* $x_1(0) = -0.6$, $x_2(0) = 0.7$, $x_3(0) = 0.8$, $z(0) = -5$.

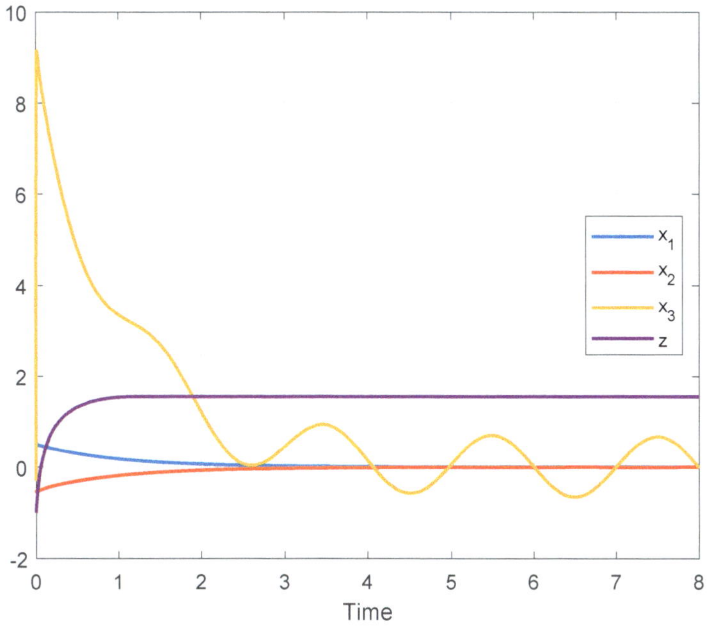

Figure 8.10. *The solution of the closed-loop system* (8.1) *with DADS controller* (8.26), (8.29), $d_1(t) = \sin(\pi t)$, $d_2(t) = 2\sin(\pi t + 0.5)$, *initial condition* $x_1(0) = 0.5$, $x_2(0) = -0.5$, $x_3(0) = -0.3$, $z(0) = -1$.

Chapter 9
Epilogue

9.1 ▪ Introduction

Future research can produce novel results that extend the DADS control scheme to interesting cases and problems. Two possible extensions of the DADS scheme are the following

- The extension to interconnected systems. To this purpose, small-gain results (like Theorem 2.7) can be very useful.
- The extension to systems where the whole state vector is not measured but only part of the components of the state are measured (a measured output). To this purpose, the DADS control scheme can be augmented by observers and filters (see Chapters 7–10 of [87]).

We are confident that some of the ideas presented in this book can give remarkable new results when applied to open/unsolved problems. In this chapter, we focus on two different issues that need further study:

- the effect of measurement errors, and
- the possible application of the DADS control scheme to *tracking* control problems for strict-feedback systems.

We next provide some preliminary ideas that will help the researcher to study these two issues.

9.2 ▪ The Effect of Measurement Errors

It is unfortunate, but this is always the case: when we apply a feedback law to a control system, we introduce a disturbance to the system. This type of disturbance is the so-called measurement error or "noise". This type of disturbance does not appear in the open-loop system and is only present in the closed-loop system, since what we actually measure is not the state $x \in \mathbb{R}^n$ of the system but

$$\hat{x} = x + e \tag{9.1}$$

where $e \in \mathbb{R}^n$ represents the measurement error or noise.

The effect of noise is difficult to analyze when the feedback law or the system is nonlinear. Very few robustness results are provided in the literature for the case where noise is present. An exception is the book [34], where the detrimental effect of noise is shown and a complete robustness analysis is provided for strict-feedback systems. A complete robustness analysis for the case where noise is present is out of the scope of the present book. However, here we would like to show the effect of measurement error to the simple scalar system that was studied in Chapter 4, namely the system

$$\begin{aligned} &\dot{x} = \theta x + u + d \\ &x \in \mathbb{R}, u \in \mathbb{R}, d \in \mathbb{R} \end{aligned} \tag{9.2}$$

where $\theta \in \mathbb{R}$ is the unknown constant parameter. The DADS controller derived in Chapter 4 is given by the equations

$$u = -\left((1+\exp(z))\frac{1+x^2}{4a} + c + \frac{1}{2c}(1+\exp(z))^2\right)x \tag{9.3}$$

$$\dot{z} = \Gamma \exp(-z)\left(\frac{1}{2}x^2 - \varepsilon\right)^+ \tag{9.4}$$

where $\varepsilon, \Gamma, c, a > 0$ are constants. It is clear that in the presence of noise the DADS controller becomes

$$u = -\left((1+\exp(z))\frac{1+(x+e)^2}{4a} + c + \frac{1}{2c}(1+\exp(z))^2\right)(x+e) \tag{9.5}$$

$$\dot{z} = \Gamma \exp(-z)\left(\frac{1}{2}(x+e)^2 - \varepsilon\right)^+ \tag{9.6}$$

The mathematical trick that one can apply for noise is to assume that $e : \mathbb{R}_+ \to \mathbb{R}$ is a bounded continuous signal with a measurable, essentially bounded derivative, i.e., $e \in W^{1,\infty}(\mathbb{R}_+)$, and not simply a function of class $L^\infty(\mathbb{R}_+)$. In this way, the closed-loop system (9.2) with (9.5), (9.6) under the transformation

$$y = x + e \tag{9.7}$$

is described by the equations

$$\dot{y} = \theta y - \left((1+\exp(z))\frac{1+y^2}{4a} + c + \frac{1}{2c}(1+\exp(z))^2\right)y + \tilde{d} \tag{9.8}$$

$$\dot{z} = \Gamma \exp(-z)\left(\frac{1}{2}y^2 - \varepsilon\right)^+ \tag{9.9}$$

where the new disturbance $\tilde{d}$ acting on the system (9.8), and including the noise and its derivative, is given by

$$\tilde{d} = d + \dot{e} - \theta e \tag{9.10}$$

System (9.8), (9.9) is exactly the closed-loop system that was studied in Chapter 4 with d replaced by $\tilde{d}$. Therefore, we can obtain exactly the same estimates, which hold for all $(y(0), z(0)) \in \mathbb{R}^2, t \geq 0, \tilde{d} \in L^\infty(\mathbb{R}_+)$ and are provided below:

$$|y(t)| \leq \exp(-ct/2)\,|y(0)| + \sqrt{\frac{2a}{c}}\left\|\tilde{d}\right\|_\infty + \sqrt{\frac{2a}{c}}\,(|\theta|-1)^+ \tag{9.11}$$

$$z(0) \leq z(t) \leq \Omega\left(\left\|\tilde{d}\right\|_\infty, |\theta|-1, |y(0)|, \exp(z(0))\right) \tag{9.12}$$

$$\limsup_{t\to+\infty}(|y(t)|) \leq \sqrt{2\varepsilon} \tag{9.13}$$

$$\limsup_{t\to+\infty}(|y(t)|) \le \sqrt{\frac{2a}{c\left(1+\exp\left(\lim\limits_{t\to+\infty}(z(t))\right)\right)}}\limsup_{t\to+\infty}\left(\left|\tilde{d}(t)\right|\right)$$
$$+\sqrt{\frac{2a}{c\left(1+\exp\left(\lim\limits_{t\to+\infty}(z(t))\right)\right)}}\left(|\theta|-1-\exp\left(\lim_{t\to+\infty}(z(t))\right)\right)^{+} \tag{9.14}$$

where $\Omega : \mathbb{R}_+ \times \mathbb{R} \times \mathbb{R}_+ \times (0,+\infty) \to \mathbb{R}$ is the continuous function defined in (4.34). Using (9.7) and (9.10) in conjunction with the triangle inequality, we obtain from (9.11), (9.12), (9.13), (9.14) the following estimates that hold for all $(x(0), z(0)) \in \mathbb{R}^2$, $t \ge 0$, $d \in L^\infty(\mathbb{R}_+)$, $e \in W^{1,\infty}(\mathbb{R}_+)$:

$$|x(t)| \le \exp(-ct/2)\,|x(0)| + \sqrt{\tfrac{2a}{c}}\left(\|d\|_\infty + \|\dot{e}\|_\infty + (|\theta|-1)^+\right)$$
$$+\left(1+|\theta|\sqrt{\tfrac{2a}{c}}+\exp(-ct/2)\right)\|e\|_\infty \tag{9.15}$$

$$z(0) \le z(t) \le \Omega\left(\|d\|_\infty + \|\dot{e}\|_\infty + |\theta|\,\|e\|_\infty,\, |\theta|-1,\, |x(0)| + \|e\|_\infty,\, \exp(z(0))\right) \tag{9.16}$$

$$\limsup_{t\to+\infty}(|x(t)|) \le \sqrt{2\varepsilon} + \limsup_{t\to+\infty}(|e(t)|) \tag{9.17}$$

$$\limsup_{t\to+\infty}(|x(t)|) \le \left(1+|\theta|\sqrt{\frac{2a}{c\left(1+\exp\left(\lim\limits_{t\to+\infty}(z(t))\right)\right)}}\right)\limsup_{t\to+\infty}(|e(t)|)$$
$$+\sqrt{\frac{2a}{c\left(1+\exp\left(\lim\limits_{t\to+\infty}(z(t))\right)\right)}}\limsup_{t\to+\infty}(|\dot{e}(t)|)$$
$$+\sqrt{\frac{2a}{c\left(1+\exp\left(\lim\limits_{t\to+\infty}(z(t))\right)\right)}}\left(\limsup_{t\to+\infty}(|d(t)|) + \left(|\theta|-1-\exp\left(\lim_{t\to+\infty}(z(t))\right)\right)^{+}\right) \tag{9.18}$$

Certain conclusions can be drawn from the estimates (9.15), (9.16), (9.17), (9.18):

1. no parameter drift, remarkably, arises in the presence of noise,
2. while suppression of the effect of d is achieved, suppression of the effect of noise is not achieved; estimate (9.17) shows that the state ultimately enters a neighborhood of radius $\sqrt{2\varepsilon} + \limsup\limits_{t\to+\infty}(|e(t)|)$.

We expect that the above conclusions are valid for all systems studied in the present book for noise of class $e \in W^{1,\infty}(\mathbb{R}_+;\mathbb{R}^n)$. However, the researcher needs to perform calculations for each specific system in order to check whether the above conclusions are valid for more general systems.

The effect of noise is shown in Figure 9.1 while the combined effect of the noise and d is shown in Figure 9.2.

A word of caution is needed here: the analysis that we presented above can be applied to many control systems and gives important results. However, the assumption that noise is a "smooth" disturbance is not realistic. Measurement noise is typically in a higher range of the frequency spectrum relative to the system's dynamics, and is for that reason often modeled as stochastic. Hence, a concerning aspect of the above analysis is that the obtained estimates depend (in general) on $\|\dot{e}\|_\infty$. While $\|e\|_\infty$ can be guaranteed to be small if the measurement device provides reliable measurements, $\|\dot{e}\|_\infty$ is not necessarily small. Consequently, a more detailed analysis (like the one provided in [34]) is needed in order to quantify the effect of the noise when noise is simply a function of class $L^\infty(\mathbb{R}_+;\mathbb{R}^n)$.

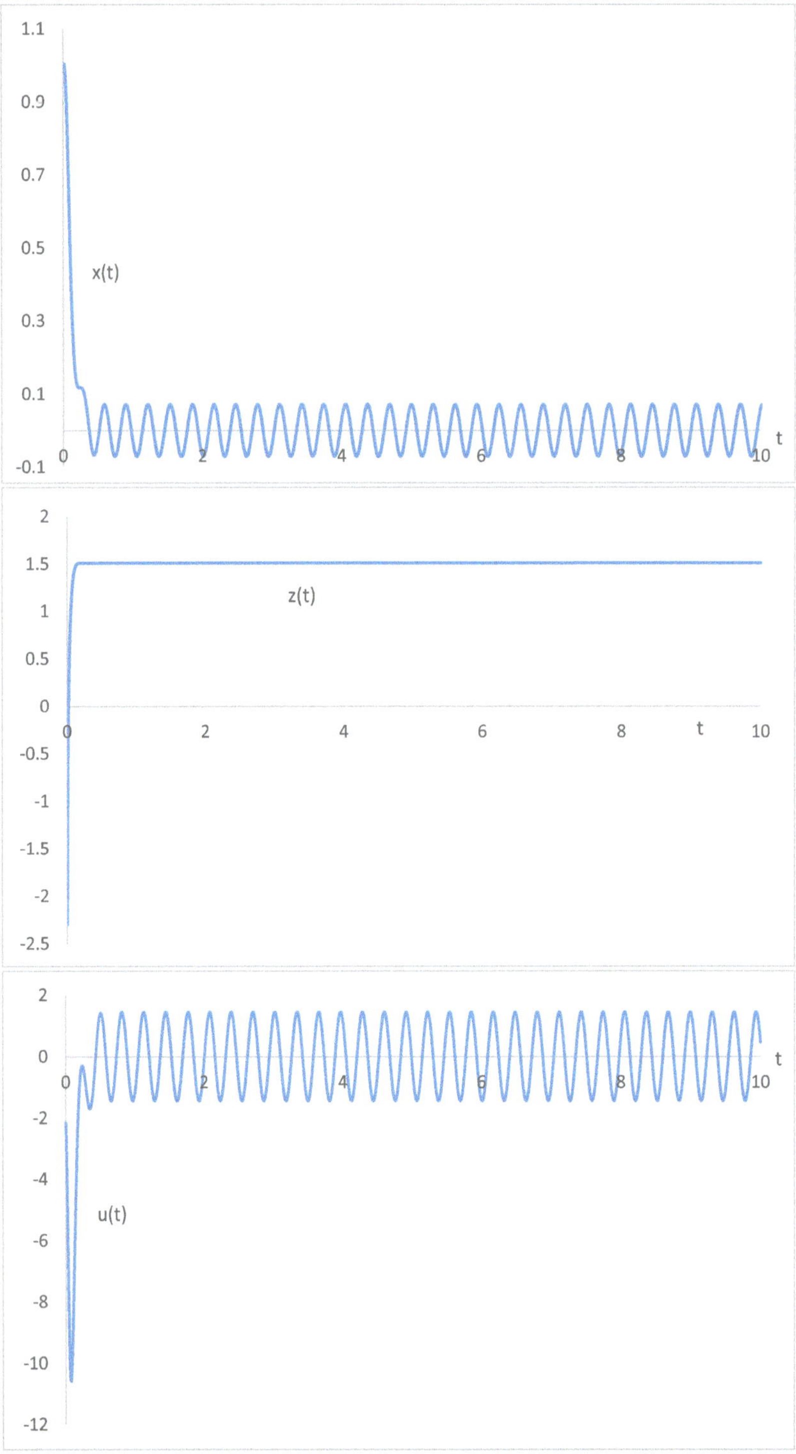

Figure 9.1. *The evolution of the state and the control input for the closed-loop system* (9.2), (9.5), (9.6) *with* $d \equiv 0$, $e(t) = 0.1\sin(20t)$, $\theta = 3$, $c = a = 1, \Gamma = 100, \varepsilon = 0.01$ *and* $x(0) = 1$, $z(0) = -\ln(10)$.

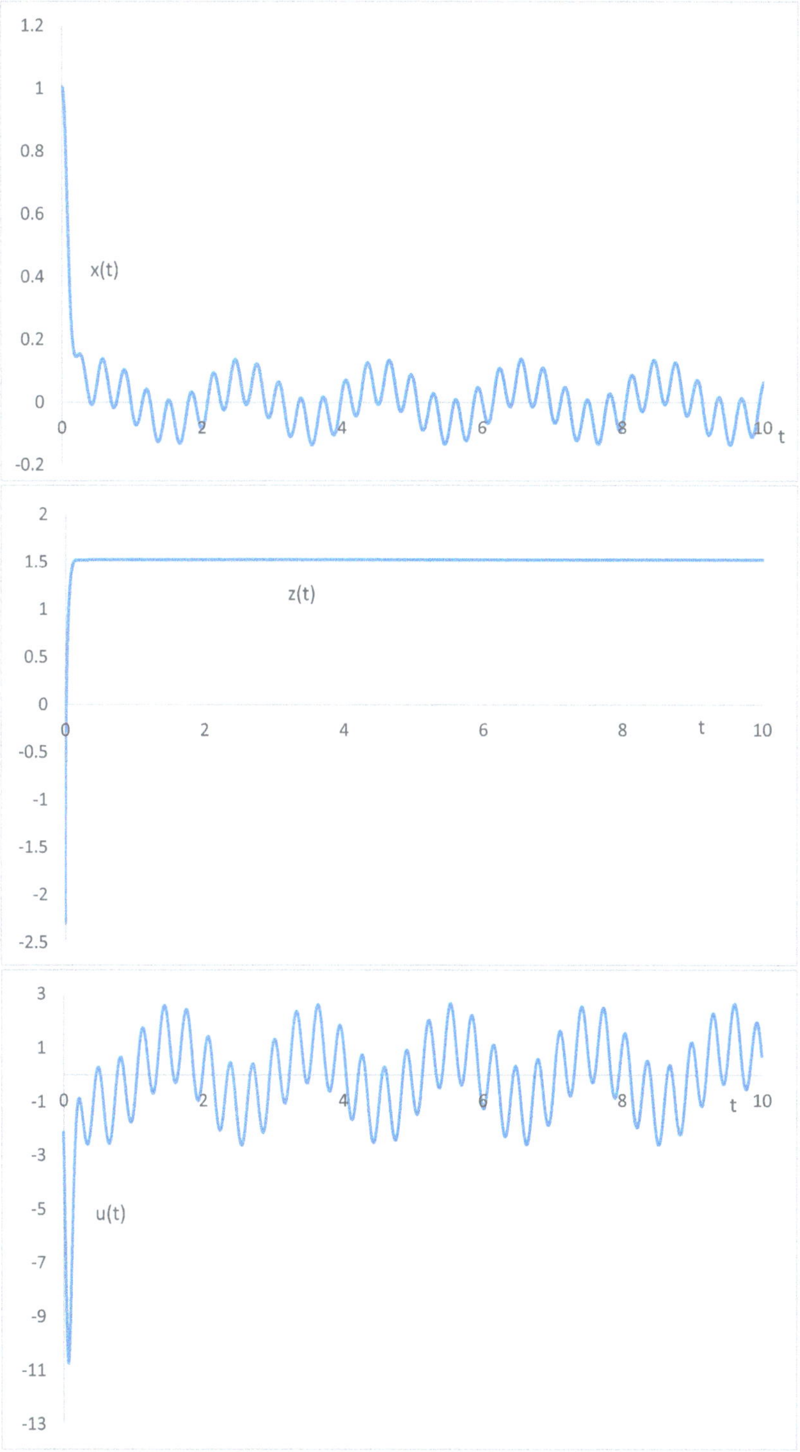

Figure 9.2. *The evolution of the state and the control input for the closed-loop system* (9.2), (9.5), (9.6) *with* $d(t) = \sin(\pi t)$, $e(t) = 0.1\sin(20t)$, $\theta = 3$, $c = a = 1, \Gamma = 100, \varepsilon = 0.01$ *and* $x(0) = 1$, $z(0) = -\ln(10)$.

9.3 ▪ Tracking Problems

The present book does not contain results for trajectory tracking problems (with the exception of the result given in Section 5.2). This does not mean that the DADS control scheme cannot be used for tracking problems. On the contrary, we expect that future research will show that tracking control problems for strict-feedback systems can be solved, robustly, by using DADS controllers.

In order to show how the DADS control scheme can be applied to strict-feedback systems for tracking control problems, we study here the benchmark system

$$\begin{aligned} &\dot{\xi}_1 = \xi_2 + \varphi(\xi_1)\,\theta + d_1 \\ &\dot{\xi}_2 = u + d_2 \\ &\xi = (\xi_1, \xi_2)' \in \mathbb{R}^2, u \in \mathbb{R}, d = (d_1, d_2)' \in \mathbb{R}^2 \end{aligned} \tag{9.19}$$

where $\varphi : \mathbb{R} \to \mathbb{R}$ is a smooth function, $\theta \in \mathbb{R}$ is an unknown constant and $d = (d_1, d_2)' \in \mathbb{R}^2$ is a disturbance. Our objective is to make the state ξ_1 track a given smooth signal $y_r : \mathbb{R}_+ \to \mathbb{R}$.

In order to bring the problem to a more familiar form, we perform the transformation

$$\begin{aligned} w_1 &= \xi_1 - y_r \\ w_2 &= \xi_2 - \dot{y}_r \\ v &= u - \ddot{y}_r \end{aligned} \tag{9.20}$$

Therefore, under the transformation (9.20), system (9.19) takes the following form:

$$\begin{aligned} \dot{w}_1 &= w_2 + \varphi(w_1 + y_r)\,\theta + d_1 \\ \dot{w}_2 &= v + d_2 \end{aligned} \tag{9.21}$$

System (9.21) is a strict-feedback system like the ones that are studied in Chapter 6, and our objective is to stabilize the output $y = w_1$. However, there are two differences with the systems that were studied in Chapter 6: (i) system (9.21) is a time-varying system (since y_r is a time-varying signal), and (ii) the coefficient of θ is not necessarily zero when $w_1 = 0$. Consequently, we next perform the disturbance transformation

$$\begin{aligned} \tilde{d}_1 &= \varphi(y_r)\,\theta + d_1 \\ \tilde{d}_2 &= d_2 \end{aligned} \tag{9.22}$$

and we obtain

$$\begin{aligned} \dot{w}_1 &= w_2 + w_1 h(y_r, w_1)\,\theta + \tilde{d}_1 \\ \dot{w}_2 &= v + \tilde{d}_2 \end{aligned} \tag{9.23}$$

where

$$h(y_r, w_1) := \int_0^1 \frac{d\varphi}{dy}(y_r + \lambda w_1)\, d\lambda \tag{9.24}$$

System (9.23) is the system to which we apply backstepping for the design of a DADS controller. The backstepping procedure that we apply is exactly in the spirit of Lemma 6.1, but not the lemma's direct application, due to the time-varying nature of model (9.23).

For the first step we consider the function

$$V_1(w_1) = w_1^2/2 \tag{9.25}$$

for the subsystem

$$\dot{w}_1 = w_2 + w_1 h(y_r, w_1)\,\theta + \tilde{d}_1 \tag{9.26}$$

Performing exactly the same steps that were used in Chapter 5, Chapter 6, and Chapter 8, we get for arbitrary constants $a, b > 0$

$$\begin{aligned}
&\dot{V}_1\left(w_1, w_2, \tilde{d}_1\right) = w_1 w_2 + \theta h\left(y_r, w_1\right) w_1^2 + \tilde{d}_1 w_1 \\
&\leq w_1 w_2 + |\theta|\left|h\left(y_r, w_1\right)\right| w_1^2 + \tfrac{a}{1+\exp(z)} \tilde{d}_1^2 + \tfrac{1+\exp(z)}{4a} w_1^2 \\
&\leq w_1 w_2 + (|\theta| - b - \exp(z))^+ \left|h\left(y_r, w_1\right)\right| w_1^2 \\
&+ (b + \exp(z))\left|h\left(y_r, w_1\right)\right| w_1^2 + \tfrac{a}{1+\exp(z)} \tilde{d}_1^2 + \tfrac{1+\exp(z)}{4a} w_1^2 \\
&\leq w_1 w_2 + \tfrac{1+\exp(z)}{4a} h^2\left(y_r, w_1\right) w_1^4 + \tfrac{1+2a+\exp(z)}{4a} w_1^2 \\
&+ \tfrac{1}{2}(b + \exp(z))^2 h^2\left(y_r, w_1\right) w_1^2 + a \tfrac{\tilde{d}_1^2 + \left((|\theta| - b - \exp(z))^+\right)^2}{1+\exp(z)}
\end{aligned} \tag{9.27}$$

Setting

$$\begin{aligned}
M\left(y_r, w_1, z\right) &:= \tfrac{1+2a+\exp(z)}{4a} + \tfrac{1+\exp(z)}{4a} h^2\left(y_r, w_1\right) w_1^2 \\
&+ c + (b + \exp(z))^2 h^2\left(y_r, w_1\right)/2
\end{aligned} \tag{9.28}$$

where $c > 0$ is an arbitrary constant, we obtain from (9.27) the following inequality for all $w = (w_1, w_2)' \in \mathbb{R}^2$, $z \in \mathbb{R}$, $\theta \in \mathbb{R}$, and $\tilde{d}_1 \in \mathbb{R}$:

$$\dot{V}_1\left(w_1, w_2, \tilde{d}_1\right) \leq -c w_1^2 + w_1\left(w_2 + M\left(y_r, w_1, z\right) w_1\right) + a \frac{\tilde{d}_1^2 + \left((|\theta| - b - \exp(z))^+\right)^2}{1 + \exp(z)} \tag{9.29}$$

For the second (and final) step, we consider the function

$$V\left(y_r, w, z\right) = w_1^2/2 + \left(w_2 + M\left(y_r, w_1, z\right) w_1\right)^2/2 \tag{9.30}$$

the update law

$$\dot{z} = \Gamma \exp(-z)\left(V\left(y_r, w, z\right) - \varepsilon\right)^+ \tag{9.31}$$

and the feedback law

$$\begin{aligned}
v = &-w_1\left(1 + \tfrac{\partial M}{\partial y_r}\left(y_r, w_1, z\right) \dot{y}_r\right) - P\left(y_r, w, z\right)\left(w_2 + M\left(y_r, w_1, z\right) w_1\right) \\
&-w_1 \tfrac{\partial M}{\partial z}\left(y_r, w_1, z\right) \Gamma \exp(-z)\left(V\left(y_r, w, z\right) - \varepsilon\right)^+ \\
&-\left(w_1 \tfrac{\partial M}{\partial w_1}\left(y_r, w_1, z\right) + M\left(y_r, w_1, z\right)\right) w_2
\end{aligned} \tag{9.32}$$

where $\Gamma, \varepsilon > 0$ are arbitrary constants where $P\left(y_r, w, z\right) > 0$ is a locally Lipschitz positive function to be selected. Using (9.25), (9.29), (9.31), and (9.32) we get

$$\begin{aligned}
&\dot{V} \leq -c w_1^2 + a \tfrac{\tilde{d}_1^2 + \left((|\theta| - b - \exp(z))^+\right)^2}{1+\exp(z)} - P\left(y_r, w, z\right)\left(w_2 + M\left(y_r, w_1, z\right) w_1\right)^2 \\
&+ \left(w_2 + M\left(y_r, w_1, z\right) w_1\right)\left(w_1 \tfrac{\partial M}{\partial w_1}\left(y_r, w_1, z\right) + M\left(y_r, w_1, z\right)\right) w_1 h\left(y_r, w_1\right) \theta \\
&+ \left(w_2 + M\left(y_r, w_1, z\right) w_1\right)\left(\tilde{d}_2 + \left(w_1 \tfrac{\partial M}{\partial w_1}\left(y_r, w_1, z\right) + M\left(y_r, w_1, z\right)\right) \tilde{d}_1\right)
\end{aligned} \tag{9.33}$$

Performing exactly the same steps that were used in Chapter 5, Chapter 6, and Chapter 8, we get from (9.33)

$$\begin{aligned}
\dot{V} &\leq -cw_1^2 + a\frac{2\tilde{d}_1^2+\tilde{d}_2^2+\left(\left(|\theta|-b-\exp(z)\right)^+\right)^2}{1+\exp(z)} - P\left(y_r,w,z\right)W^2 \\
&+|W|\left|w_1\frac{\partial M}{\partial w_1}\left(y_r,w_1,z\right)+M\left(y_r,w_1,z\right)\right|\left|w_1\right|\left|h\left(y_r,w_1\right)\right|\left(|\theta|-b-\exp(z)\right)^+ \\
&+|W|\left|w_1\frac{\partial M}{\partial w_1}\left(y_r,w_1,z\right)+M\left(y_r,w_1,z\right)\right|\left|w_1\right|\left|h\left(y_r,w_1\right)\right|\left(b+\exp(z)\right) \\
&+\frac{1+\exp(z)}{4a}\left(1+\left(w_1\frac{\partial M}{\partial w_1}\left(y_r,w_1,z\right)+M\left(y_r,w_1,z\right)\right)^2\right)W^2 \\
&\leq -\frac{c}{2}w_1^2 + 2a\frac{\tilde{d}_1^2+\tilde{d}_2^2+\left(\left(|\theta|-b-\exp(z)\right)^+\right)^2}{1+\exp(z)} - P\left(y_r,w,z\right)W^2 \\
&+\frac{1+\exp(z)}{4a}w_1^2\left(w_1\frac{\partial M}{\partial w_1}\left(y_r,w_1,z\right)+M\left(y_r,w_1,z\right)\right)^2 h^2\left(y_r,w_1\right)W^2 \\
&+\frac{b+\exp(z)}{2c}\left(w_1\frac{\partial M}{\partial w_1}\left(y_r,w_1,z\right)+M\left(y_r,w_1,z\right)\right)^2 h^2\left(y_r,w_1\right)W^2 \\
&+\frac{1+\exp(z)}{4a}\left(1+\left(w_1\frac{\partial M}{\partial w_1}\left(y_r,w_1,z\right)+M\left(y_r,w_1,z\right)\right)^2\right)W^2
\end{aligned} \tag{9.34}$$

where $W = w_2 + M\left(y_r,w_1,z\right)w_1$. Hence, by selecting

$$\begin{aligned}
P\left(y_r,w,z\right) &= \frac{1+\exp(z)}{4a}\left(1+\left(w_1\frac{\partial M}{\partial w_1}\left(y_r,w_1,z\right)+M\left(y_r,w_1,z\right)\right)^2\right) \\
&+\frac{1+\exp(z)}{4a}w_1^2\left(w_1\frac{\partial M}{\partial w_1}\left(y_r,w_1,z\right)+M\left(y_r,w_1,z\right)\right)^2 h^2\left(y_r,w_1\right) \\
&+\frac{b+\exp(z)}{2c}\left(w_1\frac{\partial M}{\partial w_1}\left(y_r,w_1,z\right)+M\left(y_r,w_1,z\right)\right)^2 h^2\left(y_r,w_1\right)+\frac{c}{2}
\end{aligned} \tag{9.35}$$

we obtain for all $w = \left(w_1,w_2\right)' \in \mathbb{R}^2$, $z \in \mathbb{R}$, $\theta \in \mathbb{R}$, and $\tilde{d} = \left(\tilde{d}_1,\tilde{d}_2\right)' \in \mathbb{R}^2$

$$\dot{V} \leq -cV\left(y_r,w,z\right) + 2a\frac{\left|\tilde{d}\right|^2+\left(\left(|\theta|-b-\exp(z)\right)^+\right)^2}{1+\exp(z)} \tag{9.36}$$

For the reader's convenience, Table 9.1 provides all the formulas that constitute the DADS controller for (9.23).

We conclude that the backstepping procedure can be applied without difficulty to the tracking problem for (9.19).

We expect that similar ideas like the ideas presented here can be used for the application of the backstepping procedure to general tracking problems for strict-feedback systems.

The derivation of stability estimates from (9.36) is a different story. The researcher needs to develop an extension of Theorem 2.8 to the time-varying case. However, this is again something doable. Here, we explain how this can be done.

First, we exploit (9.36) and (9.31) (which implies that $z(t)$ is non-decreasing) in order to obtain the following estimate for all $t \geq 0$ for which the solution of the closed-loop system (9.23) with the DADS controller exists:

$$\begin{aligned}
&V(y_r(t),w(t),z(t)) \leq \exp(-ct)V(y_r(0),w(0),z(0)) \\
&+2a\frac{\left\|\tilde{d}\right\|_\infty^2+\left(\left(|\theta|-b-\exp(z(0))\right)^+\right)^2}{c(1+\exp(z(0)))}
\end{aligned} \tag{9.37}$$

Next, we employ Lemma 4.1 in conjunction with (9.36) and (9.31) in order to obtain the estimate

$$z(0) \leq z(t) \leq \ln\left(A_1+A_2\right) \tag{9.38}$$

Table 9.1. *The DADS controller for* (9.23).

Constants	$c, a, \Gamma, \varepsilon, b > 0$
Update law	$\dot{z} = \Gamma \exp(-z)\left(V\left(y_r, w, z\right) - \varepsilon\right)^+$
Control	$v = -w_1\left(1 + \frac{\partial M}{\partial y_r}\left(y_r, w_1, z\right)\dot{y}_r\right) - P\left(y_r, w, z\right)\left(w_2 + M\left(y_r, w_1, z\right) w_1\right)$ $-w_1 \frac{\partial M}{\partial z}\left(y_r, w_1, z\right)\Gamma \exp(-z)\left(V\left(y_r, w, z\right) - \varepsilon\right)^+$ $-\left(w_1 \frac{\partial M}{\partial w_1}\left(y_r, w_1, z\right) + M\left(y_r, w_1, z\right)\right) w_2$
Auxiliary functions	$M\left(y_r, w_1, z\right) := \frac{1+2a+\rho}{4a} + \frac{1+\rho}{4a} h^2\left(y_r, w_1\right) w_1^2$ $+c + (b+\rho)^2 h^2\left(y_r, w_1\right)/2$ $V\left(y_r, w, z\right) := w_1^2/2 + \left(w_2 + M\left(y_r, w_1, z\right) w_1\right)^2/2$ $\rho := \exp(z)$ $P\left(y_r, w, z\right) := \frac{1+\rho}{4a}\left(1 + \left(w_1 \frac{\partial M}{\partial w_1}\left(y_r, w_1, z\right) + M\left(y_r, w_1, z\right)\right)^2\right)$ $+\frac{1+\rho}{4a} w_1^2\left(w_1 \frac{\partial M}{\partial w_1}\left(y_r, w_1, z\right) + M\left(y_r, w_1, z\right)\right)^2 h^2\left(y_r, w_1\right)$ $+\frac{b+\rho}{2c}\left(w_1 \frac{\partial M}{\partial w_1}\left(y_r, w_1, z\right) + M\left(y_r, w_1, z\right)\right)^2 h^2\left(y_r, w_1\right) + \frac{c}{2}$

where

$$
\begin{aligned}
A_1 &:= \max\left(\tfrac{2a}{c\varepsilon}\left(\left\|\tilde{d}\right\|_\infty^2 + \left(\left(|\theta| - b - \exp(z(0))\right)^+\right)^2\right) - 1, \exp(z(0))\right) \\
A_2 &:= \tfrac{\Gamma}{c}\left(V(y_r(0), w(0), z(0)) + \frac{2a\left(\left\|\tilde{d}\right\|_\infty^2 + \left(\left(|\theta| - b - \exp(z(0))\right)^+\right)^2\right)}{c(1+\exp(z(0)))} - \varepsilon\right)^+
\end{aligned}
\tag{9.39}
$$

At this point one big difference with the time-invariant case arises: in order to show from (9.37) and (9.38) that the solution of the closed-loop system (9.23) with the DADS controller is bounded one has to assume that the reference signal y_r is bounded, i.e., $\left\|y_r\right\|_\infty < +\infty$. Indeed, the boundedness of the closed-loop solution can be shown by exploiting the following coercivity property that is a consequence of definition (9.30):

"For every $R > 0$ the set $\left\{ (y_r, w, z) \in [-R, R] \times \mathbb{R}^2 \times [-R, R] \ : \ V(y_r, w, z) \leq R \right\}$ is bounded."

Having established that the solution of the closed-loop system (9.23) with the DADS controller is bounded as long as it exists, we can conclude that the solution exists for all $t \geq 0$. Moreover, estimates (9.37) and (9.38) hold for all $t \geq 0$.

At this point a second difference with the time-invariant case arises: in order to show from (9.37), (9.38) and Barbălat's lemma that

$$
\limsup_{t \to +\infty}\left(\frac{1}{2}\left|w_1(t)\right|^2\right) \leq \limsup_{t \to +\infty}\left(V(y_r(t), w(t), z(t))\right) \leq \varepsilon \tag{9.40}
$$

we have to show that $\frac{d}{dt}\left(V(y_r(t), w(t), z(t))\right)$ is of class $L^\infty\left(\mathbb{R}_+\right)$. Thus, we need to assume that

$$
\left\|y_r\right\|_\infty + \left\|\dot{y}_r\right\|_\infty < +\infty \tag{9.41}
$$

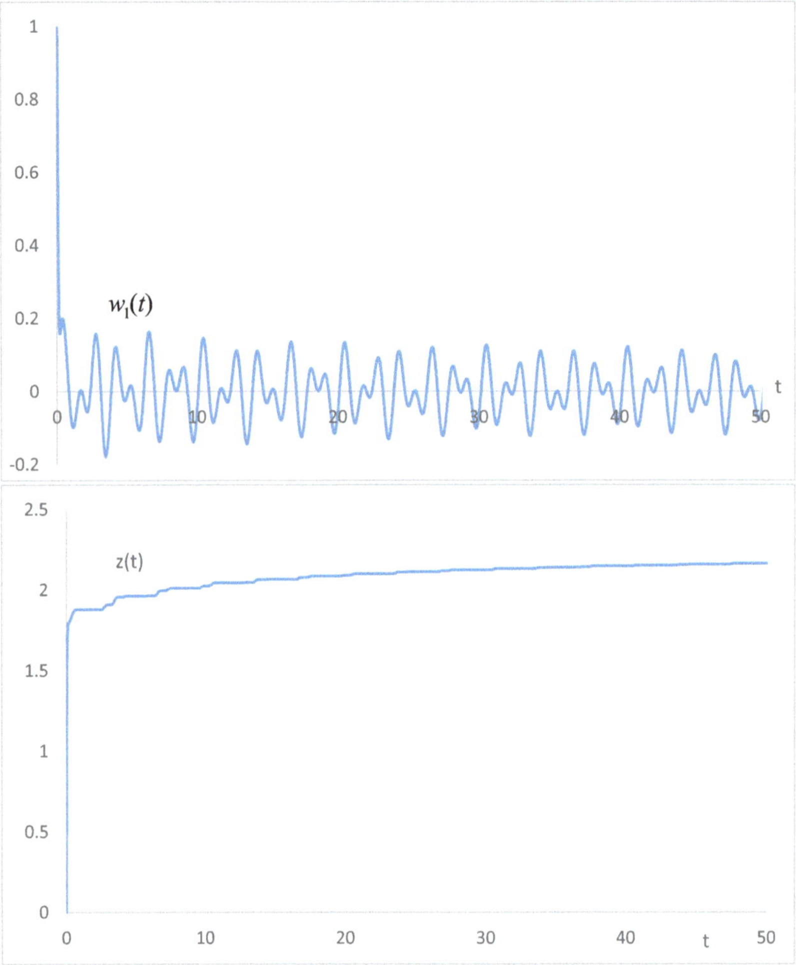

Figure 9.3. *The evolution of the state* $w_1(t)$ *and the gain* $z(t)$ *for the closed-loop system* (9.23) *with the adaptive feedback law of Table 9.1 and* $d(t) = 0, \varphi(\xi_1) = \xi_1, \theta = 3, a = b = c = 1, \varepsilon = 0.005, \Gamma = 100, w_1(0) = 0, w_2(0) = -3.5, z(0) = 0$. *The reference signal is* $y_r(t) = \sin(\pi t) + \sin(5t)$.

Let us repeat what we have achieved to show: for every $w(0) \in \mathbb{R}^2$, $\tilde{d} \in L^\infty\left(\mathbb{R}_+; \mathbb{R}^2\right)$, $\theta \in \mathbb{R}$, $y_r \in C^2\left(\mathbb{R}_+\right)$ for which (9.41) holds, the corresponding solution of the closed-loop system (9.23) with the DADS controller exists for all $t \geq 0$, is bounded and satisfies (9.40). However, taking into account the disturbance transformation (9.22) and the state transformation (9.20), we can also state our results in a different way:

"for every $\xi(0) \in \mathbb{R}^2$, $d \in L^\infty\left(\mathbb{R}_+; \mathbb{R}^2\right)$, $\theta \in \mathbb{R}$, $y_r \in C^2\left(\mathbb{R}_+\right)$ for which (9.41) holds, the corresponding solution of the closed-loop system (9.19) with the DADS controller exists for all $t \geq 0$, is bounded and satisfies $\limsup\limits_{t \to +\infty} \left(|\xi_1(t) - y_r(t)|\right) \leq \sqrt{2\varepsilon}$."

Clearly, the above result shows practical tracking with a simultaneous suppression of the effect of the disturbance. Recall that $\varepsilon > 0$ is a prespecified, but otherwise arbitrary, constant.

Figure 9.3 shows the evolution of the state $w_1(t)$ and the gain $z(t)$ for the closed-loop system (9.23) with the adaptive feedback law of Table 9.1 and $d(t) = 0, \varphi(\xi_1) = \xi_1, \theta = 3, a = b =$

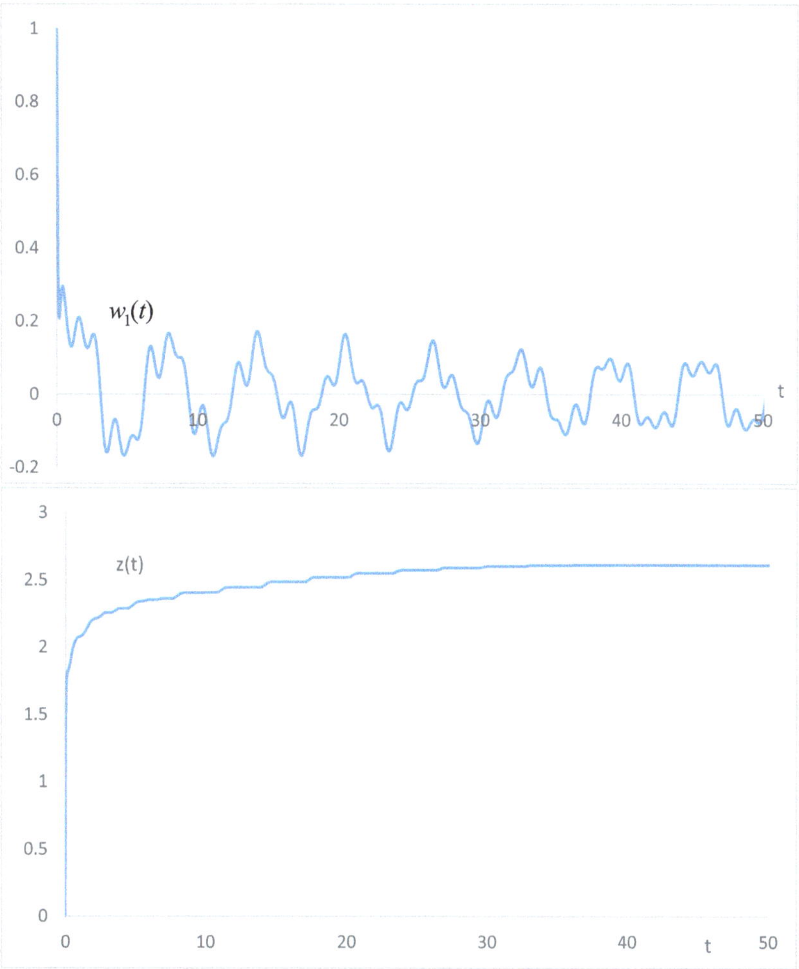

Figure 9.4. *The evolution of the state $w_1(t)$ and the gain $z(t)$ for the closed-loop system* (9.23) *with the adaptive feedback law of Table 9.1 and $d_1(t) = 10\sin(t), d_2(t) = 10\sin(2t), \varphi(\xi_1) = \xi_1, \theta = 3, a = b = c = 1, \varepsilon = 0.005, \Gamma = 100, w_1(0) = 0, w_2(0) = -3.5, z(0) = 0$. The reference signal is $y_r(t) = \sin(\pi t) + \sin(5t)$.*

$c = 1, \varepsilon = 0.005, \Gamma = 100, w_1(0) = 0, w_2(0) = -3.5, z(0) = 0$. The reference signal is $y_r(t) = \sin(\pi t) + \sin(5t)$. The case where $d_1(t) = 10\sin(t), d_2(t) = 10\sin(2t)$, and all the other data are the same is shown in Figure 9.3. In both Figure 9.3 and Figure 9.4, it is shown that the state $w_1(t)$ ultimately satisfies $|w_1(t)| \leq 0.1$, while the gain $z(t)$ remains bounded.

The analysis that is presented above can be generalized to the class of strict-feedback systems without significant problems.

9.4 ▪ Machine Learning for Adaptive Control and System Identification of Nonlinear Systems

As we have seen in Chapter 6, DADS controllers for general strict-feedback nonlinear systems become very complicated, in the sense of the algebraic expressions for the feedback law and the Lyapunov function that enters the deadzone-based law for updating the parameter $z(t)$.

Such algebraic complexity arises in all adaptive backstepping approaches [87]. It is not a consequence of the method. It is a consequence of the complexity of the strict-feedback class and, particularly, of the unmatched character of the parameter uncertainty (or of the unmatched disturbance). The higher the "uncertainty index" (the number of integrators between the input and any of the unknown parameters and disturbances in the model, the more algebraically complex the feedback and update laws. This is partly because of the symbolic differentiations of the system nonlinearities and the stabilizing and tuning functions (in the adaptive backstepping approach). The algebraic complexity goes to the extreme in adaptive control designs for stochastic nonlinear systems [26] due to the fact that the Lyapunov function, under Ito's differentiation rule, undergoes two differentiations, rather than one.

The algebraic complexity of adaptive backstepping methods has motivated many efforts for coping with the complexity. In the 1990s, symbolic software packages, in the early years of *Mathematica* and Maple, were developed for automating the design of the controller and update law expressions. These were pioneering and valuable efforts, by Reza Ghanadan and Gil Blankenship.

In the early 2000s, a mistaken view emerged that algebraic complexity of adaptive backstepping laws was caused by the method (rather than by the class of parametric strict-feedback systems). A marketing expression was coined, an "explosion of complexity" (a sensationalism bordering on intellectual vulgarity), with alternative attempts proposed to reduce the purported complexity. These attempts employed dynamic filters in time, instead of symbolic differentiations with respect to the state. They were named "dynamic surface control" and "command filtering." In these fallacious endeavors, it was overlooked (or consciously underemphasized) that the presumed "explosion of complexity" is being replaced by other "explosions:" of the dynamic order (the filter states) and of the filter gain values. Furthermore, an *implosion of the region of attraction* was introduced (unless gains are taken to extremes). Globally stabilizing and asymptotically convergent adaptive backstepping algorithms were being replaced by local practical, or at best semiglobal practical results. A methodological and theoretical regression in just about every way that matters. One has had to have been serving in an editorial role, with the associated exposure to the trends in submissions, to observe how populous a community of misled early-stage researchers can grow under the influence of misconceived ideas, accompanied by a slogan such as "explosion of complexity." Thankfully, with time, and with reviewer diligence, these efforts ultimately got understood for what they are and faded from the world of adaptive control research.

Nevertheless, adaptive backstepping designs, which achieve global stabilization and asymptotic regulation (in the disturbance-free case) are inevitably algebraically complex. And even more so are the DADS designs because they achieve more: not just global stabilization and regulation, but also global robustness to disturbances and the property of zero output asymptotic gain.

The derivation of the complicated expressions in the DADS feedback can be automated by symbolic computation. But their inherent complexity raises the questions of the cost of their real-time evaluation, of possible reduction of such complexity using approximation, and of the circumvention of the symbolic derivation for each new nonlinear system model.

It borders on obvious that a complicated function, such as a DADS feedback law for a given system, can be approximated by a neural network (NN). If such a feedback law is first symbolically derived for a given nonlinear model, and then NN-approximated, one gets an NN mapping from the plant state vector y and the gain z to the control input u. There is nothing particularly innovative in that, and one still has to perform the laborious symbolic derivation before the NN approximation.

A more interesting option, employing machine learning (ML), is one in which, for the DADS approach, a mapping from the nonlinear system's vector field (which is not a vector value but a function of the state) to the backstepping diffeomorphism, or to the control u itself, is approximated by a (deep) neural network. In this idea, the NN's input is not the state vector but the function of the system vector which characterizes the system's model (the vector field of the system). Such a mapping from the vector field, as an input function, to the control, as a scalar or a vector, is a nonlinear operator, an infinite-dimensional mapping. Such a nonlinear operator can be encoded using a neural network. To learn an approximation of such a nonlinear operator, one has to produce a data set using many samples of nonlinear vector fields, functions of the state vector. Then one has to train a deep NN to approximate this operator. The resulting neural network approximates the feedback law whose input is the system state y and the gain update z. In exactly the same way as the mapping to the control input u is learned, a mapping to the Lyapunov function V is learned.

We emphasize, we are not outlining an NN-approximation of a DADS feedback law for a given nonlinear system, but for ALL nonlinear systems in a particular class. So, a comparison with "dynamic surface control" and "command filtering," which are designs for a given system, is not even appropriate.

The idea of approximating, using machine learning, a complex model-based feedback law for a complex dynamic model has already been introduced and developed, for PDE control using backstepping, in [18], as well as in a number of other recent publications by Bhan, Shi, Krstic, Lamarque, Diagne, Qi, Wang, and coauthors. In particular, in Section VI of [18], a model-based stabilizing feedback law mapping from the system's model data and the system's state, to the system input, is learned, once and for all, for a class of hyperbolic PDE systems. What we discuss above, as a feedback mapping from the nonlinear system's vector field, and the state, to the system's input, for not one particular case of a strict-feedback system but for the entire class of strict-feedback systems, learned once and for all rather than for each new system, matches the principle introduced in Section VI of [18]. The difference is that, while in [18] generating the training set consists of solving a large set of integral equations, for DADS the training set consists of symbolic computation of a large set of DADS feedback laws for a large sample set of strict-feedback systems.

Executing such an application of ML in the context of DADS computationally, and developing the associated stability theory under the resulting deep NN approximations, is not this book's subject. We are not able to claim that such ML-approximated DADS controllers, which entail an evaluation of a neural network, offer savings in real-time computation relative to symbolically derived, numerically computed, algebraically complex DADS feedback laws. These questions, and further advances in design and analysis, represent an exciting area for future research in adaptive and robust nonlinear control. In this area, the design is model-based and the implementation is ML-based. Both a theorist and an AI specialist have a role to play in this next phase of development of the important field of robust adaptive control.

Besides encoding complex DADS feedback laws using ML, there is another domain for ML that arises from this book on robust adaptation for nonlinear systems. It is the opportunity for approximating the implementation of robust identifiers from Chapter 7. The Batch Least-Squares identifiers of Chapter 7 introduce natural nonlinear operators, as mappings from measured signals, over time intervals, to estimates of an unknown parameter vector. The sampled-data approximation of the identifier map is described as follows for a fixed integer $N > 0$:

1. Give $\sigma > 0, T > 0, \eta > 0$ and $\hat{\theta}(iT), \psi_i$.
2. Take sampled measurements of $y\,(iT + j\delta)$ and $z\,(iT + j\delta)$ for $j = 1, ..., N$, where $\delta = T/N$.

3. Use formulas (7.30), (7.31) to get $\lim_{t \to (i+1)T^-} (\chi(t))$, $\lim_{t \to (i+1)T^-} (Q(t))$, $\lim_{t \to (i+1)T^-} (G(t))$ and $\lim_{t \to (i+1)T^-} (\nu(t))$.

4. Use formula (7.25) to get $A\,(iT)$ and $B\,(iT)$.

5. Estimate $\lambda_{\text{min}}\,(B(iT))$.

6. Use (7.14), (7.15), (7.16) to get $\hat{\theta}((i+1)T)$ and ψ_{i+1}.

Thus, the sampled-data approximation of the identifier constitutes a complex nonlinear mapping, which is defined on a finite-dimensional space.

With this discussion about the robust identifier, following the earlier discussion about the DADS controller, we conclude the discussion about the vast opportunities that arise from merging the model-based, certificates-bearing results in this book, with emerging AI techniques that approximate nonlinear functions and nonlinear operators, with approximation accuracies guaranteed either by the classical approximation theorems for nonlinear functions [25] or the recent deep NN-based universal approximation theory for nonlinear operators [27].

Appendix

A.1 ▪ Introduction

In this appendix we provide the necessary mathematical background for a complete understanding of the book (including the proofs of all results). A student/researcher who studies adaptive control usually has some knowledge obtained by the books [80] and [87]. In this book we do not assume that the reader has such knowledge although prior knowledge obtained by the books [80] and [87] will certainly help the reader.

The only things that are required to be known are basic notions from a calculus course, like open and closed sets, intervals, *sup* and *inf*, (piecewise) continuous functions, limits, partial derivatives, and the chain rule. Based on these basic notions, the material that is given in this appendix will help the reader to reach the point of understanding everything.

Of course, the material in Chapter 2 (stability notions) is also a topic that the student/researcher needs to understand in order to conquer the field of adaptive control. But all notions/results given in Chapter 2 can be understood by having this useful appendix.

All results that are given below and are not attributed can be found in [20], [81].

A.2 ▪ Classes of Functions Used in the Book

This section provides all the classes of functions that are used in the present book.

Classes of functions based on continuity: Let $D \subseteq \mathbb{R}^n$ be an open set and let $S \subseteq \mathbb{R}^n$ be a set that satisfies $D \subseteq S \subseteq \bar{D}$, where $\bar{D}$ is the closure of D. By $C^0(S\ ;\ \Omega)$, we denote the class of continuous functions on S, which take values in $\Omega \subseteq \mathbb{R}^m$. By $C^k(S\ ;\ \Omega)$, where $k \geq 1$ is an integer, we denote the class of functions on $S \subseteq \mathbb{R}^n$, which take values in $\Omega \subseteq \mathbb{R}^m$ and have continuous derivatives of order k. In other words, the functions of class $C^k(S\ ;\ \Omega)$ are the functions which have continuous derivatives of order k in D that can be continued continuously to all points in $\partial D \cap S$, where ∂D is the boundary of D. When $\Omega = \mathbb{R}$ then we write $C^0(S\ )$ or $C^k(S\ )$. A function $f \in C^\infty(S\ ;\ \Omega) = \bigcap_{k=0}^{\infty} C^k(S\ ;\ \Omega)$ is called a smooth function.

Measurable functions: Let $I \subseteq \mathbb{R}$ be an interval with non-empty interior and let $D \subseteq \mathbb{R}^n$. For every $a \geq 1$, $L^a(I;D)$ denotes the set of equivalence classes of Lebesgue measurable functions $f : I \to D$ for which $\|f\|_a = \left(\int_I |f(t)|^a\, dt\right)^{1/a} < +\infty$. $L^\infty(I;D)$ denotes the set of equivalence classes of Lebesgue measurable functions $f : I \to D$ for which $\|f\|_\infty := \operatorname{ess\,sup}_{t\in I} (|f(t)|) < +\infty$. When $D = \mathbb{R}$, we write $L^a(I)$ or $L^\infty(I)$.

Note that: if the reader is not familiar with the notion of a Lebesgue measurable function, or with expressions like "a.e.=almost everywhere" and quantities like $\operatorname{ess\,sup}_{t\in I} (|f(t)|)$, the reader can still understand the theory presented in the book. All that is needed is to replace everywhere the term "Lebesgue measurable function" by the term "piecewise continuous function", the expression "a.e." by the expression "with the possible exception of a countable number of points of which there are only finitely many in every bounded interval," and the quantity $\operatorname{ess\,sup}_{t\in I}(|f(t)|)$ by the quantity $\sup_{t\in I}(|f(t)|)$. Of course, what we are suggesting is not equivalent from a mathematical point of view (there are Lebesgue measurable functions that are not piecewise continuous). However, with such a swap of notions the reader can understand the theory of the book without having to learn measure theory.

In Chapter 2 we also use the space $L^\infty_{loc}(\mathbb{R}_+; D)$. This is the space of equivalence classes of Lebesgue measurable functions $f : \mathbb{R}_+ \to D$ for which their restriction on every bounded interval $I \subseteq \mathbb{R}_+$ with non-empty interior is of class $L^\infty(I; D)$. For example, the function $f(t) = \exp(t)$ does not belong to $L^\infty(\mathbb{R}_+)$ but belongs to $L^\infty_{loc}(\mathbb{R}_+; \mathbb{R})$.

<u>Absolutely continuous functions:</u> Let $I \subseteq \mathbb{R}$ be a closed and bounded interval with non-empty interior and let $f : I \to \mathbb{R}^n$ be a continuous function. We say that f is absolutely continuous on $I \subseteq \mathbb{R}$ if there exists $g \in L^1(I; \mathbb{R}^n)$ such that $f(t) = f(\tau) + \int_\tau^t g(s)ds$ for all $t, \tau \in I$. In such a case, the function f is differentiable for $t \in I$ a.e. with $\dot{f}(t) = g(t)$ for $t \in I$ a.e.. Therefore, we can write that $\dot{f} \in L^1(I; \mathbb{R}^n)$ and $f(t) = f(\tau) + \int_\tau^t \dot{f}(s)ds$ for all $t, \tau \in I$.

When $I \subseteq \mathbb{R}$ is an interval with non-empty interior (not necessarily closed or bounded) we say that the continuous function $f : I \to \mathbb{R}^n$ is absolutely continuous on $I \subseteq \mathbb{R}$ if for every closed and bounded interval $\tilde{I} \subseteq I$ with non-empty interior the function f is absolutely continuous on $\tilde{I} \subseteq \mathbb{R}$.

If the functions $f, g : I \to \mathbb{R}$ are absolutely continuous on the interval $I \subseteq \mathbb{R}$, then the functions $f + g$, fg, and λf are absolutely continuous on I for every $\lambda \in \mathbb{R}$. Moreover, the usual rules for differentiation hold for $t \in I$ a.e.: $\frac{d}{dt}(f(t) + g(t)) = \dot{f}(t) + \dot{g}(t)$, $\frac{d}{dt}(\lambda f(t)) = \lambda \dot{f}(t)$ and $\frac{d}{dt}(f(t)g(t)) = \dot{f}(t)g(t) + f(t)\dot{g}(t)$. If $f : I \to \mathbb{R}$ is absolutely continuous on the interval $I \subseteq \mathbb{R}$, with $f(t) \neq 0$ for all $t \in I$, then $\frac{1}{f}$ is absolutely continuous on the interval I, with $\frac{d}{dt}\left(\frac{1}{f(t)}\right) = -\frac{\dot{f}(t)}{f^2(t)}$ for $t \in I$ a.e. Moreover, if $f : I \to \mathbb{R}^n$ is absolutely continuous on I and $V \in C^1(\mathbb{R}^n)$ then the function $h(t) = V(f(t))$ is absolutely continuous on I with $\frac{d}{dt}(V(f(t))) = \nabla V(f(t))\dot{f}(t)$ for $t \in I$ a.e.

In Chapter 3 (Section 3.4) and Chapter 9 (Section 9.2), we also use the Sobolev space $W^{1,\infty}(I)$, where $I \subseteq \mathbb{R}$ is an interval with non-empty interior: it is the set of absolutely continuous functions $f : \bar{I} \to \mathbb{R}$ with $f, \dot{f} \in L^\infty(I; \mathbb{R})$, where $\bar{I}$ denotes the closure of $I \subset \mathbb{R}$.

<u>Comparison functions:</u> These are classes of functions that are used frequently in stability theory (a reader who has read parts of the book [80] or parts of the more advanced books [66], [109] certainly understands this point). Chapter 2 of the present book also uses such classes of functions. By K we denote the class of increasing continuous functions $a : \mathbb{R}_+ \to \mathbb{R}_+$ with $a(0) = 0$. By K_∞ we denote the class of increasing continuous functions $a : \mathbb{R}_+ \to \mathbb{R}_+$ with $a(0) = 0$ and $\lim_{s\to+\infty}(a(s)) = +\infty$. By KL we denote the set of all continuous functions $\sigma : \mathbb{R}_+ \times \mathbb{R}_+ \to \mathbb{R}_+$ with the properties (i) for each $t \geq 0$ the mapping $\sigma(\cdot, t)$ is of class K; (ii) for each $s \geq 0$ the mapping $\sigma(s, \cdot)$ is non-increasing with $\lim_{t\to+\infty}(\sigma(s,t)) = 0$.

<u>Real functions:</u> Real functions (usually of many variables) are used throughout the book (mainly as Lyapunov-like functions). Let $S \subseteq \mathbb{R}^n$ be a non-empty set with $0 \in S$. We say that a function $V : S \to \mathbb{R}_+$ is positive definite if $V(x) > 0$ for all $x \in S$ with $x \neq 0$ and $V(0) = 0$. We say that a continuous function $V : S \to \mathbb{R}_+$ is radially unbounded if the following property holds: "for every $M > 0$ the set $\{x \in S : V(x) \leq M\}$ is compact". The term "radially unbounded" is used very much in control theory but other areas in mathematics use the term "coercive" (e.g., in numerical analysis and functional analysis) or the term "proper" (in dynamical systems).

Locally Lipschitz functions: Let $D \subseteq \mathbb{R}^p$ be a closed, non-empty set and let $f : \mathbb{R}^n \times D \to \mathbb{R}^m$ be a continuous mapping. We say that f is locally Lipschitz with respect to $x \in \mathbb{R}^n$ mapping if for every compact set $S \subset \mathbb{R}^n \times D$ there exists a constant $L > 0$ that satisfies for all $(x, d) \in S$, $(w, d) \in S$

$$|f(x,d) - f(w,d)| \leq L\,|x - w| \tag{Ap.1}$$

Notice that the fact that $f : \mathbb{R}^n \times D \to \mathbb{R}^m$ is continuous and the fact that $D \subseteq \mathbb{R}^p$ is closed guarantees that for every closed and bounded set $K \subseteq \mathbb{R}^n$ and for every $r \geq \min\{\,|d| \,:\, d \in D\}$ there exists a constant $M > 0$ such that $|f(x,d)| \leq M$ for all $x \in K$ and $d \in D$ with $|d| \leq r$. This property is important for the initial value problems (see below) because it is used in Theorem 54 on page 476 in [137].

A.3 ▪ Limits

In this book we use limits for functions $f : \mathbb{R}_+ \to \mathbb{R}$ with $t \to +\infty$. The following fact is used frequently in the book:

If $f : \mathbb{R}_+ \to \mathbb{R}$ is non-decreasing and bounded from above (i.e., $\sup\limits_{t\geq 0}(f(t)) < +\infty$) then the limit $\lim\limits_{t\to+\infty}(f(t))$ exists and satisfies $\lim\limits_{t\to+\infty}(f(t)) = \sup\limits_{t\geq 0}(f(t))$.

The above fact is equivalent to the following fact:

If $f : \mathbb{R}_+ \to \mathbb{R}$ is non-increasing and bounded from below (i.e., $\inf\limits_{t\geq 0}(f(t)) > -\infty$) then the limit $\lim\limits_{t\to+\infty}(f(t))$ exists and satisfies $\lim\limits_{t\to+\infty}(f(t)) = \inf\limits_{t\geq 0}(f(t))$.

If $f \in L^\infty(\mathbb{R}_+)$ then the function $a(T) := \operatorname*{ess\,sup}\limits_{t\geq T}(f(t))$ defined for $T \geq 0$ is non-increasing and bounded from below (by $-\|f\|_\infty$, where $\|f\|_\infty := \operatorname*{ess\,sup}\limits_{t\geq 0}(|f(t)|) < +\infty$). Consequently, the limit $\lim\limits_{T\to+\infty}(a(T))$ exists. This limit is denoted by $\limsup\limits_{t\to+\infty}(f(t))$, i.e., we have

$$\limsup_{t\to+\infty}(f(t)) = \lim_{T\to+\infty}\left(\operatorname*{ess\,sup}_{t\geq T}(f(t))\right) \tag{Ap.2}$$

Similarly, we define $\limsup\limits_{t\to+\infty}(f(t)) = \lim\limits_{T\to+\infty}\left(\sup\limits_{t\geq T}(f(t))\right)$ for all bounded functions $f : \mathbb{R}_+ \to \mathbb{R}$. It should be noticed that $\limsup\limits_{t\to+\infty}(f(t))$ exists for all bounded functions and even for functions that do not have a limit as $t \to +\infty$ (e.g., the limit $\lim\limits_{t\to+\infty}(\sin(t))$ does not exist while the limit $\limsup\limits_{t\to+\infty}(\sin(t))$ exists and is equal to 1). If the bounded function $f : \mathbb{R}_+ \to \mathbb{R}$ is non-negative (i.e., $f(t) \geq 0$ for all $t \geq 0$) and $\limsup\limits_{t\to+\infty}(f(t)) = 0$ then $\lim\limits_{t\to+\infty}(f(t)) = 0$.

If the bounded functions $f, g : \mathbb{R}_+ \to \mathbb{R}$ satisfy $f(t) \leq g(t)$ for all $t \geq T$ for some $T > 0$ then $\limsup\limits_{t\to+\infty}(f(t)) \leq \limsup\limits_{t\to+\infty}(g(t))$. For every pair of bounded functions $f, g : \mathbb{R}_+ \to \mathbb{R}$ it holds that $\limsup\limits_{t\to+\infty}(f(t) + g(t)) \leq \limsup\limits_{t\to+\infty}(f(t)) + \limsup\limits_{t\to+\infty}(g(t))$. If the bounded function $f : \mathbb{R}_+ \to \mathbb{R}$ satisfies $\lim\limits_{t\to+\infty}(f(t)) = l \in \mathbb{R}$ then $\limsup\limits_{t\to+\infty}(f(t)) = l$.

A tool that allows us to prove the existence of limits is the so-called Barbălat's Lemma (see [73], [80]): If $f : \mathbb{R}_+ \to \mathbb{R}$ is a uniformly continuous function for which the limit $\lim\limits_{T\to+\infty}\left(\int_0^T f(t)dt\right) = \int_0^{+\infty} f(t)dt$ exists and is finite then the limit $\lim\limits_{t\to+\infty}(f(t))$ exists and is equal to zero. Barbălat's lemma is used in Chapter 2, Chapter 4, and Chapter 6 of the book.

A.4 ▪ Initial Value Problems

In this book, we study initial value problems of the form

$$\dot{x} = f(x, d)\ ,\ x \in \mathbb{R}^n\ ,\ d \in D \tag{Ap.3}$$

$$x(0) = x_0 \tag{Ap.4}$$

where $D \subseteq \mathbb{R}^p$ is a given closed set and $x_0 \in \mathbb{R}^n$, $d \in L^\infty_{loc}(\mathbb{R}_+; D)$ are given data. Theorem 54 on page 476 in [137] and Proposition C.3.6 on page 481 in [137] imply that the initial value problem (Ap.3), (Ap.4) is uniquely solvable when f is locally Lipschitz with respect to $x \in \mathbb{R}^n$. More specifically, if f is locally Lipschitz with respect to $x \in \mathbb{R}^n$ then there exists $t_{\max} \in (0, +\infty]$ and a unique absolutely continuous function $x : [0, t_{\max}) \to \mathbb{R}^n$ that satisfies (Ap.4) as well as the system of differential equations $\dot{x}(t) = f(x(t), d(t))$ for $t \in [0, t_{\max})$ a.e.. Furthermore, if $t_{\max} < +\infty$ then $\limsup\limits_{t \to t^-_{\max}} (|x(t)|) = +\infty$. This function $x : [0, t_{\max}) \to \mathbb{R}^n$ is called the unique solution of the initial value problem (Ap.3), (Ap.4).

There are some important properties of the solutions of initial value problems that a researcher should know. The first property is the "Newtonian" causality property, which says that the solution $x(T)$ of the initial value problem (Ap.3), (Ap.4) at a time $T \geq 0$ depends only on the initial condition $x_0 \in \mathbb{R}^n$ and the values of the input $d \in L^\infty_{loc}(\mathbb{R}_+; D)$ up to time $T \geq 0$. In other words, the future values of the input $d(t)$ for $t > T$ play no role. This property has some significant consequences. For example, suppose that there exists a function $P : \mathbb{R}_+ \times \mathbb{R}^n \times \mathbb{R}^n \times \mathbb{R}_+$ for which we can establish the inequality

$$P\left(t, x(t), x_0, \|d\|_\infty\right) \leq 0 \tag{Ap.5}$$

for all inputs $d \in L^\infty(\mathbb{R}_+; D)$ and for certain $t \geq 0$, $x_0 \in \mathbb{R}^n$. Then we can say that the inequality

$$P\left(t, x(t), x_0, \operatorname*{ess\,sup}_{0 \leq s \leq t}(|d(s)|)\right) \leq 0 \tag{Ap.6}$$

holds for all inputs $d \in L^\infty_{loc}(\mathbb{R}_+; D)$. The two statements are equivalent. This explains why ISS and IOS are sometimes formulated with $\sup\limits_{0 \leq s \leq t}(|d(s)|)$ (which actually means $\operatorname*{ess\,sup}\limits_{0 \leq s \leq t}(|d(s)|)$, e.g., in [66], [80]) instead of $\|d\|_\infty$ (as we do in Chapter 2 of the present book).

The second property of the solutions of initial value problems that a researcher should know is a very far-reaching property. It is the so-called "semigroup property", which says that the solution $x(T)$ of the initial value problem (Ap.3), (Ap.4) at a time $T > 0$ is the same with the solution $z(T - t)$ of the initial value problem $\dot{z} = f(z, \tilde{d})$, $z(0) = x(t)$ with the "shifted" input $\tilde{d} \in L^\infty_{loc}(\mathbb{R}_+; D)$ defined by $\tilde{d}(\tau) := d(t + \tau)$ for $\tau \geq 0$. In the previous statement, the time $T > 0$ is a time for which $x(T)$ exists and $t \in [0, T]$ is arbitrary.

The semigroup property has important consequences. For example, suppose that there exists a function $P : \mathbb{R}_+ \times \mathbb{R}^n \times \mathbb{R}^n \times \mathbb{R}_+$ for which we can establish the inequality

$$P\left(t, x(t), x_0, \operatorname*{ess\,sup}_{0 \leq s \leq t}(|d(s)|)\right) \leq 0 \tag{Ap.7}$$

for all inputs $d \in L^\infty_{loc}(\mathbb{R}_+; D)$ and for all $t \geq 0$, $x_0 \in \mathbb{R}^n$. Then the following inequality also holds for all $t_0 \in [0, t]$:

$$P\left(t - t_0, x(t), x(t_0), \operatorname*{ess\,sup}_{t_0 \leq s \leq t}(|d(s)|)\right) \leq 0 \tag{Ap.8}$$

Similarly, if there exist functions $h : \mathbb{R}^n \to \mathbb{R}^k$, $\gamma \in K$ and $\beta \in KL$ for which the inequality

$$|h(x(t))| \leq \beta(|x_0|, t) + \gamma(\|d\|_\infty) \tag{Ap.9}$$

holds for all inputs $d \in L^\infty(\mathbb{R}_+; D)$ and for all $t \geq 0$, $x_0 \in \mathbb{R}^n$ then by causality we obtain the inequality

$$|h(x(t))| \leq \beta(|x_0|, t) + \gamma\left(\operatorname*{ess\,sup}_{0 \leq s \leq t}(|d(s)|)\right) \tag{Ap.10}$$

for all inputs $d \in L^\infty_{loc}(\mathbb{R}_+; D)$ and for all $t \geq 0$, $x_0 \in \mathbb{R}^n$. Then the semigroup property implies the inequality

$$|h(x(t))| \leq \beta(|x(t_0)|, t - t_0) + \gamma\left(\operatorname*{ess\,sup}_{t_0 \leq s \leq t}(|d(s)|)\right) \tag{Ap.11}$$

for all inputs $d \in L^\infty_{loc}(\mathbb{R}_+; D)$, for all $t \geq 0$, $x_0 \in \mathbb{R}^n$, and for all $t_0 \in [0, t]$. Restricting our attention to a specific (but arbitrary) input $d \in L^\infty(\mathbb{R}_+; D)$ and since for every $\varepsilon > 0$ there exists $t_0(\varepsilon) > 0$ with $\operatorname*{ess\,sup}_{t_0(\varepsilon) \leq s \leq t}(|d(s)|) < \limsup_{\tau \to +\infty}(|d(\tau)|) + \varepsilon$, we get the inequality

$$|h(x(t))| \leq \beta(|x(t_0)|, t - t_0(\varepsilon)) + \gamma\left(\limsup_{s \to +\infty}(|d(s)|) + \varepsilon\right) \tag{Ap.12}$$

for all $t \geq t_0(\varepsilon)$. Since $\lim_{t \to +\infty}(\beta(|x(t_0)|, t - t_0(\varepsilon))) = 0$, we obtain the estimate

$$\limsup_{t \to +\infty}(|h(x(t))|) \leq \gamma\left(\limsup_{t \to +\infty}(|d(t)|) + \varepsilon\right) \tag{Ap.13}$$

for all $\varepsilon > 0$. The previous estimate (by continuity of $\gamma \in K$) gives

$$\limsup_{t \to +\infty}(|h(x(t))|) \leq \gamma\left(\limsup_{t \to +\infty}(|d(t)|)\right) \tag{Ap.14}$$

and, since $d \in L^\infty(\mathbb{R}_+; D)$ and $x_0 \in \mathbb{R}^n$ are arbitrary, we conclude that the above estimate (output asymptotic gain property) holds for the solution of (Ap.3), (Ap.4) for all $d \in L^\infty(\mathbb{R}_+; D)$, $x_0 \in \mathbb{R}^n$.

A.5 ▪ Obtaining Estimates

The whole book deals with estimates for the solutions of initial value problems. Estimates are obtained by means of inequalities. Here we provide many inequalities that are used in the present book.

Young's inequality: For every $a, b \geq 0$, $p, q > 1$ with $\frac{1}{p} + \frac{1}{q} = 1$ and for every $\varepsilon > 0$ it holds that

$$ab \leq \frac{\varepsilon}{p} a^p + \frac{1}{q\varepsilon^{q/p}} b^q \tag{Ap.15}$$

In this book, we mainly use inequality (Ap.15) with $p = q = 2$.

"Plus" inequality: For every $x, a \in \mathbb{R}$, the inequality $x \leq a$ implies the inequality $x^+ \leq a^+$.

"Product" inequality: For every $a \in K_\infty$ there exists $\kappa \in K_\infty$ such that $a(rs) \leq \kappa(r)\kappa(s)$ for all $r, s \geq 0$. See Corollary 10 in [139].

Upper bound for continuous functions: For every $V \in C^0(\mathbb{R}^n)$ with $V(0) = 0$ there exists $\zeta \in K_\infty$ such that $|V(x)| \le \zeta(|x|)$ for all $x \in \mathbb{R}^n$. Simply define

$$a(s) := \max\{\, |V(x)| \,:\, x \in \mathbb{R}^n, |x| \le s \,\} \tag{Ap.16}$$

and apply Lemma 2.4 on page 65 in [66].

Upper bound for smooth mappings: For every $h \in C^\infty(\mathbb{R}^n \times \mathbb{R}^k; \mathbb{R}^m)$ with $h(0,z) = 0$ for all $z \in \mathbb{R}^k$ there exists $r \in C^\infty(\mathbb{R}^n \times \mathbb{R}^k; \mathbb{R}_+)$ such that $|h(x,z)| \le r(x,z)\,|x|$ for all $(x,z) \in \mathbb{R}^n \times \mathbb{R}^k$. First assume that $m = 1$ and write

$$h(x,z) = \left(\int_0^1 \frac{\partial h}{\partial x}(\lambda x, z) d\lambda\right) x \tag{Ap.17}$$

which gives

$$\begin{aligned} |h(x,z)| &= \left|\int_0^1 \tfrac{\partial h}{\partial x}(\lambda x, z) d\lambda\right| |x| \le \left(\int_0^1 \left|\tfrac{\partial h}{\partial x}(\lambda x, z)\right| d\lambda\right) |x| \\ &\le \left(\int_0^1 \left(1 + \left|\tfrac{\partial h}{\partial x}(\lambda x, z)\right|^2\right) d\lambda\right) |x| \end{aligned} \tag{Ap.18}$$

If $m > 1$ then we apply the above estimate in a componentwise manner and we add all estimates.

Lower bound for positive definite functions: For every positive definite function $V \in C^0(\mathbb{R}^n)$ there exists a positive definite function $\rho \in C^0(\mathbb{R}_+)$ such that $V(x) \ge \rho(|x|)$ for all $x \in \mathbb{R}^n$. See Proposition 2.2 on page 107 in [66].

Lower bound for positive definite and radially unbounded functions: For every positive definite and radially unbounded function $V \in C^0(\mathbb{R}^n)$ there exists $\zeta \in K_\infty$ such that $V(x) \ge \zeta(|x|)$ for all $x \in \mathbb{R}^n$. See Proposition 2.2 on page 107 in [66].

Differential inequality with nonlinear dissipation: For every $\rho \in K_\infty$ there exist functions $\sigma \in KL$, $\gamma \in K_\infty$ with the following property: for every $T \in (0, +\infty]$, $\mu \ge 0$ and for every absolutely continuous function $y : [0,T) \to \mathbb{R}_+$ for which the following inequality holds for $t \in [0,T)$ a.e.:

$$\dot{y}(t) \le -\rho(y(t)) + \mu \tag{Ap.19}$$

the following estimate holds for all $t \in [0,T)$:

$$y(t) \le \sigma(y(0), t) + \gamma(\mu) \tag{Ap.20}$$

The existence of functions $\sigma \in KL$, $\gamma \in K_\infty$ follows from Lemma 2.14 on page 82 in [66] and by noticing that (Ap.19) shows that the implication

$$y(t) \ge \rho^{-1}(2\mu) \Rightarrow \dot{y}(t) \le -\frac{1}{2}\rho(y(t)) \tag{Ap.21}$$

holds for $t \in [0,T)$ a.e.

Linear differential inequality: Let $c \in \mathbb{R}$, $T \in (0, +\infty]$ be given and let $f \in L^1([0,T))$ be a given function. Then for every absolutely continuous function $y : [0,T) \to \mathbb{R}$ for which the following inequality holds for $t \in [0,T)$ a.e.:

$$\dot{y}(t) \le -c\,y(t) + f(t) \tag{Ap.22}$$

the following estimate holds for all $t \in [0, T)$:

$$y(t) \leq \exp(-ct)y(0) + \int_0^t \exp\left(-c(t-s)\right) f(s)ds \tag{Ap.23}$$

If in addition $c > 0$ and $f \in L^\infty\left([0,T)\right)$ with $\|f\|_\infty \leq \mu$, where $\mu \geq 0$ is a constant, then the following estimate holds for all $t \in [0, T)$:

$$y(t) \leq \exp(-ct)y(0) + \mu/c \tag{Ap.24}$$

The proof of this fact follows from noticing that the function

$$z(t) = \exp(ct)y(t) - \int_0^t \exp\left(cs\right) f(s)ds \tag{Ap.25}$$

is absolutely continuous and satisfies for $t \in [0, T)$ a.e. the inequality $\dot{z}(t) \leq 0$.

Bibliography

[1] V. Adetola and M. Guay, *Finite-Time Parameter Estimation in Adaptive Control of Nonlinear Systems*, IEEE Transactions on Automatic Control, 53 (2008), pp. 807–811. (Cited on pp. 2, 57)

[2] V. Adetola and M. Guay, *Performance Improvement in Adaptive Control of Linearly Parameterized Nonlinear Systems*, IEEE Transactions on Automatic Control, 55 (2010), pp. 2182–2186. (Cited on pp. 2, 57)

[3] F. Allgower, J. Ashman and A. Ilchmann, *High-Gain Adaptive λ-Tracking for Nonlinear Systems*, Automatica, 33 (1997), pp. 881–888. (Cited on p. 3)

[4] B. Andrievsky, E. Kudryashova, N. Kuznetsov and O. Kuznetsova, *Aircraft Wing Rock Oscillations Suppression by Simple Adaptive Control*, Aerospace Science and Technology, 105 (2020), 106049. (Cited on p. 129)

[5] B. Andrievsky, E. Kudryashova, N. Kuznetsov, O. Kuznetsova, T. Mokaev and S. Tomashevich, *Simple Adaptive Control of Aircraft Roll Angle, Suppressing the Wing Rock Oscillations*, Mathematics in Science, Engineering, and Aerospace, 10 (2019), pp. 373–386. (Cited on p. 129)

[6] D. Angeli and E. D. Sontag, *Forward Completeness, Unboundedness Observability, and their Lyapunov Characterizations*, Systems and Control Letters, 38 (1999), pp. 209–217. (Cited on pp. 11, 20)

[7] G. Arslan and T. Basar, *Disturbance Attenuating Controller Design for Strict-Feedback Systems with Structurally Unknown Dynamics*, Automatica, 37 (2001), pp. 1175–1188. (Cited on p. 3)

[8] A. Astolfi and R. Ortega, *Immersion and invariance: a new tool for stabilization and adaptive control of nonlinear systems*, IEEE Transactions on Automatic Control, 48 (2003), pp. 590–606. (Cited on p. 4)

[9] M. Bagheri, I. Karafyllis, P. Naseradinmousavi and M. Krstic, *Adaptive Control of a Two-Link Robot Using Batch Least-Squares Identifier*, IEEE/CCA Journal of Automatica Sinica, 8 (2021), pp. 86–93. (Cited on pp. 2, 4, 44)

[10] C. P. Bechlioulis and G. A. Rovithakis, *Robust Adaptive Control of Feedback Linearizable MIMO Nonlinear Systems with Prescribed Performance*, IEEE Transactions on Automatic Control, 53 (2008), pp. 2090–2099. (Cited on p. 5)

[11] C. P. Bechlioulis and G. A. Rovithakis, *Adaptive Control with Guaranteed Transient and Steady State Tracking Error Bounds for Strict Feedback Systems*, Automatica, 45 (2009), pp. 532–538. (Cited on p. 5)

[12] C. P. Bechlioulis and G. A. Rovithakis, *Prescribed Performance Adaptive Control for Multi-Input Multi-Output Affine in the Control Nonlinear Systems*, IEEE Transactions on Automatic Control, 55 (2010), pp. 1220–1226. (Cited on p. 5)

[13] C. P. BECHLIOULIS AND G. A. ROVITHAKIS, *Robust Partial-State Feedback Prescribed Performance Control of Cascade Systems with Unknown Nonlinearities*, IEEE Transactions on Automatic Control, 56 (2011), pp. 2224–2230. (Cited on p. 5)

[14] C. P. BECHLIOULIS AND G. A. ROVITHAKIS, *A Low-Complexity Global Approximation-Free Control Scheme with Prescribed Performance for Unknown Pure Feedback System*, Automatica, 50 (2014), pp. 1217–1226. (Cited on p. 5)

[15] T. BERGER, A. ILCHMANN AND E. P. RYAN, *Funnel Control of Nonlinear Systems*, Mathematics of Control, Signals, and Systems, 33 (2021), pp. 151–194. (Cited on p. 5)

[16] T. BERGER, H. H. LE AND T. REIS, *Funnel Control for Nonlinear Systems with Known Strict Relative Degree*, Automatica, 87 (2018), pp. 345–357. (Cited on p. 5)

[17] T. BERGER, A. ILCHMANN AND E. P. RYAN, *Funnel Control—A Survey*, arXiv:2310.03449[math.OC]. (Cited on p. 5)

[18] L. BHAN, Y. SHI AND M. KRSTIC, *Neural Operators for Bypassing Gain and Control Computations in PDE Backstepping*, IEEE Transactions on Automatic Control, 69 (2024), pp. 5310–5325. (Cited on p. 157)

[19] S. BOULOUMA, S. LABIOD AND H. BOUBERTAKH, *Linear Adaptive Actuator Failure Compensation for Wing Rock Motion Control*, Aerospace Science and Technology, 67 (2017), pp. 155–168. (Cited on p. 129)

[20] H. BREZIS, *Functional Analysis, Sobolev Spaces and Partial Differential Equations*, Springer, 2011. (Cited on p. 159)

[21] E. BULLINGER AND F. ALLGÖWER, *Adaptive λ-Tracking for Nonlinear Higher Relative Degree Systems*, Automatica, 41 (2005), pp. 1191–1200. (Cited on p. 3)

[22] E. CAPELLO, G. GUGLIERI AND D. SARTORI, *Performance Evaluation of an L1 Adaptive Controller for Wing-Body Rock Suppression*, Journal of Guidance, Control, and Dynamics, 35 (2012), pp. 1702–1708. (Cited on p. 129)

[23] G. CHOWDHARY AND E. JOHNSON, *Least Squares Based Modification for Adaptive Control*, in Proceedings of the IEEE Conference on Decision Control, Atlanta, GA, USA, 2010, pp. 1767–1772. (Cited on pp. 2, 57)

[24] G. CHOWDHARY, M. MULHEGG AND E. JOHNSON, *Exponential Parameter and Tracking Error Convergence Guarantees for Adaptive Controllers*, International Journal of Control, 87 (2014), pp. 1583–1603. (Cited on p. 4)

[25] G. CYBENKO, *Approximation by Superpositions of a Sigmoidal Function*, Mathematics of Control, Signals and Systems, 2 (1989), pp. 303–314. (Cited on p. 158)

[26] H. DENG, M. KRSTIC AND R. J. WILLIAMS, *Stabilization of Stochastic Nonlinear Systems Driven by Noise of Unknown Covariance*, IEEE Transactions on Automatic Control, 46 (2001), pp. 1237–1253. (Cited on p. 156)

[27] B. DENG, Y. SHIN, L. LU, Z. ZHANG AND G. E. KARNIADAKIS, *Approximation Rates of DeepONets for Learning Operators Arising from Advection-Diffusion-Equations*, Neural Networks, 153 (2022), pp. 411–426. (Cited on p. 158)

[28] G. DIDINSKY AND T. BASAR, *Design of Minimax Controllers for Linear Systems with Nonzero Initial States under Specified Information Structures*, International Journal of Robust and Nonlinear Control, 2 (1992), pp. 1–30. (Cited on p. 3)

[29] G. Didinsky and T. Basar, *Minimax Adaptive Control of Uncertain Plants*, in Proceedings of the 33rd IEEE Conference on Decision and Control, 1994, pp. 2839–2844. (Cited on p. 3)

[30] G. Didinsky, T. Basar and P. Bernhard, *Structural Properties of Minimax Controllers for a Class of Differential Games Arising in Nonlinear Control*, Systems and Control Letters, 21 (1993), pp. 433–441. (Cited on p. 3)

[31] G. Didinsky, Z. Pan and T. Basar, *Parameter Identification for Uncertain Plants Using H^∞ Methods*, Automatica, 31 (1995), pp. 1227–1250. (Cited on pp. 119, 128)

[32] Y. Dong, Z. Shi, K. Chen, S. Tong and D. Wei, *The Suppression of Flying-Wing Roll Oscillations with Open and Closed-Loop Spanwise Blowing*, Aerospace Science and Technology, 99 (2020), 105766. (Cited on p. 129)

[33] R. Engel and G. Kreisselmeier, *A Continuous Time Observer which Converges in Finite Time*, IEEE Transactions on Automatic Control, 47 (2002), pp. 1202–1204. (Cited on p. 57)

[34] R. A. Freeman and P. V. Kokotovic, *Robust Nonlinear Control Design. State-Space and Lyapunov Techniques*, Birkhäuser, 1996. (Cited on pp. 145, 147)

[35] P. Georgieva, A. Ilchmann and M.-F. Weirig, *Modelling and Adaptive Control of Aerobic Continuous Stirred Tank Reactors*, European Journal of Control, 7 (2001), pp. 476–491. (Cited on p. 3)

[36] G. C. Goodwin and K. S. Sin, *Adaptive Filtering Prediction and Control*, Prentice-Hall, 1984. (Cited on p. 1)

[37] G. Guglieri and F. B. Quagliotti, *Analytical and Experimental Analysis of Wing Rock*, Nonlinear Dynamics, 24 (2001), pp. 129–146. (Cited on p. 129)

[38] G. Guglieri, *A Comprehensive Analysis of Wing Rock Dynamics for Slender Delta Wing Configurations*, Nonlinear Dynamics, 67 (2012), pp. 1559–1575. (Cited on p. 129)

[39] G. Guglieri and D. Sartori, *Design of a Sliding Mode Control for Wing Rock Suppression in Highly-Swept Wing Aircraft*, International Journal of Aerospace Sciences, 2 (2013), pp. 1–10. (Cited on p. 129)

[40] L. Guo, *On Critical Stability of Discrete-Time Adaptive Nonlinear Control*, IEEE Transactions on Automatic Control, 42 (1997), pp. 1488–1499. (Cited on pp. 2, 57)

[41] L. Guo, *Feedback and Uncertainty: Some Basic Problems and Results*, Annual Reviews in Control, 49 (2020), pp. 27–36. (Cited on pp. 2, 57)

[42] C. M. Hackl, N. Hopfe, A. Ilchmann, M. Mueller and S. Trenn, *Funnel Control for Systems with Relative Degree Two*, SIAM Journal on Control and Optimization, 51 (2013), pp. 965–995. (Cited on p. 5)

[43] M. Hartman, N. Bauer and A. R. Teel, *Robust Finite-Time Parameter Estimation Using a Hybrid Systems Framework*, IEEE Transactions on Automatic Control, 57 (2012), pp. 2956–2962. (Cited on pp. 2, 57)

[44] J. P. Hespanha and A. S. Morse, *Certainty Equivalence Implies Detectability*, Systems and Control Letters, 36 (1999), pp. 1–13. (Cited on p. 2)

[45] J. P. Hespanha, D. Liberzon and A. S. Morse, *Supervision of Integral-Input-to-State Stabilizing Controllers*, Automatica, 38 (2002), pp. 1327–1335. (Cited on pp. 2, 3, 57, 87)

[46] J. P. Hespanha, D. Liberzon and A. S. Morse, *Hysteresis-Based Switching Algorithms for Supervisory Control of Uncertain Systems*, Automatica, 39 (2003), pp. 263–272. (Cited on pp. 2, 3, 57)

[47] J. P. HESPANHA, D. LIBERZON AND A. S. MORSE, *Overcoming the Limitations of Adaptive Control by Means of Logic-Based Switching*, Systems and Control Letters, 49 (2003), pp. 49–65. (Cited on p. 2)

[48] C. HUA, P. NING AND K. LI, *Adaptive Prescribed-Time Control for a Class of Uncertain Nonlinear Systems*, IEEE Transactions on Automatic Control, 67 (2022), pp. 6159–6166. (Cited on p. 5)

[49] R. HUANG, J. ZHANG AND X. ZHANG, *Adaptive Tracking Control of Uncertain Switched Non-Linear Systems with Application to Aircraft Wing Rock*, IET Control Theory & Applications, 10 (2016), pp. 1755–1762. (Cited on p. 129)

[50] F. IKHOUANE AND M. KRSTIC, *Robustness of the tuning functions adaptive backstepping design for linear systems*, IEEE Transactions on Automatic Control, 43 (1998), pp. 431–437. (Cited on p. 3)

[51] F. IKHOUANE AND M. KRSTIC, *Adaptive backstepping with parameter projection: robustness and asymptotic performance*, Automatica, 34 (1998), pp. 429–435. (Cited on p. 3)

[52] A. ILCHMANN, *Non-Identifier-Based Adaptive Control of Dynamical Systems: A Survey*, IMA Journal of Mathematical Control and Information, 8 (1991), pp. 321–366. (Cited on p. 3)

[53] A. ILCHMANN, *Adaptive λ-Tracking for Polynomial Minimum Phase Systems*, Dynamics and Stability of Systems, 13 (1998), pp. 341–371. (Cited on p. 3)

[54] A. ILCHMANN AND M. PAHL, *Adaptive Multivariable Regulation of a Biogas Tower Reactor*, European Journal of Control, 4 (1998), pp. 116–131. (Cited on p. 3)

[55] A. ILCHMANN AND E. P. RYAN, *Universal λ-Tracking for Nonlinearly-Perturbed Systems in the Presence of Noise*, Automatica, 30 (1994), pp. 337–346. (Cited on p. 3)

[56] A. ILCHMANN, E. P. RYAN AND C. J. SANGWIN, *Tracking with Prescribed Transient Behaviour*, ESAIM Control, Optimisation and Calculus of Variations, 7 (2002), pp. 471–493. (Cited on p. 5)

[57] A. ILCHMANN AND E. P. RYAN, *High-Gain Control without Identification: A Survey*, GAMM-Mitteilungen, 31 (2008), pp. 115–125. (Cited on pp. 3, 87)

[58] P. A. IOANNOU AND J. SUN, *Robust Adaptive Control*, Prentice Hall, 1996. (Cited on pp. 1, 2, 3, 34, 57, 73, 87)

[59] P. A. IOANNOU AND B. FIDAN, *Adaptive Control Tutorial*, SIAM, 2006. (Cited on pp. 2, 3, 34, 57, 73, 87)

[60] H. ISHII AND T. BASAR, *Quantization in Parameter Identification*, IEEE Transactions on Automatic Control, 53 (2008), pp. 2186–2192. (Cited on p. 119)

[61] Z.-P. JIANG, A. TEEL AND L. PRALY, *Small-Gain Theorem for ISS Systems and Applications*, Mathematics of Control, Signals and Systems, 7 (1994), pp. 95-120. (Cited on p. 40)

[62] Z.-P. JIANG AND L. PRALY, *Design of Robust Adaptive Controllers for Nonlinear Systems with Dynamic Uncertainties*, Automatica, 34 (1998), pp. 825–840. (Cited on p. 87)

[63] S. JOSHI, A. G. SREENATHA AND J. CHANDRASEKHAR, *Suppression of Wing Rock of Slender Delta Wings Using a Single Neuron Controller*, IEEE Transactions on Control Systems Technology, 6 (1998), pp. 671–677. (Cited on p. 129)

[64] I. KARAFYLLIS, *Finite-Time Global Stabilization by Means of Time-Varying Distributed Delay Feedback*, SIAM Journal Control and Optimization, 45 (2006), pp. 320–342. (Cited on p. 57)

[65] I. KARAFYLLIS, *Global Stabilization by Means of Discrete-Delay Static Output Feedback*, Systems and Control Letters, 57 (2008), pp. 987–995. (Cited on p. 57)

[66] I. KARAFYLLIS AND Z.-P. JIANG, *Stability and Stabilization of Nonlinear Systems*, Springer-Verlag (Series: Communications and Control Engineering), 2011. (Cited on pp. 10, 15, 22, 23, 24, 26, 84, 90, 160, 162, 164)

[67] I. KARAFYLLIS AND L. GRUNE, *Feedback Stabilization Methods for the Numerical Solution of Systems of Ordinary Differential Equations*, Discrete and Continuous Dynamical Systems: Series B, 16 (2011), pp. 283–317. (Cited on pp. 74, 77)

[68] I. KARAFYLLIS AND Z.-P. JIANG, *Hybrid Dead-Beat Observers for a Class of Nonlinear Systems*, Systems and Control Letters, 60 (2011), pp. 608–617. (Cited on pp. 2, 57)

[69] I. KARAFYLLIS AND Z.-P. JIANG, *Reduced Order Dead-Beat Observers for the Chemostat*, Nonlinear Analysis Real World Applications, 14 (2013), pp. 340–351. (Cited on pp. 2, 57)

[70] I. KARAFYLLIS AND M. KRSTIC, *Adaptive Certainty-Equivalence Control with Regulation-Triggered Finite-Time Least-Squares Identification*, IEEE Transactions on Automatic Control, 63 (2018), pp. 3261–3275. (Cited on pp. 2, 4, 44, 45, 47, 119, 126, 127)

[71] I. KARAFYLLIS, M. KRSTIC AND K. CHRYSAFI, *Adaptive Boundary Control of Constant-Parameter Reaction-Diffusion PDEs Using Regulation-Triggered Finite-Time Identification*, Automatica, 103 (2019), pp. 166–179. (Cited on p. 44)

[72] I. KARAFYLLIS, M. KONTORINAKI AND M. KRSTIC, *Adaptive Control by Regulation-Triggered Batch Least-Squares*, IEEE Transactions on Automatic Control, 65 (2020), pp. 2842–2855. (Cited on pp. 2, 4, 44, 45, 47, 119, 126, 127)

[73] I. KARAFYLLIS AND A. CHAILLET, *Lyapunov Conditions for Uniform Asymptotic Output Stability and a Relaxation of Barbălat's Lemma*, Automatica, 132 (2021), 109792. (Cited on pp. 4, 10, 12, 15, 29, 32, 65, 87, 161)

[74] I. KARAFYLLIS, M. KRSTIC AND A. ASLANIDIS, *On Disturbance-to-State Adaptive Stabilization without Parameter Bound by Nonlinear Feedback of Delayed State and Input*, SIAM Journal on Control and Optimization, 61 (2023), pp. 3584–3607. (Cited on pp. 5, 50, 53, 54, 67, 119, 123)

[75] I. KARAFYLLIS, A. ASLANIDIS AND M. KRSTIC, *Adaptive Regulation with Global KL Guarantees*, IEEE Transactions on Automatic Control, 69 (2024), pp. 2–15. (Cited on pp. 2, 4, 47, 87)

[76] I. KARAFYLLIS AND M. KRSTIC, *Deadzone-Adapted Disturbance Suppression Control for Global Practical IOS and Zero Asymptotic Gain to Matched Uncertainties*, Systems and Control Letters, 185 (2024), 105746. (Cited on pp. 6, 67, 86)

[77] I. KARAFYLLIS, D. THEODOSIS AND M. PAPAGEORGIOU, *Forward Completeness in Open Sets and Applications to Control of Automated Vehicles*, IEEE Transactions on Automatic Control, 70 (2025), pp. 2406–2421. (Cited on p. 11)

[78] I. KARAFYLLIS, M. KRSTIC AND A. ASLANIDIS, *Deadzone-Adapted Disturbance Suppression Control for Strict-Feedback Systems*, Automatica, 171 (2025), 111986. (Cited on pp. 6, 40, 90)

[79] I. KARAFYLLIS AND M. KRSTIC, *Global Stability Notions to Enhance the Rigor and Robustness of Adaptive Control*, arXiv:2410.16691[math.OC]. (Cited on p. 40)

[80] H. K. KHALIL, *Nonlinear Systems*, 2nd Edition, Prentice-Hall, 1996. (Cited on pp. 6, 7, 21, 29, 31, 32, 159, 160, 161, 162)

[81] A. N. KOLMOGOROV AND S. V. FOMIN, *Introductory Real Analysis*, Dover, 1975. (Cited on p. 159)

[82] P. KRISHNAMURTHY AND F. KHORRAMI, *Dynamic High-Gain Scaling: State and Output Feedback with Application to Systems with ISS Appended Dynamics Driven by All States*, IEEE Transactions on Automatic Control, 49 (2004), pp. 2219–2239. (Cited on p. 87)

[83] P. KRISHNAMURTHY, F. KHORRAMI AND M. KRSTIC, *Adaptive Output-Feedback Stabilization in Prescribed Time for Nonlinear Systems with Unknown Parameters Coupled with Unmeasured States*, International Journal of Adaptive Control and Signal Processing, 35 (2021), pp. 184–202. (Cited on p. 4)

[84] M. KRSTIC, *On Using Least-Squares Updates Without Regressor Filtering in Identification and Adaptive Control of Nonlinear Systems*, Automatica, 45 (2009), pp. 731–735. (Cited on pp. 2, 57)

[85] M. KRSTIC, I. KANELLAKOPOULOS AND P. KOKOTOVIC, *Adaptive Nonlinear Control Without Overparametrization*, Systems and Control Letters, 19 (1992), pp. 177–185. (Cited on pp. 1, 2, 100)

[86] M. KRSTIC AND P. KOKOTOVIC, *Transient performance improvement with a new class of adaptive controllers*, Systems Control Letters, 21 (1993), pp. 451–461. (Cited on p. 4)

[87] M. KRSTIC, I. KANELLAKOPOULOS AND P. KOKOTOVIC, *Nonlinear and Adaptive Control Design*, Wiley, 1995. (Cited on pp. 1, 2, 3, 4, 7, 16, 33, 47, 57, 87, 89, 90, 112, 129, 138, 145, 156, 159)

[88] M. KRSTIC AND P. KOKOTOVIC, *Adaptive Nonlinear Design with Controller-Identifier Separation and Swapping*, IEEE Transactions on Automatic Control, 40 (1995), pp. 426–440. (Cited on p. 2)

[89] M. KRSTIC AND P. KOKOTOVIC, *Control Lyapunov Functions for Adaptive Nonlinear Stabilization*, Systems and Control Letters, 26 (1995), pp. 17–23. (Cited on pp. 2, 93, 100)

[90] M. KRSTIC AND P. KOKOTOVIC, *Modular Approach to Adaptive Nonlinear Stabilization*, Automatica, 32 (1996), pp. 625–629. (Cited on pp. 2, 87)

[91] M. KRSTIC, *Invariant Manifolds and Asymptotic Properties of Adaptive Nonlinear Stabilizers*, IEEE Transactions on Automatic Control, 41 (1996), pp. 817–829. (Cited on p. 4)

[92] V. LAKSHMIKANTHAM AND S. LEELA, *Differential and Integral Inequalities*, Volume 1, Academic Press, 1969. (Cited on p. 37)

[93] V. LAKSHMIKANTHAM AND L. SALVADORI, *On Massera Type Converse Theorem in Terms of two Different Measures*, Bollettino dell' Unione Matematica Italiana, 13 (1976), pp. 293–301. (Cited on p. 37)

[94] E. LAVRETSKY AND K. WISE, *Robust and Adaptive Control: With Aerospace Applications*, 2nd Edition, Springer, 2024. (Cited on p. 5)

[95] Z.-H. LI AND M. KRSTIC, *Optimal Design of Adaptive Tracking Controllers for Non-linear Systems*, Automatica, 33 (1997), pp. 1459–1473. (Cited on pp. 93, 100)

[96] X. LIU, R. ORTEGA, H. SU AND J. CHU, *Immersion and Invariance Adaptive Control of Nonlinearly Parameterized Nonlinear Systems*, IEEE Transactions on Automatic Control, 55 (2010), pp. 2209–2214. (Cited on p. 4)

[97] A. LORIA, R. KELLY AND A. R. TEEL, *Uniform Parametric Convergence in the Adaptive Control of Mechanical Systems*, European Journal of Control, 11 (2005), pp. 87–100. (Cited on p. 4)

[98] S. LU AND T. BASAR, *Robust Nonlinear System Identification Using Neural-Network Models*, IEEE Transactions on Neural Networks, 9 (1998), pp. 407–429. (Cited on p. 119)

[99] B.-F. MA, X.-Y. DENG, Z. RONG AND B. WANG, *The Self-Excited Rolling Oscillations Induced by Fore-Body Vortices*, Aerospace Science and Technology, 47 (2015), pp. 299–313. (Cited on p. 129)

[100] I. M. Y. MAREELS, *A Simple Selftuning Controller for Stably Invertible Systems*, Systems and Control Letters, 4 (1984), pp. 5–16. (Cited on p. 37)

[101] R. MARINO AND P. TOMEI, *Robust Stabilization of Feedback Linearizable Time-Varying Uncertain Nonlinear Systems*, Automatica, 29 (1993), pp. 181–189. (Cited on p. 87)

[102] F. MAZENC, M. DE QUEIROZ AND M. MALISOFF, *Uniform Global Asymptotic Stability of a Class of Adaptively Controlled Nonlinear Systems*, IEEE Transactions on Automatic Control, 54 (2009), pp. 1152–1158. (Cited on p. 4)

[103] F. MAZENC, M. MALISOFF AND M. DE QUEIROZ, *Uniform Global Asymptotic Stability of Adaptive Cascaded Nonlinear Systems with Unknown High-Frequency Gains*, Nonlinear Analysis, Theory, Methods and Applications, 74 (2011), pp. 1132–1145. (Cited on p. 4)

[104] J. W. MELODY, T. BASAR, W. R. PERKINS AND P. G. VOULGARIS, *Parameter Identification for Inflight Detection and Characterization of Aircraft Icing*, Control Engineering Practice, 8 (2000), pp. 985–1001. (Cited on p. 119)

[105] D. E. MILLER AND E. J. DAVISON, *An Adaptive Controller Which Provides Lyapunov Stability*, IEEE Transactions on Automatic Control, 34 (1989), pp. 599–609. (Cited on p. 2)

[106] D. E. MILLER AND E. J. DAVISON, *An Adaptive Controller Which Provides an Arbitrarily Good Transient and Steady-State Response*, IEEE Transactions on Automatic Control, 36 (1991), pp. 68–81. (Cited on pp. 2, 3, 57)

[107] D. E. MILLER, *Near Optimal LQR Performance for a Compact Set of Plants*, IEEE Transactions on Automatic Control, 51 (2006), pp. 1423–1439. (Cited on p. 2)

[108] D. E. MILLER AND J. R. VALE, *Pole Placement Adaptive Control with Persistent Jumps in the Plant Parameters*, Mathematics of Control, Signals, and Systems, 26 (2014), pp. 177–214. (Cited on p. 2)

[109] A. MIRONCHENKO, *Input-to-State Stability. Theory and Applications*, Springer, 2023. (Cited on p. 160)

[110] A. MIRONCHENKO, *Lyapunov Criteria for Robust Forward Completeness of Distributed Parameter Systems*, Systems and Control Letters, 180 (2023), 105618. (Cited on pp. 11, 20)

[111] A. S. MORSE, *Recent Problems in Parameter Adaptive Control*, in I. D. Landau (editor), Outils et Modéles Mathématiques pour l'Automatique, l'Analyse de Systémes et le Traitment du Signal, Birkhäuser, 1983, pp. 733–740. (Cited on p. 37)

[112] A. S. MORSE, *Supervisory Control of Families of Linear Set-Point Controllers-Part 1: Exact Matching*, IEEE Transactions on Automatic Control, 41 (1996), pp. 1413–1431. (Cited on p. 2)

[113] A. S. MORSE, *Supervisory Control of Families of Linear Set-Point Controllers-Part 2: Robustness*, IEEE Transactions on Automatic Control, 42 (1997), pp. 1500–1515. (Cited on p. 2)

[114] A. A. MOVCHAN, *Stability of Processes with Respect to two Metrics*, Journal of Applied Mathematics and Mechanics, 24 (1960), pp. 1506–1524. (Cited on p. 37)

[115] K. S. NARENDRA AND A. M. ANNASWAMY, *Stable Adaptive Systems*, Dover, 2005. (Cited on pp. 1, 2, 57, 73, 87)

[116] N. NGUYEN, *Least-Squares Model-Reference Adaptive Control with Chebyshev Orthogonal Polynomial Approximation*, Journal of Aerospace Information Systems, 10 (2013), pp. 268–286. (Cited on pp. 2, 57)

[117] S. I. NICULESCU, *Delay Effects on Stability, A Robust Control Approach*, Springer-Verlag, 2001. (Cited on p. 57)

[118] J. ORLOWSKI, A. CHAILLET AND M. SIGALOTTI, *Counterexample to a Lyapunov Condition for Uniform Asymptotic Partial Stability*, IEEE Control Systems Letters, 4 (2020), pp. 397–401. (Cited on pp. 14, 15)

[119] J. Orlowski, A. Chaillet, A. Destexhe and M. Sigalotti, *Adaptive Control of Lipschitz Time-Delay Systems by Sigma Modification with Application to Neuronal Population Dynamics*, Systems and Control Letters, 159 (2022), 105082. (Cited on p. 34)

[120] Z. Pan and T. Basar, *Parameter Identification for Uncertain Linear systems with Partial State Measurements Under an H^∞ criterion*, IEEE Transactions on Automatic Control, 41 (1996), pp. 1295–1311. (Cited on p. 119)

[121] Z. Pan and T. Basar, *Adaptive Controller Design for Tracking and Disturbance Attenuation in Parametric Strict-Feedback Nonlinear Systems*, IEEE Transactions on Automatic Control, 43 (1998), pp. 1066–1083. (Cited on p. 3)

[122] Z. Pan and T. Basar, *Adaptive Controller Design and Disturbance Attenuation for Minimum Phase MIMO Linear Systems with Noisy Output Measurements and with Measured Disturbances*, International Journal of Adaptive Control and Signal Processing, 37 (2023), pp. 2577–2626. (Cited on p. 3)

[123] B. B. Peterson and K. S. Narendra, *Bounded Error Adaptive Control*, IEEE Transactions on Automatic Control, 27 (1982), pp. 1161–1168. (Cited on p. 87)

[124] Polycarpou, M. M. and P. A. Ioannou, *A robust adaptive nonlinear control design*, in Proceedings of the 1993 American Control Conference, San Francisco, CA, 1993, pp. 1365–1369. (Cited on p. 3)

[125] L. Praly, G. Bastin, J.-B. Pomet and Z.-P. Jiang, *Adaptive Stabilization of Nonlinear Systems*, in P. V. Kokotovic (editor), Foundations of Adaptive Control, Springer-Verlag, 1991, pp. 347–434. (Cited on pp. 1, 2)

[126] L. Praly, *Adaptive Regulation: Lyapunov Design with a Growth Condition*, International Journal of Adaptive Control and Signal Processing, 6 (1992), pp. 329–351. (Cited on p. 2)

[127] L. Praly and Z.-P. Jiang, *Linear Output Feedback with Dynamic High Gain for Nonlinear Systems*, Systems and Control Letters, 53 (2004), pp. 107–116. (Cited on p. 87)

[128] H.-J. Rong, S. Han, J.-M. Bai and Y.-Q. Liang, *Improved Adaptive Control for Wing Rock via Fuzzy Neural Network with Randomly Assigned Fuzzy Membership Function Parameters*, Aerospace Science and Technology, 39 (2014), pp. 614–627. (Cited on p. 129)

[129] J. G. Rueda-Escobedo and J. A. Moreno, *Strong Lyapunov Functions for two Classical Problems in Adaptive Control*, Automatica, 124 (2021), 109250. (Cited on p. 4)

[130] S. Sastry and M. Bodson, *Adaptive Control: Stability, Convergence, and Robustness*, Dover, 2011. (Cited on pp. 1, 2, 57, 74, 87)

[131] H. R. Schwarz, *Numerical Analysis. A Comprehensive Introduction*, Wiley, 1989. (Cited on p. 125)

[132] Y. Song, K. Zhao and M. Krstic, *Adaptive Backstepping with Exponential Regulation in the Absence of Persistent Excitation*, in Proceedings of the American Control Conference, 2016, pp. 739–744. (Cited on p. 4)

[133] Y. Song, K. Zhao and M. Krstic, *Adaptive Control with Exponential Regulation in the Absence of Persistent Excitation*, IEEE Transactions on Automatic Control, 62 (2017), pp. 2589–2596. (Cited on p. 4)

[134] E. D. Sontag, *Smooth Stabilization Implies Coprime Factorization*, IEEE Transactions on Automatic Control, 34 (1989), pp. 435–443. (Cited on p. 40)

[135] E. D. Sontag and Y. Wang, *On Characterizations of the Input-to-State Stability Property*, Systems and Control Letters, 24 (1995), pp. 351–359. (Cited on p. 40)

[136] E. D. SONTAG AND Y. WANG, *New Characterizations of the Input-to-State Stability*, IEEE Transactions on Automatic Control, 41 (1996), pp. 1283–1294. (Cited on p. 40)

[137] E. D. SONTAG, *Mathematical Control Theory. Deterministic Finite Dimensional Systems*, 2nd Edition, Springer, New York, 1998. (Cited on pp. 161, 162)

[138] E. D. SONTAG, *Comments on Integral Variants of ISS*, Systems and Control Letters, 34 (1998), pp. 93–100. (Cited on p. 40)

[139] E. D. SONTAG AND Y. WANG, *Notions of Input to Output Stability*, Systems and Control Letters, 38 (1999), pp. 235–248. (Cited on pp. 15, 40, 163)

[140] E. SONTAG AND Y. WANG, *Lyapunov Characterizations of Input-to-Output Stability*, SIAM Journal on Control and Optimization, 39 (2000), pp. 226–249. (Cited on pp. 15, 16, 40)

[141] E. D. SONTAG, *Input-to-State Stability: Basic Concepts and Results*, in P. Nistri and G. Stefani (editors) Nonlinear and Optimal Control Theory. Lectures Given at the C.I.M.E. Summer School Held in Cetraro, Italy, June 19–29, 2004, Vol. 1932, Lecture Notes in Mathematics, Springer, 2008, pp. 163–220. (Cited on p. 40)

[142] H. TAGHAVIFAR, Y. QIN AND C. HU, *Adaptive Immersion and Invariance Induced Optimal Robust Control of Unmanned Surface Vessels with Structured/Unstructured Uncertainties*, Ocean Engineering, 239 (2021), 109792. (Cited on p. 4)

[143] G. TAO, *Adaptive Control Design and Analysis*, Wiley, 2003. (Cited on p. 3)

[144] A. R. TEEL AND L. PRALY, *A Smooth Lyapunov Function from a Class KL Estimate Involving two Positive Semidefinite Functions*, ESAIM Control, Optimisation and Calculus of Variations, 5 (2002), pp. 313–367. (Cited on p. 37)

[145] I. E. TEZCAN AND T. BASAR, *Disturbance Attenuating Adaptive Controllers for Parametric Strict Feedback Nonlinear Systems with Output Measurements*, Journal of Dynamic Systems, Measurement, and Control, 121 (1999), pp. 48–57. (Cited on p. 3)

[146] K. TSAKALIS AND P. IOANNOU, *Linear Time-Varying Systems: Control and Adaptation*, Prentice Hall, 1993. (Cited on p. 3)

[147] V. I. VOROTNIKOV, *Partial Stability and Control*, Birkhäuser, Boston, 1998. (Cited on p. 37)

[148] J. WANG AND M. DIAGNE, *Delay-Adaptive Boundary Control of Coupled Hyperbolic PDE-ODE Cascade Systems*, arXiv:2301.12372[math.OC]. (Cited on p. 44)

[149] J. WANG AND M. KRSTIC, *Regulation-Triggered Adaptive Control of a Hyperbolic PDE-ODE Model with Boundary Interconnections*, International Journal of Adaptive Control and Signal Processing, 35 (2021), pp. 1513–1543. (Cited on p. 44)

[150] L. WANG AND C. M. KELLETT, *Adaptive Tracking Control via Immersion and Invariance: An (i)ISS perspective*, in Proceedings of the IEEE 58th Conference on Decision and Control, 2019, pp. 7019–7024. (Cited on p. 4)

[151] Z. WANG AND D. R. LIU, *Data-Based Output Feedback Control Using Least Squares Estimation Method for a Class of Nonlinear Systems*, International Journal of Robust and Nonlinear Control, 24 (2014), pp. 3061–3075. (Cited on pp. 2, 57)

[152] J. C. WILLEMS AND C. I. BYRNES, *Global Adaptive Stabilization in the Absence of Information on the Sign of the High Frequency Gain*, in A. Bensoussan and J. L. Lions (editors), Analysis and Optimization of Systems, in Proceedings of the 6th INRIA Conference, Nice, France, Springer-Verlag, 1984, pp. 49–57. (Cited on p. 37)

[153] W. YANG, G. CUI, Z. LI AND C. TAO, *Adaptive Practical Fixed-Time Control for a Class of Nonlinear Systems with Input Saturation*, in Proceedings of the 39th Chinese Control Conference, 2020, pp. 440–445. (Cited on p. 4)

[154] YAO, B. AND M. TOMIZUKA, *Robust adaptive nonlinear control with guaranteed transient performance*, in Proceedings of the 1995 American Control Conference, Seattle, WA, 1995, pp. 2500–2504. (Cited on p. 3)

[155] A. YARZA, V. SANTIBANEZ AND J. MORENO-VALENZUELA, *An Adaptive Output Feedback Motion Tracking Controller for Robot Manipulators: Uniform Global Asymptotic Stability and Experimentation*, International Journal of Applied Mathematics and Computer Science, 23 (2013), pp. 599–611. (Cited on p. 4)

[156] H. YOON, B. E. BATEMAN AND B. N. AGRAWAL, *Laser Beam Jitter Control Using Recursive-Least-Squares Adaptive Filters*, Journal of Dynamic Systems, Measurement, and Control, 133 (2011), 041001. (Cited on pp. 2, 57)

[157] T. YOSHIZAWA, *Stability Theory by Liapunov's Second Method*, Publications of the Mathematical Society of Japan, 1966. (Cited on p. 40)

[158] J. YU, A. STANCU, Z. WU AND Y. WU, *Adaptive Finite/Fixed-Time Stabilizing Control for Nonlinear Systems with Parametric Uncertainty*, International Journal of Robust and Nonlinear Control, 33 (2023), pp. 1513–1530. (Cited on p. 4)

[159] Z. ZANG AND R.R. BITMEAD, *Transient bounds for adaptive control systems*, in Proceedings of the 29th IEEE Conference on Decision and Control, Honolulu, HI, 1990, pp. 2724–2729. (Cited on p. 4)

[160] ZHANG, Y. AND P. A. IOANNOU, *A robust modification of a new class of adaptive controllers*, in Proceedings of the 1995 American Control Conference, Seattle, WA, 1995, pp. 2510–2514. (Cited on p. 3)

[161] ZHANG, Y. AND P. A. IOANNOU, *Stability and performance of nonlinear robust adaptive control*, in Proceedings of the 34th IEEE Conference on Decision and Control, New Orleans, LA, USA, 1995, pp. 3941–3946. (Cited on p. 3)

[162] K. ZIMENKO, D. EFIMOV AND A. POLYAKOV, *Adaptive Finite-Time and Fixed-Time Control Design Using Output Stability Conditions*, International Journal of Robust and Nonlinear Control, 32 (2022), pp. 6361–6378. (Cited on p. 4)

[163] M. ZRIBI, S. ALSHAMALI AND M. AL-KENDARI, *Suppression of the Wing-Rock Phenomenon Using Nonlinear Controllers*, Nonlinear Dynamics, 71 (2013), pp. 313–322. (Cited on p. 129)

Index